The
Plant
Kingdom

The Plant Kingdom

EVOLUTION AND FORM

SAMUEL R. RUSHFORTH

Brigham Young University

PRENTICE-HALL, INC., Englewood Cliffs, New Jersey

Library of Congress Cataloging in Publication Data

RUSHFORTH, SAMUEL R.
   The plant kingdom.

   Includes bibliographies and index.
   1. Botany. I. Title.
QK47.R913   581   75–23463
ISBN 0-13-680405-5

© 1976 by PRENTICE-HALL, INC.
Englewood Cliffs, New Jersey

10 9 8 7 6 5 4 3 2 1

Printed in the United States of America

PRENTICE-HALL INTERNATIONAL, INC., *London*
PRENTICE-HALL OF AUSTRALIA, PTY. LTD., *Sydney*
PRENTICE-HALL OF CANADA, LTD., *Toronto*
PRENTICE-HALL OF INDIA PRIVATE LIMITED, *New Delhi*
PRENTICE-HALL OF JAPAN, INC., *Tokyo*
PRENTICE-HALL OF SOUTHEAST ASIA (PTE.) LTD., *Singapore*

Contents

Preface

A young man once told me that he had attended a recent concert without enjoying it very much. We discussed the reason and decided that he had not enjoyed the music because he did not understand it. He knew nothing of the life of the composer, of his style, or of what he was trying to say with his music. After some further discussion, we decided that in order to really enjoy such music, the listener must learn something about it.

Similarly, it takes understanding and awareness to fully enjoy the natural world. A person who is not familiar with plants and animals, and their interactions and habitats cannot appreciate nature or life as much as someone who is. Consequently, it is important for us to strive to understand nature with its infinite patterns and variations, and to study the relationships between plants and animals and find where we fit into the pattern. When we come to understand our role in the natural world, we can live a happier, more rewarding life. Thus, moments of confusion and disillusionment may be softened by an appreciation of natural things. Literally beauty may be sought and found in our own backyard. No one need despair for want of wonder since each of us is constantly surrounded by it.

It is the aim of this book to make the readers more aware of the plants about them. If they can walk through the forest after reading this book and gain pleasure from recognizing a moss or fern, and know something of its life style, the goal of this book will have been reached. Beauty is in the eye of the understander—it is often not enough to be simply a beholder.

A word is in order about the format of this book. I have followed a rather rigid pedagogy throughout, since I believe that in an introductory work, the student should concentrate on the conceptual material and not struggle with organizational changes in the various chapters. I have been supported in this project by many of my colleagues who have used the laboratory manual *Plants and Man* (Burgess Pub.) prepared by William Tidwell and me, and who have urged me to write a text along the same organizational lines.

In most cases the factual material treated in this book is not new. That is, most of it may be obtained from a good senior- or graduate-level morphology book such as Bold's *Plant Morphology*. However, the level of this book and the approach are unique. The book is aimed squarely at the beginning student in a plant kingdom or lower-level plant morphology course. I have had several students read portions of my manuscript as it has been completed, and have benefited from their advice on level and format. I hope I have succeeded since I have long despaired of texts written "above and beyond" the audience they were intended for.

Many people have aided in the development of this book, and I wish to thank them all. Somehow the production of any book is a team effort with help coming from former mentors, friends, and family as well as from colleagues. Naomi Hebbert and Paul Smith are particularly to be thanked for their fine drawings. I am also grateful to several colleagues for their critical review and help in improving the manuscript.

Finally, in the words of Mark Twain, whom I admire greatly, ". . . and so there ain't nothing more to write about, and I'm rotten glad of it, because if I'd a knowed what a trouble it was to make a book I wouldn't a tackled it and I ain't agoing to no more."

SAMUEL R. RUSHFORTH

The
Plant
Kingdom

1 The Science of Botany

This Red Delicious apple symbolizes our dependency upon plants for food. We are likewise dependent upon plants for many other equally important reasons. (Courtesy of Sheril Burton, Brigham Young University.)

Figure 1-1. Green plants often form very dense covers of vegetation on the surface of the earth. This photograph of an aspen forest shows that essentially no bare ground is present.

Figure 1-2. Plants also may be very dense in the sea. These brown algae have washed up onto the shore forming "windrows" of plant material. A dense underwater forest of algae occurs a few hundred yards offshore from where this photograph was taken. Microscopic plants are even more important in the sea than these larger forms. (Courtesy of Jack Brotherson, Brigham Young University.)

A BRIEF INTRODUCTION

Growing children gradually become aware of the natural things surrounding them. They first develop an immense interest in moving creatures, manifested in the collection and often critical examination of beetles, butterflies, lizards, or anything else that moves. As children continue to grow and mature, they often become aware of plants. They may also begin to discover the great beauty of many of these plants, and the interdependence of plants and animals. Unfortunately, some children lose their interest in the natural world as they grow older. However, many never suffer this loss, and some become biologists. Biologists who study plants primarily are known as botanists. Those who study animals are zoologists, and those concentrating on microorganisms are microbiologists. Even so, the sciences of botany, zoology, and microbiology cannot be separated easily since living organisms are interrelated and interdependent in nature.

Green plants are all about us. They cover much of the surface of the land (Fig. 1-1), and they occur in tremendous numbers in the seas, ponds, and rivers of the world (Fig. 1-2). In fact, green plants made life as we now know it possible on earth and are responsible for allowing it to continue. They provide our primary source of food, produce the oxygen we breathe, provide most of our building materials and paper pulp (as well as countless other products), and are a constant source of beauty and contentment to millions of people. No educated person should be unfamiliar with the plant world and its basic importance to life.

Some people study botany because of the great practical importance of plants. Others realize that something very aesthetic and peaceful may be gleaned from an awareness of plants. Untold beauty awaits all people willing to put forth the effort to really "see." For instance, a student approached me recently and asked if we could study only flowering plants in our plant kingdom course. This student reasoned that they are "the only ones we ever see anyway." I suggested that perhaps what was needed was a deeper awareness and sharper interest on his part. We walked about campus together for fifteen minutes, and in that time we found plants representing most of the various divisions of the plant kingdom. He was amazed at the variety and beauty of these plants and went on to develop a keen interest and appreciation for the plant world.

Another reason for studying botany is to develop an environmental awareness in order to make sound conclusions regarding the development and preservation of our natural resources. For instance, understanding how aquatic plants are affected by increasing organic pollution may provide the basis for rational decisions that are necessary to insure that we, our children, and our grandchildren may have a habitable world now and in the future. It is our responsibility to be informed about conservation issues and to be knowledgeable about the best solutions to these problems based on sound biological principles.

A SHORT HISTORY

The study of plants is not new. People throughout history have been botanists of a sort for many reasons. Primitive people were primarily interested in plants because of their important use as food. Even early hunting cultures apparently often relied heavily on plants and thus were forced by necessity to be aware of desirable and poisonous or unpalatable species.

As people began to domesticate plants for food, it became increasingly important that they be aware of the characteristics of the plants they wished to grow. Thus, early agriculturists continually selected new plants for themselves and their domesticated animals and disregarded or discarded less desirable ones. As people became more efficient at cultivating foods, they became aware of many things they had not noticed before. For example, they found that diseases often ravaged their crops, and that these diseases were prevalent under certain environmental conditions. They noted that their crops often grew better if they spread ashes or certain minerals on their field before crops were planted. They saw that their crops grew better in certain types of soil than in others, and that some of their crops were better suited for growth in certain localities than in others. In other words, earliest agricultural people became interested in plants and their environment for very practical reasons. They needed to eat and to build shelters.

As people became more adept at growing crops, the land was able to support more and more humans. Likewise, leisure time developed as people became able to grow more than enough food for themselves and their immediate families. As these two important trends continued, some people became interested in plants for other than purely practical reasons. Thus, some became interested in flowers for their aesthetic values. Others grew or gathered plants for their unique flavors, unusual smells, hallucinogenic properties, or medicinal values.

Figure 1-3. This reproduction from a tomb at Saqqara dated about 2400 S.C. shows the harvesting of grain and its subsequent binding and loading onto donkeys. (Reprinted by permission from Charles Singer, et al., eds. *A History of Technology*, Oxford University Press, Cambridge, 1954).

Figure 1-4. This relief shows the "Winged Genius" pollinating the sacred tree. Together with the previous figure it indicates that the level of agricultural attainment by certain groups of early people were fairly significant. (Reproduced by permission of the Museum of Fine Arts, Boston.)

Figure 1-5. Anton van Leeuwenhoek was a Dutch merchant who became interested in lenses. He was largely responsible for the invention of the microscope. This invention has proven to be one of the most important in all of biology. (Courtesy of the Rijks-Museum, Amsterdam.)

The early Chinese were perhaps the first to become particularly adept in both agrarian and nonagrarian uses of plants. Prior to 3000 B.C. the Chinese had developed a rather extensive agriculture based on soy beans, rice, barley, oranges, and other crops. They were (and still are) also skilled at using plants medicinally. For instance, they used such plants as marijuana and the opium poppy for the relief of pain. *Ephedra* was used for the control of asthma and ginseng and pomegranate roots for weakness and other infirmities. These plants are still used today by Chinese and other cultures. One of the first books ever written was authored some 5000 years ago by the Chinese botanist Sheng Nung. The first of many herbals written throughout history, this book described several Chinese plants and their uses in medicinal practices.

Many other early civilizations also used plants both medicinally and for food. Long before the time of Christ, the Egyptians had developed intensive cultivation of food crops including cereal grains and fruits (Figs. 1-3 and 1-4). The level of early Egyptian civilization was apparently quite high. This was mainly because of efficient agriculture, which allowed many citizens abundant leisure time for education and development of the arts.

The Greeks were very aware of agriculture and biology in general, and made important contributions to these and other sciences. Perhaps the greatest early Greek biologist was Aristotle, who lived from 384–322 B.C. Aristotle was interested in all forms of life and wrote extensively of his skilled studies of both plants and animals. Unfortunately, many of his botanical writings have been lost. Perhaps the most important contribution attributed to Aristotle was his development of sound scientific methods for making accurate observations and descriptions.

The Greek botanist Theophrastus, a student of Aristotle, has been called by many the *father of botany*. He was interested in plants scientifically as well as for their practical uses. Theophrastus wrote several books, but perhaps his most significant was *Historia Plantarum* or *Inquiry Into Plants*. In it he discussed the classification, reproduction, form, and ecology of plants as well as their medicinal values. *Inquiry Into Plants* represents a magnificent early botanical treatise.

A Roman, Pliny the Elder, was active in biological studies during the first century A.D. He scoured the books and manuscripts of previous biologists and compiled the book, *Historia Naturalis*. This work is not only a discussion of biological science, but of folklore, the arts, and contemporary discoveries of the time as well. Pliny treated many species of plants, but all are discussed from a purely economic standpoint.

The Roman surgeon, Dioscorides, made significant contributions to botany at about the same time as Pliny the Elder. He published a book on medicinal botany in which he discussed about 600 different plants. His discussion of these plants was usually very accurate, and his book had a great effect on the later development of botanical science.

Just as in many other fields of science, very little of importance was added to botanical knowledge during the Dark Ages. The Arabic and Persian cultures were the only ones that contributed significantly during this period. However, these cultures were more interested in the physical sciences, and little botanical work was done—or at least recorded. Certainly the most important event of the Dark Ages was the development of universities that were destined to become the seats of great stimulation in all areas of human endeavor, including the study of plants.

The Renaissance witnessed a rebirth of all learning and study, and reawakened an interest in botany. Many herbals were written during this time, and people began to be more interested in plants medicinally, scientifically, and aesthetically. Several botanists were active from the mid 1600's to the mid 1700's and were responsible for many scientific developments.

During this time, an unlikely candidate for making history, Anton van Leeuwenhoek (Fig. 1-5), a Dutch cloth merchant and surveyor, became interested in grinding lenses. This became his intense hobby, and it eventually resulted in the manufacture of many fine microscopes (Fig. 1-6). Leeuwenhoek became so famous for his work with the microscope that he was elected to the *Royal Society of London,* and was visited by many important people of the day, including the Queen of England. Many prominent scientists corresponded with him concerning problems and uses of the microscope. The development of the microscope (Fig. 1-7) represents a milestone in the history of

Figure 1-6. After a brief period of experimentation, early microscopists succeeded in producing high quality microscopes. Even though this replica of one of the early microscopes manufactured by van Leeuwenhoek looks crude, he was capable of observing bacteria. (Courtesy of the Armed Forces Institute of Pathology.)

(A)

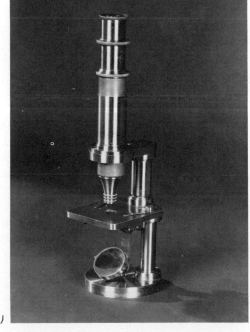

(B)

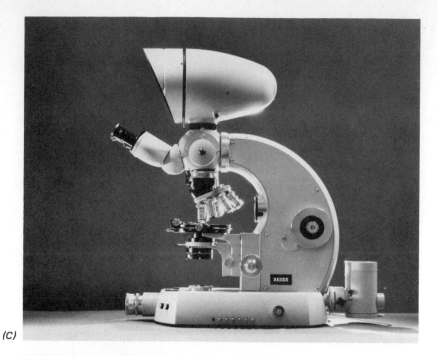

(C)

(D)

Figure 1-7. Microscopes have changed in appearance a good deal since they were first developed. These microscopes graphically show those changes. *A* is a replica of a Culpeper microscope in use during the early 1700's. *B* is a Zeiss microscope manufactured in about 1850. *C* is a modern photomicroscope by Zeiss, which combines a fine microscope with photographic equipment. *D* is the new Axiomat by Zeiss, which is a square modular system allowing great adaptability. (Courtesy of Carl Zeiss, Inc.)

botany since so much of our modern botanical knowledge has been directly dependent on the study of microscopic plants and plant parts. Even though Leeuwenhoek made few direct contributions to botany, he is considered important to botanists because his perfection of the microscope was critical for the future development of the plant sciences.

John Ray (1627–1705) was an intense and accurate observer who was interested in the classification of plants and animals and who devoted his life to their study. Perhaps Ray's greatest work was *Historia Generalis Plantarum* in which he described over 18,000 species of plants, representing nearly all the plants known at that time.

Carolus Linnaeus (1707–1778) (Fig. 1-8) was certainly one of the most influential botanists who ever lived. Linnaeus published hundreds of accurate articles and books, perhaps the most important of which is *Species Plantarum* (Fig. 1-9). This book laid the foundation for modern plant classification. The Linnaen classification system is a binomial system. That is, every plant described by Linnaeus was given two names. The first is the name of the *genus* of the plant and the second is the name of the *species*. Several closely related plants may have the same generic name, but each is given a unique species name. Thus, the genus name for apple trees is *Malus*. However, this genus contains several types of apple trees, each of which is given a separate species name. To name a few, *Malus sylvestrus* is the common cultivated apple tree, and *Malus floribunda* is the showy flowering crab apple tree. The Linnaen system of classification is in international use today, and the names given to many plants by Linnaeus in the eighteenth century still endure.

Linnaeus popularized the study of plants. As a result of his efforts, many people began to be interested in the beauty of plants and in the science of botanical classification. Europeans, as a whole, are still more interested in plants than Americans. It is common for European mail carriers, store clerks, home makers, and others from all walks of life to belong to botanical societies and to devote much of their spare time to collecting, growing, and studying plants.

Two other botanists must be named to complete this thumbnail botanical history. Charles Darwin (1809–1882) (Fig. 1-10), a medical student turned naturalist, was a shrewd observer who was widely read. His knowledge of geology and especially his acceptance of the rather startling nineteenth century theory that the earth was many millions of years old were particularly important. If the planet was this old, Darwin reasoned that living organisms have had a good deal of time to change by the process that we now popularly know of as *evolution*. Following this reasoning, Darwin set about to determine the methods whereby species change and developed the theory of *natural selection*. Briefly, this theory states that each species produces more offspring than can survive in any given year, and that only the most fit of the offspring live to reproduce and carry on the species. This theory revolutionized thinking in the natural sciences and has come a long way toward providing a common denominator for all of the sciences. Natural selection will be discussed further in Chap. 3.

One thing that remained a puzzle to Darwin was the means whereby variation is introduced into a species. He knew that this introduction was a critical part of his theory but readily admitted that the methods for organisms obtaining variation were beyond his knowledge. It is interesting that a contemporary of Darwin, a German monk named

Figure 1-8. Linnaeus is considered by modern biologists to be the father of plant classification. (Courtesy of the American Museum of Natural History.)

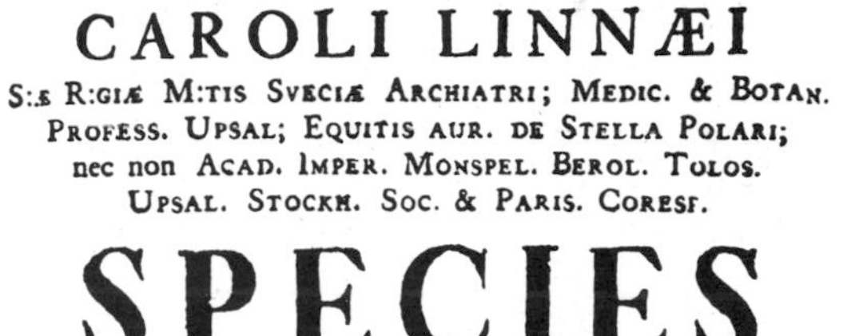

CAROLI LINNÆI

S:ᴀ R:ɢɪᴀ M:ᴛɪꜱ Sᴠᴇᴄɪᴀ Aʀᴄʜɪᴀᴛʀɪ; Mᴇᴅɪᴄ. & Bᴏᴛᴀɴ. Pʀᴏꜰᴇꜱꜱ. Uᴘꜱᴀʟ; Eǫᴜɪᴛɪꜱ ᴀᴜʀ. ᴅᴇ Sᴛᴇʟʟᴀ Pᴏʟᴀʀɪ; ɴᴇᴄ ɴᴏɴ Aᴄᴀᴅ. Iᴍᴘᴇʀ. Mᴏɴꜱᴘᴇʟ. Bᴇʀᴏʟ. Tᴏʟᴏꜱ. Uᴘꜱᴀʟ. Sᴛᴏᴄᴋʜ. Sᴏᴄ. & Pᴀʀɪꜱ. Cᴏʀᴇꜱꜰ.

SPECIES PLANTARUM,

EXHIBENTES

PLANTAS RITE COGNITAS,

ᴀ ᴅ

GENERA RELATAS,

ᴄᴜᴍ

Dɪꜰꜰᴇʀᴇɴᴛɪɪꜱ Sᴘᴇᴄɪꜰɪᴄɪꜱ, Nᴏᴍɪɴɪʙᴜꜱ Tʀɪᴠɪᴀʟɪʙᴜꜱ, Sʏɴᴏɴʏᴍɪꜱ Sᴇʟᴇᴄᴛɪꜱ, Lᴏᴄɪꜱ Nᴀᴛᴀʟɪʙᴜꜱ, Sᴇᴄᴜɴᴅᴜᴍ

SYSTEMA SEXUALE

ᴅɪɢᴇꜱᴛᴀꜱ.

Tᴏᴍᴜꜱ I.

Cum Privilegio S. R. M:tis Sueciæ & S. R. M:tis Polonicæ ac Electoris Saxon.

HOLMIÆ,
Iᴍᴘᴇɴꜱɪꜱ LAURENTII SALVII.
1753.

Figure 1-9. This figure shows the title page from the important book by Linnaeus, *Species Plantarum*, 1753.

Figure 1-10. Charles Darwin was one of the most influential biologists who ever lived. His theories have caused profound changes in biology during the past century and have caused people to look at themselves in a new light. (Courtesy of the American Museum of Natural History.)

Figure 1-11. Gregor Mendel was a German monk who performed several important experiments with garden peas. He founded the science of heredity. His work was an important addition to the work of Darwin since it explained one of the most important methods of accounting for variability in offspring. (Courtesy of the American Museum of Natural History.)

Gregor Mendel (1822–1884) (Fig. 1-11) was performing experiments on plant breeding at the same time Darwin was writing the results of his studies. Mendel was a careful researcher who is often referred to as the *father of genetics.* Among other things, genetics is the study of the sexuality and inheritance patterns of organisms. Mendel demonstrated ways by which the offspring of two parent plants may differ significantly from either parent. Thus, variation in successive generations was shown to be a reality, and the work of Mendel added beautifully to that of Darwin.

Many other famous researchers have added to the work of Darwin and Mendel during the twentieth century. Evolution and genetics are now seen to be inseparably linked and are the topic of research in thousands of laboratories throughout the world.

Significant contributions to botany have multiplied since the pioneering work of the men mentioned in this chapter, and the frontiers of botanical science are being continuously pushed forward. Today many specialized fields of botany are studied. *Cytology* is the study of cells; *ecology* is the study of relationships between the plant and its environment; *physiology* is the study of the physical and chemical processes of the plant; *anatomy* is the study of plant tissues and organs; *morphology* is the study of form and reproduction in plants; and *taxonomy* is the science of classification.

A recent discovery of note for aiding botanical study is the electron microscope (Fig. 1-12). By using this instrument, botanists have been able to investigate aspects of cells and tissues that were previously unknown. Such studies have demonstrated that even the simplest cell is tremendously complex.

Botany is one of the most exciting and rapidly expanding of all sciences. Many doors remain to be opened, and many exciting discoveries lie ahead. Some of the most demanding world problems, such as the need for increased food production and the development of cures for dread diseases, may eventually be solved by future plant scientists.

FURTHER READING

ADAMS, A., *The Eternal Quest: The Story of the Great Naturalists,* G. P. Putnam Sons, New York, 1969.

ASBERG, MARIE and WILLIAM STEARN, "Linnaeus's Oland and Gotland Journey, 1741," *Biological Journal of the Linnaen Society,* 5(1):1–107, 1973.

BURROUGHS, R. D., "The Lewis and Clark Expedition's Botanical Discoveries," *Natural History,* 75(1):56–62, 1966.

DARLINGTON, C. D., "The Origins of Agriculture," *Natural History,* 79(5):46–57, 1970.

DEBEER, GAXIN, "Darwin's Origin Today," *Natural History,* 75(7):62–71, 1966.

DIOSCORIDES, *The Greek Herbal,* Hafner Press, New York, about 70 A.D., reprinted 1968.

Figure 1-12. One of the recent important inventions that has led to a greater understanding of all the biological sciences is the electron microscope. This Zeiss EM-10 electron microscope allows examination of cells and tissues at very high magnification and resolution. (Courtesy of Carl Zeiss, Inc.)

EISLEY, LOREN, "Charles Darwin," *Scientific American,* 194(2):62–72, 1956.

EWAN, JOSEPH (Ed.), *A Short History of Botany in the United States,* Hafner Press, New York, 1969.

GALSTON, ARTHUR, "Botanist Charles Darwin," *Natural History,* 82(10):85–93, 1973.

GARDNER, ELDON, *History of Biology,* Third edition, Burgess Publishing Co., Minneapolis, 1972.

HEISER, CHARLES, JR., *Seed to Civilization,* W. H. Freeman and Company, San Francisco, 1973.

HOSKIN, MICHAEL, "A Chat with Charles Darwin," *Natural History,* 81(6):12–20, 1972.

KLEIN, RICHARD, "The Botanists Library," *Natural History,* 77(4):66–69, 1968.

LINN, ALAN, "Corn, the New World's Secret Weapon and the Builder of its Civilization," *Smithsonian,* 4(5):58–65, 1973.

MIALL, L., *The Early Naturalists: Their Lives and Work,* Hafner Press, New York, 1912, reprinted 1969.

THOMAS, PHILIP, "Nights in Pliny's Garden," *Natural History,* 81(3):70–77, 1972.

2 Cells, Reproduction, and Life Cycle

Blossoms of a flowering almond tree.

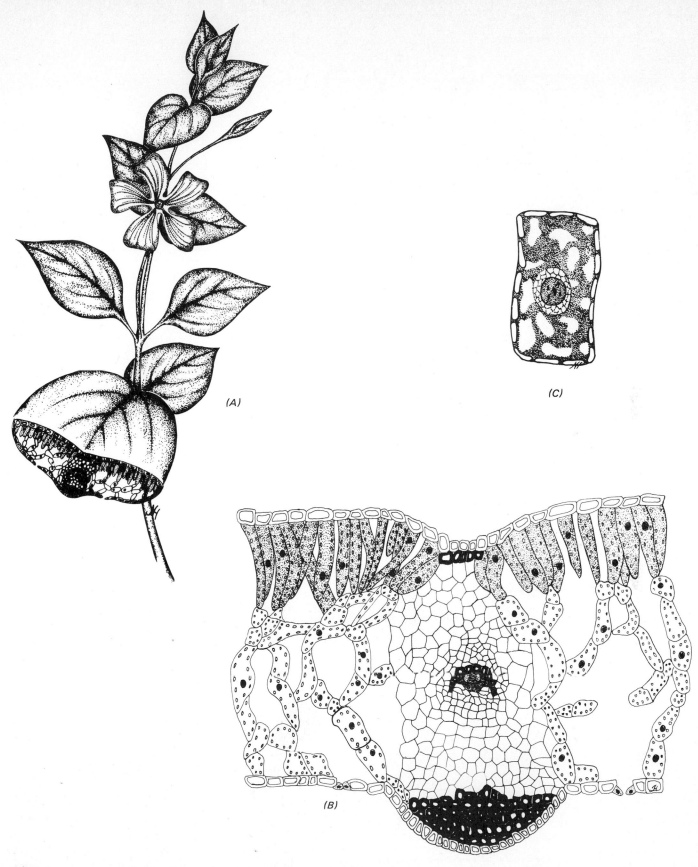

Figure 2-1. One of the important discoveries brought about by the invention of the microscope was that plants are composed of cells. *A* shows a flowering plant with a cross section cut through a leaf. *B* is an enlargement of the leaf section, and *C* is a single cell of the leaf shown at still higher magnification.

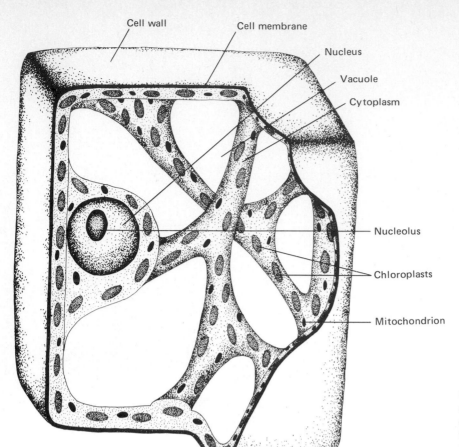

Cell wall

Cell membrane

Nucleus

Vacuole

Cytoplasm

Nucleolus

Chloroplasts

Mitochondrion

Figure 2-2. This diagramatic representation of a plant cell may help you to visualize a cell as a three dimensional structure. Various parts of the cell are labeled and may be compared to photographs later in this chapter.

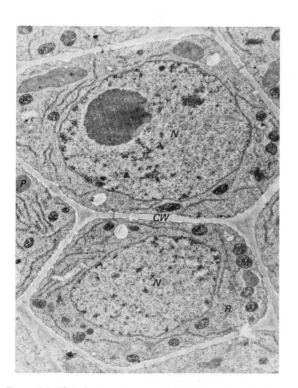

Figure 2-3. This electron micrograph of alfalfa root cells demonstrates several conspicuous cellular structures including cell walls (CW), mitochondria (M), nuclei (N), ribosomes (R), immature plastids (P), and others. (Courtesy of B. A. Tait, Southern Utah State College.)

THE PLANT CELL

All plants are composed of cells. Some are unicellular while others are composed of millions of cells (Fig. 2-1). Several characteristics are shared by all plant cells. These are important to understand in order to provide a foundation for discussing similarities and differences among the various divisions of the plant kingdom.

Plant and animal cells are alike in many ways, but differ in certain significant features. For instance, plant cells generally have an external *cell wall* (Figs. 2-2, 2-3, and 2-4) whereas animal cells lack such a wall. This cell wall may be composed of one or more of many chemical substances. It varies widely among plant divisions, both in composition and structure. The cell walls of most plants are composed largely of the complex carbohydrate, *cellulose*. Cellulose molecules are arranged into long strands known as *microfibrils* which overlap to form the basic structure of the wall (Fig. 2-5). This wall is porous and allows water and most dissolved substances to pass through without restriction, but it is also rigid, imparting a definite three-dimensional shape to the cell. Other substances found in the cell wall of some plants include hemicellulose, suberin, lignin, proteins, chitin, silicon, and pectic compounds.

Both plant and animal cells have a *cell membrane* or *plasmalemma* (Fig. 2-4). This membrane allows free exchange of water between the cell and its external environment, but unlike the cell wall it does not allow the free passage of most dissolved substances. Because of this, the cell membrane is known as a *differentially permeable membrane.* Since most dissolved substances do not pass freely into the cell, energy must be expended by the cell in order to obtain nutrients necessary for cell growth and metabolism. This process is complex and much research is currently underway to determine the exact methods by which it occurs.

Just as any other substance, water always passes from an area where it is in higher concentration to an area where it is in lesser concentration. This process is known as *diffusion.* In the cell, when diffusion occurs through the plasma membrane, the process is referred to as *osmosis.* Since the amount of water inside a cell is generally less than the amount directly outside the cell, water passes into the cell until the cell swells and the plasma membrane is pushed outward against the cell wall. This outward pressure is known as *turgor pressure,* and a cell exhibiting turgor pressure is a *turgid* cell. Under normal conditions plant cells are turgid. However, if the concentration of water inside the cell becomes greater than outside, water leaves the cell, the membrane shrinks away from the wall, and the cell is said to be *plasmolyzed.* Thus, if a carrot root is left dry, it wilts. This occurs since the cells of the root lose water to the atmosphere because the concentration of water within the carrot cells is greater than that of the atmosphere of the room. Similarly, if a water fern or other aquatic plant is removed from its habitat for a period of time, the cells plasmolyze, ultimately causing the death of the organism. Likewise, if a terrestrial plant is uprooted from the soil, the plant loses its source of water and the cells soon plasmolyze and die.

The study of the contents within the plasmalemma of a cell is an exciting and rewarding field. Many early microscopists were aware that plants are composed of cells and that cells contain smaller structures. As the microscope was refined and improved, scientists noted that many cells contain from one to several large conspicuous bodies. Later, it became evident that even a simple cell contained a host of tiny structures of uncertain function.

Early cytologists were also aware of a semiliquid substance within the plasma membrane. This material has been given various names, but is most often known as *groundplasm.* This material is chemically complex and often changes physical states from semisolid to liquid. Numerous cellular components or *organelles* are imbedded in this groundplasm. These organelles together with the plasma membrane and groundplasm of a cell form a *protoplast.* The protoplast is the living, functioning unit of the cell. The term *cytoplasm* is often used to discuss the living portions of a cell exclusive of the nucleus.

The largest and most conspicuous organelle of most cells is the *nucleus* (Figs. 2-2, 2-3, and 2-6) which is re-

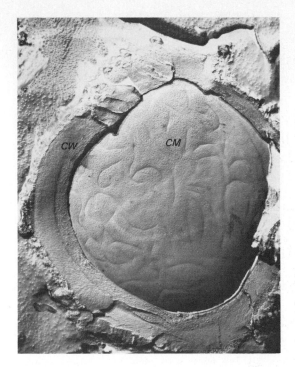

Figure 2-4. This barley smut spore was fractured especially to show the cell membrane (CM) and cell wall (CW). (Electron micrograph courtesy of W. M. Hess, Brigham Young University.)

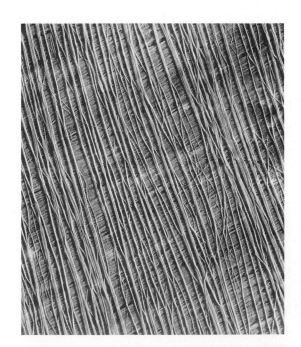

Figure 2-5. The cell walls of many plants are composed of cellulose microfibrils which overlap to create a rigid, permeable structure. The electron micrograph of the cell wall of an alga illustrated here clearly demonstrates the pattern of microfibrils. (Courtesy of R. D. Preston, University of Leeds.)

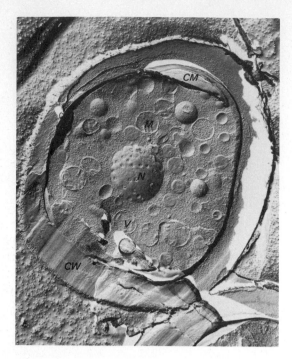

Figure 2-6. This barley smut spore was fractured through the middle of the cell to demonstrate the cell wall (CW) cell membrane (CM), nucleus (N), vacuoles (V), and mitochondria (M). (Electron micrograph courtesy of W. M. Hess, Brigham Young University.)

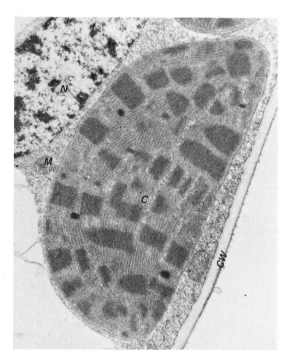

Figure 2-7. This electron micrograph of a portion of a sugar cane leaf cell demonstrates a large prominent chloroplast (C). It also shows the cell wall (CW), a portion of the nucleus (N), and a mitochondrion (M). (Courtesy of W. M. Hess, Brigham Young University.)

sponsible for controlling most of the cellular functions. The nucleus is a complex structure, containing semiliquid *nucleoplasm,* RNA (ribonucleic acid) molecules, protein, and *chromatin material.* This chromatin is also complex and always carries genes composed of DNA (deoxyribonucleic acid). Each gene is slightly different in structure from all other genes and is responsible for the production of a specific molecule of RNA. This specific molecule of RNA then is responsible for the production of a specific enzyme. Enzymes are extremely important in the cell since they are responsible for controlling and regulating all cellular processes. Thus, division, growth, life, and death of the cell are all dependent on enzymatic reactions initiated by genes in the chromatin material of the nucleus.

Plant cells may contain three basic types of *plastids* that are generally absent from animal cells. Plastids are organelles with various functions in the cell and are bound by double membranes. The *chromoplast* is a plastid that contains one to several colored pigments. The most common type of chromoplast is the *chloroplast* (Fig. 2-7), which contains chlorophyll and functions in *photosynthesis.* Chloroplasts vary widely in shape and size. Some are rather small and disk-shaped while others are long and ribbon-like. The configuration of the chloroplast is often useful in classifying plants.

Photosynthesis is the process wherein the sun's energy is used to transform the carbon in carbon dioxide molecules into the carbon of simple sugar molecules. In other words, the radiant energy of the sun is transformed into chemical energy which is needed by the cell to carry out its life processes. Chemically, the equation for this conversion is

$$6CO_2 + 12H_2O \xrightarrow[\text{chlorophyll}]{\text{light}} C_6H_{12}O_6 + 6H_2O + 6O_2$$

Photosynthesis is certainly one of the most important processes occurring on earth. If this energy converting process did not occur, life as we now know it could not exist. It has been estimated that a fantastic 200,000,000,-000 tons of carbon are transferred from carbon dioxide to simple sugars each year by photosynthesis! The simple sugars formed by this process are the basis for all life on earth with very few exceptions, since green plants ultimately are the source of energy for all animals.

Photosynthesis is remarkable for more than the sugars produced. Green plants produce large amounts of oxygen as a byproduct of photosynthesis. Oxygen is used by both plants and animals in their life processes. Without green plants to generate oxygen, life on earth for all but a very small group of chemosynthetic organisms would be impossible. So, from the standpoint of both sugars produced and oxygen liberated, photosynthesis is the most important chemical process occurring on earth.

Chromoplasts also may contain such pigments as brownish *xanthophylls* and yellow or orange *carotenes.* The role of these pigments is apparently twofold. First, they aid in photosynthesis by absorbing light energy and transferring it to chlorophyll. Second, they protect the chlorophyll from oxidation by direct sunlight. If these pigments are present in large amounts, the color of the chlorophyll may be obscured. The kinds and amounts of xanthophylls, carotenes, and chlorophylls differ among plants of the various plant divisions. This is particularly true of the primitive green plants known as *algae.* These variations are reflected in the names of the different divisions of algae, such as *Chlorophyta* or green algae, *Phaeophyta* or brown algae, *Chrysophyta* or golden algae, and so forth.

Many botanists are presently studying chloroplasts and the photosynthetic process. Much exciting new information is becoming known through this research, and perhaps at some future time people will be able to synthesize sugars by photosynthesis in an artificial system outside of a green plant cell.

Two other common noncolored plastids are *amyloplasts,* which function in producing and storing starch grains (Fig. 2-8), and *elioplasts,* which are oil-storage plastids. These plastids may be present or absent in plant cells depending on the species of the plant, location of the cell within the plant, and the specific ecological conditions under which the plant grows.

Mitochondria are present in most plant cells (Figs. 2-3, 2-6, and 2-9). These organelles are bound by a double membrane and are often compartmented into many chambers. Respiration occurs within mitochondria. This vital process is the conversion of the chemical energy stored in food products, such as simple sugars produced by photosynthesis, into a form of chemical energy used by the cell for all energy-requiring functions.

Several other organelles are common in plant cells. These include *Golgi bodies,* which function in cell-wall formation, *lysosomes,* which secrete enzymes to degrade the protoplast at the time of death of the cell, *ribosomes,* which are responsible for protein synthesis, and the *endoplasmic reticulum* (Fig. 2-3), which functions in intracellular communication and segregating chemicals in the protoplasm. Much new information dealing with the structure and function of these organelles is becoming available through electron miscroscopic and biochemical studies. Students interested in further details concerning the cell and its contents may consult any modern cytology book.

The simplest plants are unicellular; that is, their entire plant body is composed of but a single cell. On the other hand, more complex plants are multicellular. Multicellular plants may be composed of cells essentially alike or of widely divergent cell types with different functions. Thus, an algal plant may be a filament of a small number of

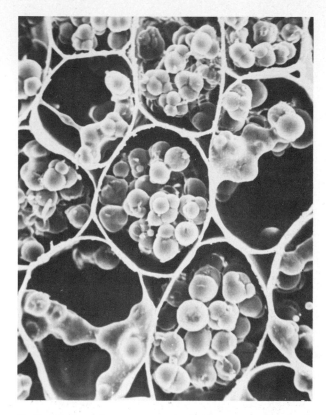

Figure 2-8. Starch is one very common food storage product that is produced by many plants. It is often deposited in amyloplasts and certain storage cells, such as those illustrated in this scanning electron micrograph of buttercup root that has become nearly filled with large starch grains. (Courtesy of E. Parsons and B. Bole, University of Bristol. Used by permission of the *Journal of Microscopy.*)

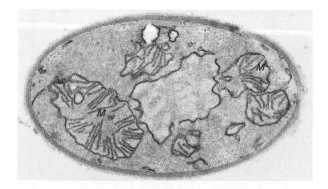

Figure 2-9. Mitochondria are double membrane-bound organelles that function in respiration. This electron micrograph of a yeast cell shows several prominent mitochondria (M). (Courtesy of M. Osumi, Japan Women's University.)

identical cells, whereas a flowering plant may be composed of millions of cells that differ widely in both appearance and function.

MITOSIS AND CYTOKINESIS

Cells divide to form new cells. The division often occurs in two steps. The first is *mitosis* or nuclear division and the second is *cytokinesis* or cytoplasmic division. Mitosis and cytokinesis are the processes of vegetative cell division. All but the most primitive plants and animals grow by these processes. New cells, or *daughter cells,* produced by cell division are essentially alike and are like the *mother cell* that divided to produce them.

Mitosis proceeds by an orderly sequence of steps or phases. *Interphase* is the longest phase of the mitotic

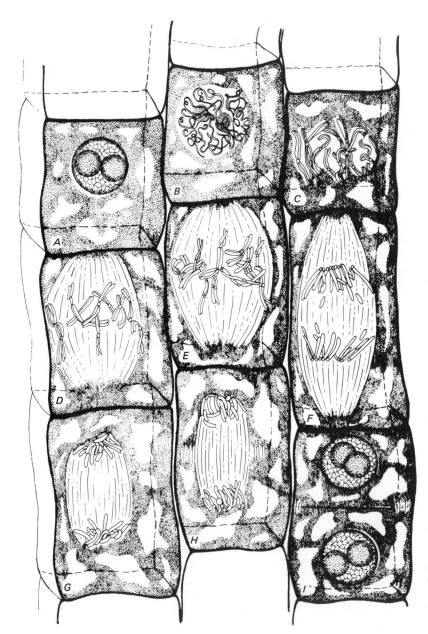

Figure 2-10. This diagram of mitosis in plant cells shows the various stages in this type of division. Interphase is shown at A. B and C are stages of prophase. D and E show the chromosomes aligning at the metaphase plate and F, G, and H show progessive stages of anaphase. Telophase is shown at I. This figure was redrawn from J. D. Dodd, *Form and Function in Plants.* (Used with the permission of Iowa State University Press.)

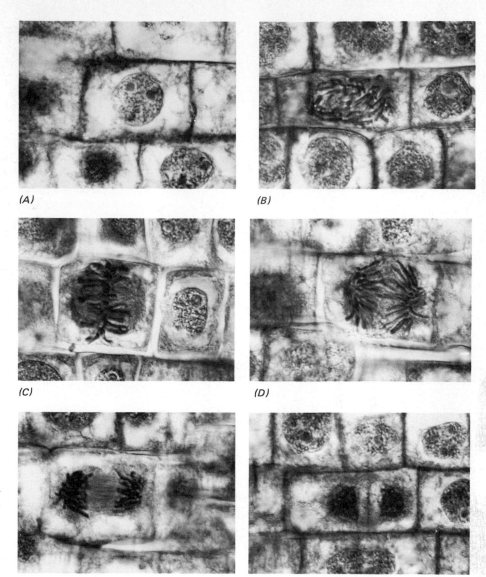

(A)

(B)

(C)

(D)

(E)

(F)

Figure 2-11. This sequence of photographs demonstrates mitosis in the root cells of onion. The plant was killed and the cells were stained prior to obtaining these photos. *A* shows a cell with its nucleus in interphase. *B* shows prophase and *C* is metaphase. Two stages in anaphase are shown at *D* and *E* and telophase is illustrated at *F*.

process (Fig. 2-10*A*). Distinct, separate chromosomes are not evident during interphase. The nucleus is in its typical membrane-bound, dark-staining state, and it carries on all of its normal functions of directing the activities of the cell. The chromatin (elongated, uncoiled chromosomes) exactly duplicates itself during this phase.

Prophase (Figs. 2-10*B*, 2-10*C*, 2-11*B*, and 2-12*A*) is begun when the chromatin material becomes arranged into tightly coiled, orderly structures known as chromosomes. As prophase proceeds, the nuclear membrane is lost. Since the chromatin material is replicated during interphase, each prophase chromosome is composed of two equal halves known as *chromatids*.

During *metaphase* the chromosomes become aligned near the center of the cell on the *metaphase* or *equatorial plate* (Figs. 2-10*D*, 2-10*E*, 2-11*C*, and 2-12*B*). When a cell in metaphase is examined in side view, the chromosomes appear to be aligned along a line through the center of the cell. If the same cell is examined from top or polar view, the chromosomes often are arranged into a ring-like configuration.

Chromatids are separated from each other during *anaphase* and move apart rather rapidly (Figs. 2-10*F*, 2-10*G*, 2-10*H*, 2-11*D*, 2-11*E*, 2-12*C*, 2-12*D*, and 2-12*E*). Each chromatid is now considered to be a daughter chromosome. During the last stage or *telophase*, the daughter chromosomes become aggregated at opposite ends of the cell (Figs. 2-10*I*, 2-11*F*, and 2-12*F*) and a nuclear membrane develops around each new nucleus. Cytokinesis follows when a new cell wall develops separating the newly-formed nuclei. Figures 2-11*F* and 2-12*F* show early stages in the development of this wall.

The phase terminology commonly used to describe mitosis creates a tendency to think of this process as a series of discontinuous, mildly-related stages where the chromosomes "jump" from one phase to another. This concept is inaccurate since mitosis is a very-smooth, continuous process. Scientists have created phase terminology merely to better describe this phenomenon.

Interphase is the longest of the phases. Its duration depends on the species and the specific, local environment of the cell, but it normally lasts several times as long as

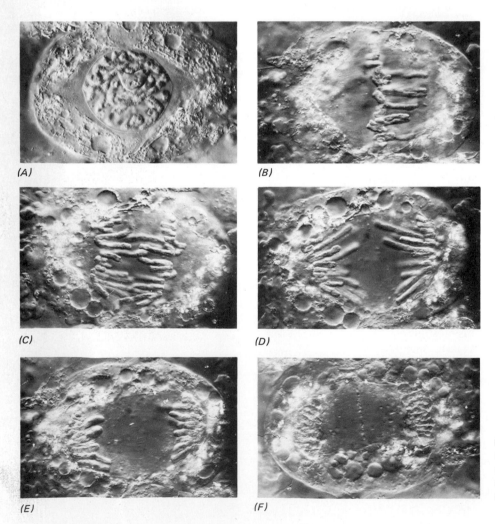

(A)

(B)

(C)

(D)

(E)

(F)

Figure 2-12. These photographs were taken with Nomarski microscopy and show living cells undergoing mitosis. Prophase is shown at A; metaphase at B; three successive stages of anaphase at C–E; and telophase at F. (Courtesy of A. Bajer, University of Oregon.)

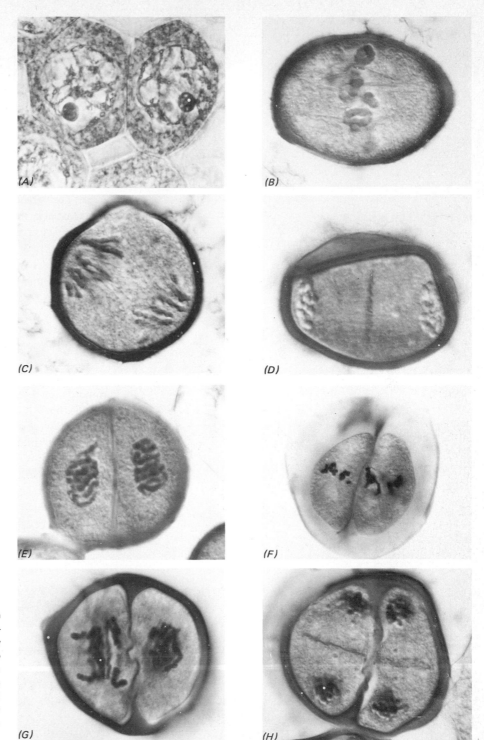

Figure 2-13. The meiotic division process (reduction division) is illustrated in these photographs of developing lily pollen grains. *A* shows two cells in early prophase I. *B* shows the chromosomes aligned at the metaphase plate in metaphase I. Anaphase I is shown at *C* and *D* depicts telophase I. Prophase II may be seen at *E* and metaphase II at *F*. *G* shows anaphase II and *H* depicts telophase II. Note that four daughter cells are produced by meiosis rather than two as in mitosis.

the other phases combined. Once prophase begins, it takes from about 15 min to 7 hr to complete nuclear division.

MEIOSIS

Certain specialized cells of many plants and animals divide by a second type of division known as *meiosis*. This process is often known as reduction division since the number of chromosomes in each daughter cell produced by meiosis is only one half as many as that of the mother cell. Meiosis occurs in *diploid* cells. A diploid cell contains two complete sets of chromosomes and each chromosome has a duplicate or *homolog*. A diploid cell is often noted as a 2N cell. A cell with only one set of chromosomes is *haploid*. This condition often is abbreviated as the 1N nuclear state. In other words, N is an algebraic symbol that may represent any number. If a cell of a haploid fern plant contained 50 chromosomes ($N = 50$), a diploid cell of that plant would contain 100 chromosomes ($2N = 100$).

The actual process of meiotic division is similar in many ways to mitosis. It proceeds by an orderly sequence of phases similar to those of mitosis. However, the nucleus passes through two sets of phases rather than one. Thus, instead of a single prophase, a nucleus dividing meiotically passes through *prophase I* and *prophase II*. Prior to *prophase I,* the chromatin material duplicates itself just as in mitosis. Thus, each chromosome at meiosis is likewise composed of two chromatids.

The greatest difference between mitosis and meiosis occurs in *prophase I* (Fig. 2-13*A*). In meiosis each chromosome pairs with its homolog at this stage. In other words, each chromosome in the cell has an exact duplicate, and these duplicate chromosomes are attracted to each other and become paired. This pairing up process never occurs in mitosis since the chromosomes remain completely independent.

At meiotic *metaphase I* (Fig. 2-13*B*) the homologous chromosome pairs become aligned on the metaphase plate. Each of these chromosome pairs contains four chromatids.

During *anaphase I* (Fig. 2-13*C*), entire chromosomes of the homologous pairs separate and move to opposite poles of the cell. Each such chromosome is still composed of two chromatids, and each new cell has only one half as many chromosomes as the mother cell.

Telophase I (Fig. 2-13*D*) occurs with the characteristic "disappearance" of chromosomes through uncoiling, and often with the accompanying formation of a new wall by cytokinesis.

The daughter cells or nuclei may or may not enter *interphase II*. If they do, it is often for a short time, and chromatin material is never duplicated during this phase.

Prophase II (Fig. 2-13*E*) is identified when the chromosomes recoil and again become evident. This phase is followed rapidly by *metaphase II* (Fig. 2-13*F*) when the separate chromosomes of each daughter cell align themselves on the metaphase plate. This process is generally synchronous in the two daughter cells.

The chromatids of each chromosome separate during *anaphase II* and move toward the poles of their cells. Thus, this second anaphase is essentially like the anaphase of mitosis. *Telophase II* (Fig. 2-13*H*) follows this anaphase. It also resembles a mitotic telophase since the chromatids soon lose their evident identity and a nuclear membrane develops around each newly formed nucleus. Cytokinesis follows and four new haploid daughter cells result.

In summary, a few important differences exist between mitosis and meiosis (Table 2-1). First, either a haploid or diploid nucleus may divide by mitosis, whereas only a diploid (or polyploid) nucleus may divide by meiosis. Second, chromosomes remain independent (unpaired) throughout mitosis, while homologous chromosomes pair during *prophase I* of meiosis. In turn, chromosomes are aligned on the metaphase plate unpaired in mitosis, whereas pairs of homologous chromosomes are aligned in meiosis. Third, mitosis results in the production of two daughter cells, each with the same chromosome number as the mother cell, while meiosis results in four daughter cells, each with half as many chromosomes as the mother cell.

REPRODUCTION OF PLANTS

Mitosis and meiosis are both methods of cellular reproduction. Plants are able to reproduce as organisms because their cells divide. Plant reproduction occurs by a wide variety of methods. One method widely used by simple plants is *fragmentation*. Reproduction by fragmentation will occur if plant parts are separated somehow and the fragments begin to grow independently. This simple type of reproduction based on cell division is the only method used by some plants. Many of these plants have developed elaborate methods to aid fragmentation.

Other plants produce specialized cells or structures that are released and grow to produce a new plant. Reproduction of this type without the production or union of gametes is *asexual reproduction*. All cells or structures functioning in asexual reproduction are produced by mitosis.

TABLE 2-1. A Comparison Between Mitosis and Meiosis.

	Mitosis	*Meiosis*
Number of daughter cells produced	2	4
Chromosome number of daughter cell nuclei	Mother Cell Daughter Cell $1N \longrightarrow 1N$ $2N \longrightarrow 2N$	Mother Cell Daughter Cell $2N \longrightarrow 1N$
Events at *Prophase I*	Chromosomes remain unpaired	Homologous chromosomes pair
Events at *Anaphase I*	Chromatids separate	Homologous chromosomes separate
Events at *Anaphase II*	Does not occur	Chromatids separate

Most plants also reproduce by *sexual reproduction*. This process is similar to reproduction in animals. *Gametes* are produced by sexually-reproducing plants, and two gametes unite by *syngamy* (fertilization) to form a *zygote*. Gametes are haploid in chromosome number, and the zygote is diploid.

Sexual reproduction is an extremely significant biological process. The offspring produced by this process have genes from both parent plants in new combinations. This process of genetic recombination may produce highly-fit progeny that are actually more successful than either parent plant. This is one important basis for plant evolution. People use this principle continually for developing new types of plants and improving already existing species.

THE LIFE CYCLE OF PLANTS

All species of plants and animals exhibit a *life cycle* or *life history*. This cycle is the pattern of events that leads from a mature individual organism of one generation to a new individual organism of the next. Thus, the life cycle of humans always contains meiosis and syngamy, and could be diagrammed as in the chart shown on the left.

Note from this life cycle diagram that a mature *Homo sapiens* is 2N in chromosome number and produces haploid gametes directly by meiosis. This life cycle is typical of most higher animals.

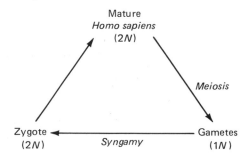

The life cycle of most plants differs considerably from that of most animals. One fundamental difference is that the diploid plant of most species does not produce gametes directly. It produces *meiospores* by meiosis rather than gametes. A meiospore germinates and grows to produce a haploid *gametophyte*. This haploid plant is a separate and distinct plant which functions in producing gametes. The gametes produced by a gametophyte unite to form a zygote that germinates to produce the meiospore producing plant or *sporophyte*. Thus, the generalized life cycle of a typical sexually-reproducing plant may be illustrated in the following diagram.

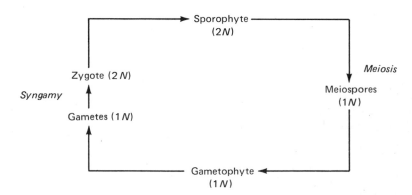

It is important to understand this generalized life cycle. With few exceptions, all the life cycles to be studied in later chapters will conform broadly to this one. Bear in mind

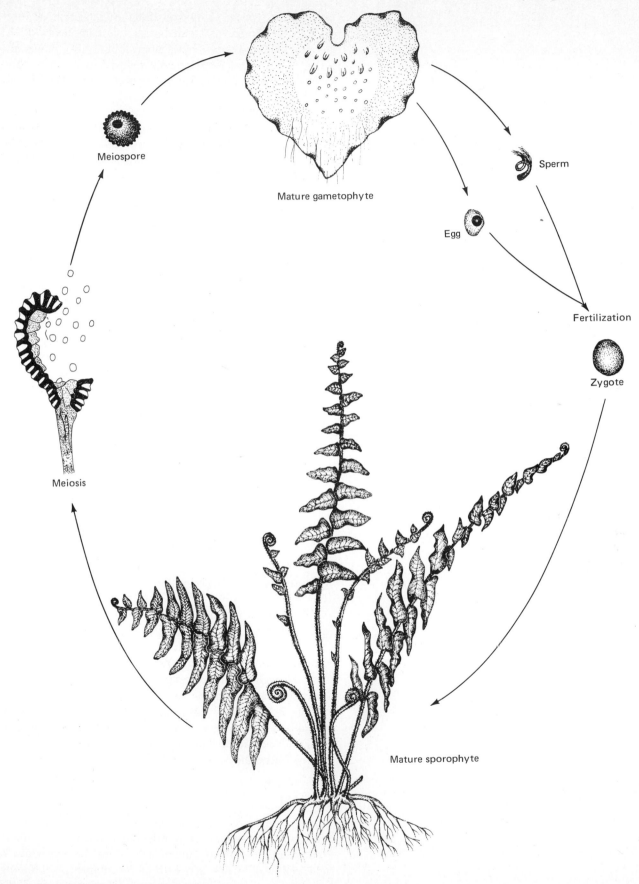

Meiospore

Mature gametophyte

Sperm

Egg

Fertilization

Meiosis

Zygote

Mature sporophyte

Figure 2-14. Simplified life cycle diagram of a fern.

that various species of plants differ in the form or morphology, the longevity, and the relative complexity of the gametophyte and sporophyte generations. Also, plants differ in the kinds of gametes and meiospores produced and in the specific details of sexual reproduction. The various ways in which these generations differ, and the ways by which they are produced will be studied in later chapters.

Since the life cycle of most plants exhibits two distinct plants or generations with different functions alternating with each other, plants are said to demonstrate an *alternation of generations*. This is an important concept that needs to be well understood.

Most people are familiar with ferns. These plants occur rather commonly, and are especially prevalent in moist climates. The large green plant common on the forest floor and often grown as a house plant is the sporophyte. Many people are not familiar with the fern gametophyte which is a small, inconspicuous plant. This plant is often about the size of your fingernail or smaller, and functions in producing gametes. Two gametes unite to produce a zygote as in all sexually-reproducing plants. The zygote then divides by mitosis and cytokinesis to develop into the mature sporophyte plant. Thus, the life cycle of a fern fits the generalized life cycle of a sexually-reproducing plant as follows:

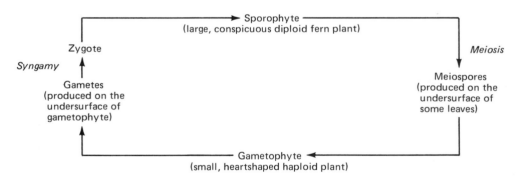

Throughout this book, you will study how plants differ from each other, how these differences evolved, and how they are used in classification. Since you will study and compare many life cycle diagrams, it is important to understand the cyclic nature of sexual reproduction. A simplified life cycle diagram of the fern discussed above is outlined in Fig. 2-14. Study this diagram, compare it with the generalized life cycle of a sexually-reproducing plant, and be certain you understand the important concept of alternation of generations.

SUMMARY

All plants are composed of cells, although some are unicellular and others are multicellular and very complex. Plant cells differ from animal cells since most produce a rigid cell wall. Cells are extremely complex, containing many organelles including the nucleus, mitochondria, and plastids. Photosynthesis occurs in the chloroplasts of plant cells. If this process did not occur, life as we know it on earth could not exist.

Plant cells divide. When the division produces two daughter cells, each with the same chromosome number as the mother cell, the cell has divided by mitosis. When the result of division is four daughter cells, each with only one half as many chromosomes as the mother cell, reduction division or meiosis has occurred.

Plants have the capacity to reproduce. This reproduction is asexual if no union of gametes occurs, or sexual if fertilization and reduction division are involved. Many plants possess both types of reproduction.

Two distinct generations alternate with each other in the life cycle of most sexually-reproducing plants. The haploid gametophyte produces gametes and the diploid sporophyte produces meiospores. The form, longevity, and complexity of these generations are significant in the classification of plants.

FURTHER READING

BERRILL, NORMAN, *Sex and the Nature of Things,* Apollo Books, New York, 1953.

BONNER, JOHN T., *Morphogenesis: An Essay on Development,* Athenum Books, New York, 1963.

BRACHET, JEAN, "The Living Cell," *Scientific American* 205(3):50–62, 1961.

DeROBERTIS, E. D., W. W. NOWINSKI and F. A. SAEZ, *Cell Biology,* Fifth edition, W. B. Saunders Co., Philadelphia, 1970.

ESAU, KATHERINE, *Plant Anatomy,* Second edition, John Wiley and Sons, New York, 1965.

FAHN, A., *Plant Anatomy,* Second edition, Pergamon Press, New York, 1974.

HARTMAN, PHILIP and S. SUSKIND, *Gene Action,* Prentice-Hall, Inc., Englewood Cliffs, New Jersey, 1965.

LEVINE, R. P., "The Mechanism of Photosynthesis," *Scientific American,* 221(6):58–70, 1969.

MAZIA, DANIEL, "How Cells Divide," *Scientific American,* 205(3):100–120, 1961.

MAZIA, DANIEL, "The Cell Cycle," *Scientific American,* 230(3):55–64, 1974.

MIRSKY, ALFRED, "The Discovery of DNA," *Scientific American,* 218(6):78–88, 1968.

3 Evolution and Classification of Plants

New plant species are constantly evolving. The fruit (top) and leaf shown in the center of this photo were taken from a *Purshia glandulosa* plant which evolved from crosses between *Cowania stansburiana* (left) and *Purshia tridentata* (right). (Courtesy of Howard Stutz, Brigham Young University.)

If we were to examine representatives of all plant species presently occurring on earth, we would be impressed immediately with their tremendous diversity. Some are extremely large and complex, with roots, stems, leaves, and flowers. Others are very small and inconspicuous, requiring the microscope for their examination. Many of the smallest plants are composed of but a single cell. The study of the origin and development of this great diversity is one of the most exciting and difficult fields in all biology.

In general, small, simple plants are primitive forms, and larger plants that are more complex are advanced forms. It is thought that in the past, primitive forms gave rise to the more complex, advanced forms through a series of changes. This process of species change through time is known as *evolution*. Even now plants and animals are constantly changing or evolving. There are several ways by which plant or animal species become modified. One of the foremost methods is known as *natural selection*.

Our present knowledge of natural selection rests on the foundation work of two famous, highly observant nineteenth century naturalists, Charles Darwin and Alfred R. Wallace. Briefly, natural selection is the process whereby unfit organisms are eliminated from populations through time since the only ones to reproduce are those that do well in their environment. Weak or otherwise unfit organisms often die before they reproduce, or else they produce inviable offspring. The result of this process is the natural improvement of species so that they become better able to fit their environment. This occurs because the genes of weak individuals are eliminated from the pool of genes available to the species.

Sexual reproduction in combination with natural selection is very important in evolution. It is by sexual reproduction that new, untried combinations of genes arise in the offspring of plants and animals. These new combinations may produce individuals weaker than the parents and thus be eliminated by natural selection. However, often they produce individuals that are in some way superior to the parents. These new individuals are then selected for by natural selection.

To illustrate this process, suppose a natural population of plants occurs in a mountain range at the edge of a desert. Further suppose that by sexual reproduction new combinations of genes are produced and a few of the offspring are able to live at slightly higher temperatures than most plants of the population. Assume that at this time, local weather patterns begin to change slightly, causing both the average annual temperature and the yearly temperature extremes to increase. As this occurs, the plants that cannot tolerate the higher temperatures will not grow well and hence will not reproduce or will reproduce only sporadically. However, the newly produced plants that can tolerate higher temperature will reproduce and flourish. The result is that those plants able to withstand the higher temperatures will become more common, and the plants not able to withstand the higher temperatures will be eliminated. Thus, by the weeding out of unfit individuals through natural selection, the population as a whole gradually will become able to exist in the higher temperatures of the changing environment.

The plants of the newly-modified population may have much the same appearance as those prior to the environmental changes, or they may appear substantially different. Suppose, for instance, that the plants able to tolerate the higher temperatures also have slightly larger flowers of a different color than the heat-sensitive plants. This difference also will be assimilated into the new population. Thus, by natural selection the plants of this new population will appear significantly different from their ancestral counterparts.

Natural selection, together with other evolutionary processes, occurs constantly in all populations, and largely accounts for the great diversity of plants and animals on earth. One of the major attempts of this book is to point out several evolutionary trends in the plant kingdom, and to correlate these with the natural classification of plants.

CLASSIFICATION OF PLANTS

Since such a large number of diverse plants occurs on earth, it has been necessary to develop a method for classifying them. Many such schemes were developed and used throughout early botanical history. However, early classification systems were at best clumsy and were often completely unusable. Such systems were often based on characteristics of external plant parts such as leaf size, stem shape, flower color, height of the plant and so forth. The habitat of the plant was even occasionally used for classification. Most of these features are extremely variable and change according to the environment. For instance, even leaves on the same plant may differ greatly in size and shape since those developing in the shade are often much larger. Another problem with these early classification systems is that only economically-significant plants were considered important enough to classify. Little thought was given to other plants of the plant kingdom. These systems were usually very confusing, and plants that were obviously not related were often placed in the same or closely-related category. Such systems are said to be *artificial*. To compound these problems, no universal name for a plant was used. Thus one botanist often gave a plant one name while another gave the same plant a completely different name. Great difficulty arose when trying to communicate the names of such plants.

Considering these problems, Carolus Linnaeus developed a functional system of plant classification for at least two critical reasons. First, he was concerned with developing a natural system of classification where only closely related plants were placed in the same genus. Second, Linnaeus proposed and used a binomial system of nomenclature where each plant was given a genus and species name. This nomenclature was to be universally accepted.

After studying hundreds of different plants, Linnaeus noted that their reproductive structures tend to remain constant even under differing ecological conditions. Also the reproductive structures of closely-related plants are very similar and so tend to show natural relationships. Thus, much of the classification system developed and made popular by Linnaeus is based on the reproductive structures of plants. This method of classification has proved to be very satisfactory. It is still used today although many modifications and improvements of the original systems used by Linnaeus have been made.

Linnaeus was also concerned that a single plant be known by one name, rather than by a variety of names. Thus, he proposed that every plant be given a formal Latinized genus and species name. Later botanists formalized the suggestions of Linnaeus. When a new plant is discovered, its discoverer or an authority names the plant and describes it in Latin in a publication which is readily available to botanists. This then is the name by which all other botanists in the world recognize that plant. This method of naming and describing plants has been very useful in the development of plant classification. It allows botanists throughout the world to communicate without becoming confused as to what plant is being discussed.

The discovery of the microscope was a significant step in the development of plant classification. Biologists began to observe and study plants that were unknown prior to that time. For instance, in a single drop of pond water many small green plants could be observed. Later, many non-green microscopic plants were observed. All of these plants were subsequently studied, named, and described according to the system developed by Linnaeus.

Several methods of arranging similar plants together in classification systems have been tried in the past. It was common for early botanists to recognize three distinct plant classes: herbs, shrubs and trees. Later botanists found this system unsatisfactory because a plant often is a shrub under certain environmental conditions and a tree under others.

Today, plants are placed into broad groups based on several criteria. These criteria include pigments in the cells, morphology of the reproductive structures, the presence or absence of an embryo and conductive tissue, and so forth. Most botanists now place similar plants together into broad groups known as *divisions*. Thus, the division Cyanophyta contains plants commonly known as blue-green algae. These plants all contain characteristic blue and green pigments in their cells. Likewise they lack roots, stems, and leaves as well as organized nuclei and plastids. In addition these plants have a multilayered cell

Division: Anthphyta
Class: Dicotyledonae
Order: Ranales
Family: Ranunculaceae
Genus: *Ranunculus*
Species: *occidentalis*

Figure 3-1. Linnaeus made important contributions to the science of plant classification. In particular, he suggested that all plants should have both a genus and species name. Also plants can be placed into higher levels of classification. Thus, similar genera may be placed into a family, similar families may be placed into an order, etc. This concept is illustrated for this wild buttercup.

Division: Chlorophyta
Class: Chlorophyceae
Order: Zygnematales
Family: Zygnemataceae
Genus: *Spirogyra*
Species: *insignis*

Figure 3-2. The classification scheme illustrated for a flowering plant in Figure 3-1 also applies to primitive plants such as the alga shown in this figure.

wall and they do not reproduce sexually. No other group of plants in the entire plant kingdom exhibits all of these characteristics. Each division of the plant kingdom likewise exhibits a distinctive set of characteristics which separates it from all other divisions.

It is often possible to place plants into other broad classification categories within divisions. For instance, four such categories or *classes* are recognized with the division Chrysophyta. Classes are composed of one or more plant *orders* which in turn contain closely-related *families* of plants. Each family is composed of closely allied genera which in turn are made up of related species.

As an example of this classification scheme, suppose a small wildflower has been collected in a damp meadow during a mountain hike. On consulting a reference book we find that this plant has been given the Latinized name *Ranunculus occidentalis.* The generic name for this plant is *Ranunculus,* and the name for this particular species is *occidentalis. Ranunculus* and several other genera are placed together into the family Ranunculacae which is placed in turn into the order Ranales. The order Ranales is placed into the class Dicotyledonae of the division Anthophyta (Figs. 3-1 and 3-2).

Even though botanists have made great progress in plant classification during the last 200 years, no mutual agreement yet exists on the position of plants in classification schemes above the level of genus. Particular disagreement exists in the case of classes and divisions. Some botanists prefer to recognize only a few divisions, each containing several to many classes. Others recognize several divisions containing fewer classes. Almost all presently recognize more plant divisions than the few used in the past. Nineteen divisions of plants are recognized and discussed in this book. These divisions are briefly characterized in Table 3-1, and a representative of each is illustrated in Fig. 3-3.

Accurate estimates of the number of plant species contained in the various divisions are difficult to make. In fact, many botanists feel that perhaps only about one half to two thirds of the plants occurring on earth have been presently discovered and named. Some estimate that ultimately more than 500,000 species of plants will be described. Currently described plants include around 37,000 species of algae, 80,000 species of fungi, 20,000 species of mosses and liverworts, 17,000 species of "lower" vascular plants such as ferns, cycads, and conifers, and over 250,000 species of flowering plants.

EVOLUTIONARY TENDENCIES AMONG PLANTS

In general terms, *thallophytes* (algae and fungi) are considered to be "lower" or primitive plants, while embryo-producing plants are considered to be "higher" or more

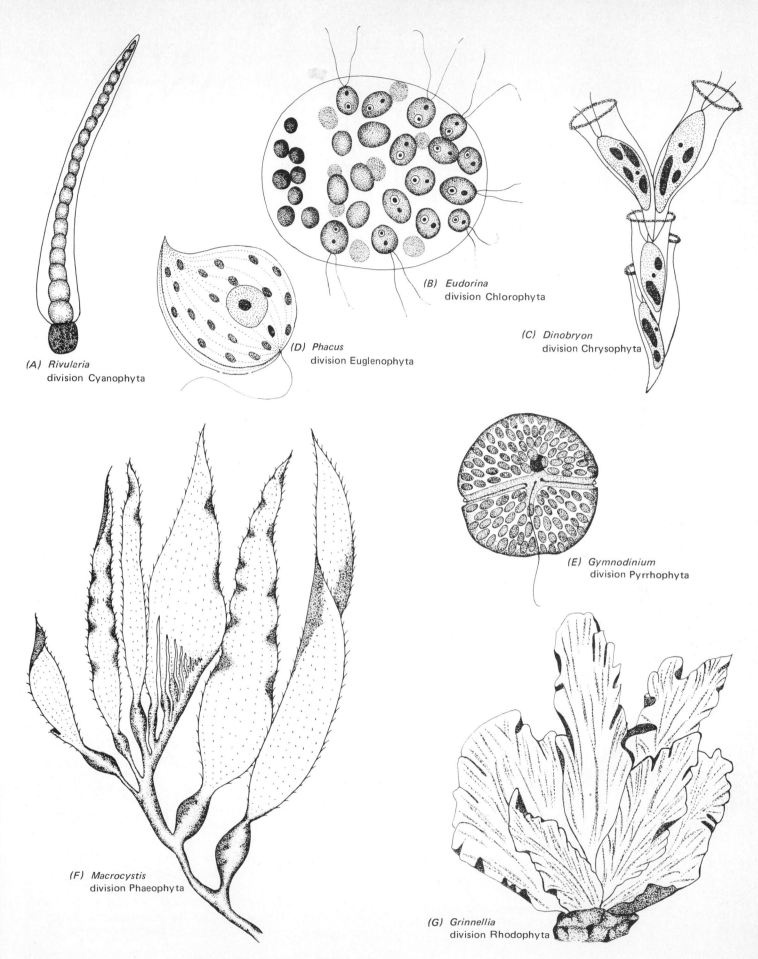

(A) *Rivularia*
division Cyanophyta

(B) *Eudorina*
division Chlorophyta

(C) *Dinobryon*
division Chrysophyta

(D) *Phacus*
division Euglenophyta

(E) *Gymnodinium*
division Pyrrhophyta

(F) *Macrocystis*
division Phaeophyta

(G) *Grinnellia*
division Rhodophyta

Figure 3-3. Illustration of a representative plant from each division of the plant kingdom.

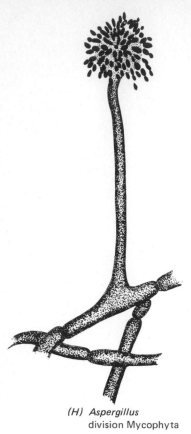

(H) *Aspergillus*
division Mycophyta

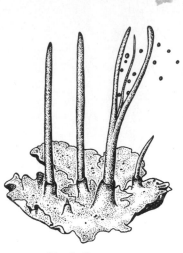

(I) *Anthoceros*
division Bryophyta

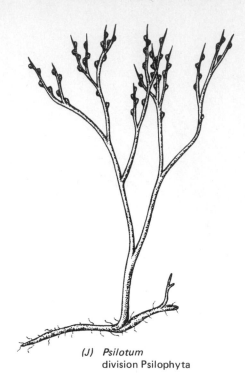

(J) *Psilotum*
division Psilophyta

(K) *Lycopodium*
division Lycophyta

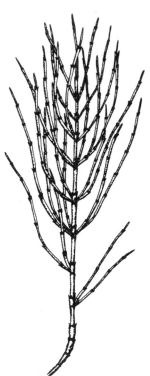

(L) *Equisetum*
division Equisetophyta

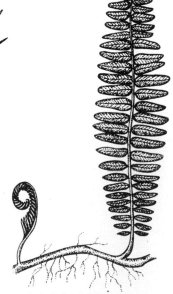

(M) *Polypodium*
division Filicophyta

Figure 3-3. (continued)

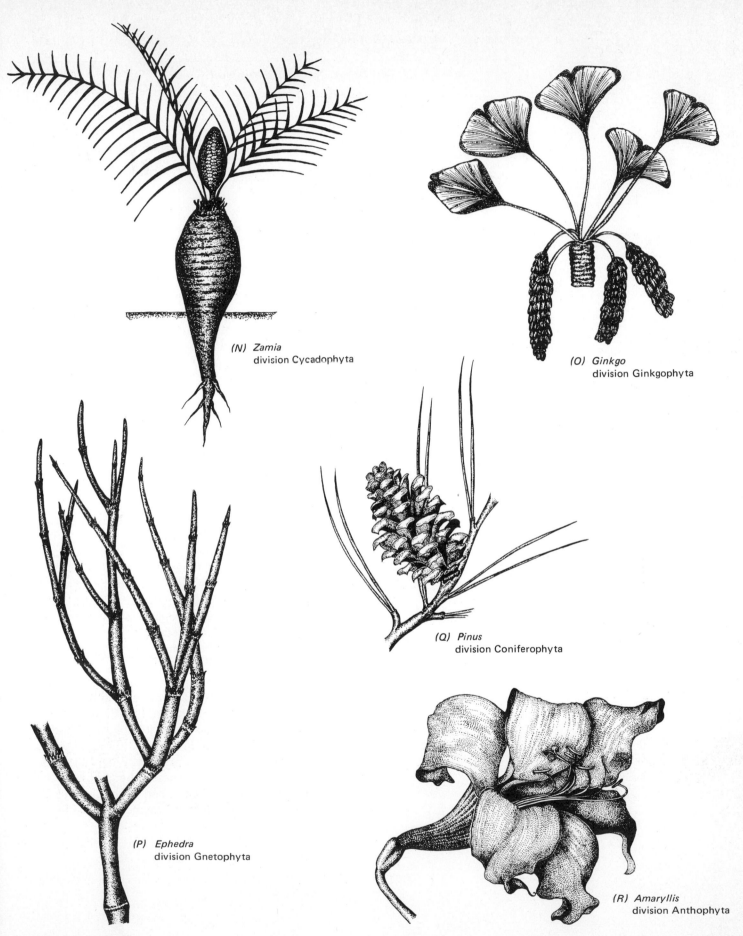

(N) *Zamia*
division Cycadophyta

(O) *Ginkgo*
division Ginkgophyta

(Q) *Pinus*
division Coniferophyta

(P) *Ephedra*
division Gnetophyta

(R) *Amaryllis*
division Anthophyta

Figure 3-3. (continued)

TABLE 3-1.   The Divisions of Plants in the Plant Kingdom.

THALLOPHYTES: plants lacking roots, stems and leaves; the zygote does not develop into an embryo	SCHIZOPHYTA — the blue-green algae CYANOPHYTA — the green algae CHRYSOPHYTA — the golden algae EUGLENOPHYTA — the green flagellates PYRRHOPHYTA — the dinoflagellates PHAEOPHYTA — the brown algae RHODOPHYTA — the red algae MYCOPHYTA — the fungi	**NONVASCULAR PLANTS:** plants lacking specialized conductive tissue
EMBRYOPHYTES: plants nearly always with roots, stems and leaves; the zygote develops into an embryo	BRYOPHYTA — the mosses and liverworts	
	PSILOPHYTA — the psilophytes LYCOPHYTA — the lycopods EQUISETOPHYTA — *Equisetum* FILICOPHYTA — the ferns GINKGOPHYTA — *Ginkgo* GNETOPHYTA — *Ephedra* CYCADOPHYTA — the cycads CONIFEROPHYTA — the conifers ANTHOPHYTA — the flowering plants	**VASCULAR PLANTS:** plants containing specialized tissue for transporting water, nutrients and photosynthetic products

advanced. Some lower plants are thought to be the ancestors of higher plants through evolution.

With this brief introduction, several evolutionary trends observed in the plant kingdom may be discussed. First, lower plants generally use more of their tissues in reproduction. In many algae and fungi, it is common for each cell of the entire plant to enter into reproduction. Second, the gametophyte generation of a lower plant is usually the predominant plant while the sporophyte is usually greatly reduced and is often represented only by the zygote. This trend is reversed among higher plants since the sporophyte plant predominates and the gametophyte is highly reduced. Third, the complexity of lower plants is almost always far less than that of higher plants. Lower plants never produce organs such as roots, stems, or leaves. Fourth, most lower plants are restricted to aquatic or moist habitats or at least require water for sexual reproduction. Most higher plants are terrestrial, or at least are not directly dependent on water for sexual reproduction. These general trends and many specific evolutionary trends and pathways will be discussed in more detail in various chapters of this book.

FURTHER READING

BAILEY, L. S., *How Plants Get Their Names,* Dover Publications, New York, 1963.

BANKS, H., *Evolution and Plants of the Past,* Wadsworth Publishing Co., Belmont, California, 1970.

BECKER, HERMAN, "Flowers, Insects and Evolution," *Natural History,* 74(2):38–45, 1965.

BENSON, L., *Plant Classification,* D. C. Heath and Company, Lexington, Mass., 1957.

EHRLICH, PAUL and PETER RAVEN, "Butterflies and Plants," *Scientific American,* 216(6):104–113, 1967.

ORGEL, L. E., *The Origins of Life: Molecules and Natural Selection,* John Wiley and Sons, New York, 1973.

STEBBINS, G. L., *Processes of Organic Evolution,* Second edition, Prentice-Hall, Inc., Englewood Cliffs, New Jersey, 1971.

Division Schizophyta: the Bacteria

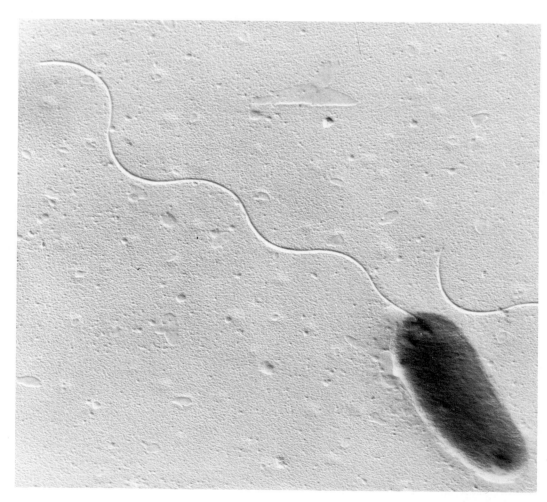

An individual cell of the bacterium *Pseudomonas* sp. clearly showing a flagellum. (Electron micrograph courtesy of James V. Allen, Brigham Young University.)

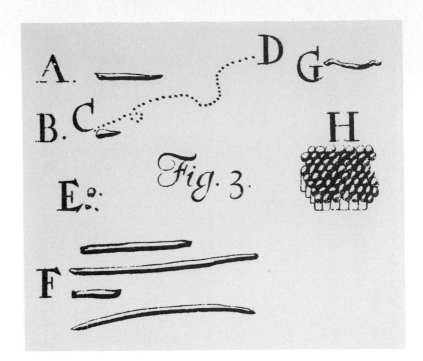

Figure 4-1. van Leeuwenhoek designed and built the earliest microscopes and used them in observing microscopic life. These drawings of bacteria by van Leeuwenhoek were published in 1864 and are the first drawings of bacteria ever published.

Anton van Leeuwenhoek, the famous Dutch microscope builder of the middle 1600's, was the first scientist to examine and describe bacteria. He discovered these tiny "animalcules," as he called them, by accident while examining a pepper water preparation for the "sharp hooks" on pepper which he supposed caused the mouth to sting. Leeuwenhoek later found bacteria in preparations made by scraping his cheeks and teeth. He made the first drawings of bacteria and sent them in a letter to the Royal Society of London (Fig. 4-1). From such humble beginnings, the study of bacteria has become one of the most significant of all scientific endeavors because of the extreme direct importance of bacteria to man.

GENERAL CHARACTERISTICS

The bacterial cell is rather simple in construction when compared to that of the higher plants and animals. It is bordered on the outside by a definite cell wall that is rigid and therefore, responsible for the shape of the cell (Fig. 4-2). This wall is unusual in its chemical construction, and certain of the compounds comprising the wall are found nowhere else in the plant kingdom. All bacterial walls contain modified sugar molecules and the basic molecular units of proteins known as *amino acids*. In addition, several other chemical substances such as fats and proteins may be present according to the species.

Cell walls of bacteria are constructed according to two different patterns. The first type is composed of tightly interconnected molecules that form a very rigid and strong three-dimensional wall. The walls of such bacteria allow the cell to react with a stain developed in 1884 by the Danish physician Hans Christian Gram and are thus known

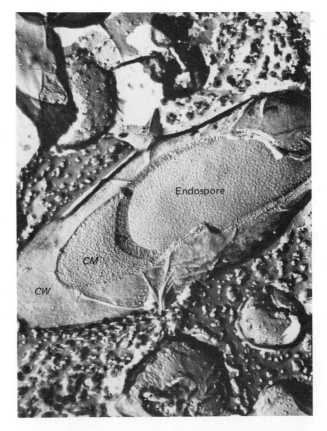

Figure 4-2. This electron micrograph of the bacterium *Bacillus stearothermophilus* shows several levels within the cell since it was fractured prior to photographing. Thus, the cell wall (CW), cell membrane (CM), and cytoplasm (CP) are all evident. In addition, the layer exposed near the center of this cell is the cell wall of a spore which is forming within. Such internally produced bacterial spores are known as endospores. (Courtesy of Balzers Artiengesellschaft.)

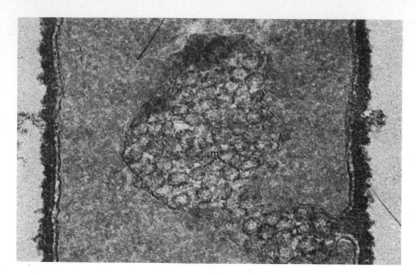

Figure 4-3. This electron micrograph of a cell of *Bacillus* clearly shows a mesosome (m) attached to the cell membrane. (Courtesy of W. van Iterson.)

as *gram-positive bacteria*. The second major wall-type is more fragile and contains an extra protein layer lacking in gram-positive bacteria. These walls do not allow the cell to react with the stain developed by Gram and are therefore *gram-negative bacteria*. The wall formation of bacteria is interesting because several of the most effective recent antibiotics attack the bacterial wall. Penicillin, for instance, interferes with the synthesis of a particular cell-wall component that is present in much higher concentration in gram-positive bacteria. Thus, penicillin is much more effective in controlling gram-positive bacteria and other antibiotics are required to control gram-negative species.

In addition to the cell wall proper, many bacterial cells are surrounded by a sheath of gelatinous material. This sheath is rather variable in both composition and thickness, and it is occasionally useful in differentiating among bacterial species. The value of this sheath to the bacterium is speculative, although some bacteriologists have suggested it may be a protective mechanism against viral pathogens and environmental changes.

As expected, the protoplast of the cell is bordered by a cell membrane (Fig. 4-2). The bacterial membrane has the same chemical and physical properties as the membrane of higher plants and animals. It is semipermeable, thereby allowing the free passage of water. In addition, the membrane actively takes up and releases chemical substances needed or discarded by the cell.

Bacterial cells lack membrance-bound organelles. However, the cell membrane often forms complex infolded membrane bundles. These are known as *mesosomes* (Fig. 4-3) and are thought to function in the synthesis of materials needed for the cell including wall materials and DNA. Since nearly all chemical reactions occurring within a cell must take place on a membrane, the bacterial cell membrane has many of the same functions as a mitochondrion of a higher cell.

Cytoplasm of a bacterial cell often appears granulate under the electron microscope (Fig. 4-4). Even so, it is

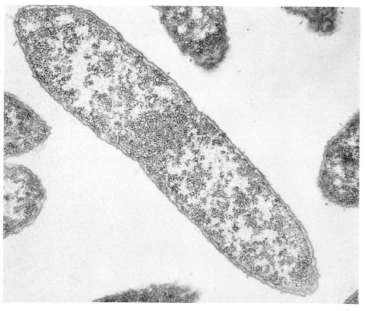

Figure 4-4. This high resolution electron micrograph of a bacterial cell (*Erwinia amylovora*) clearly shows the cell wall, cell membrane, ribosomes, and fibrous nuclear material. (Photograph courtesy of P. Huang and R. N. Goodman, University of Missouri. Used by permission of the American Society for Microbiology.)

devoid of the large organelles characteristic of cells of higher plants and animals. The granular nature of the cytoplasm is caused by a heavy concentration of *ribosomes*. These are responsible for synthesizing proteins and are numerous to allow for rapid protein production. This in turn allows a bacterium to divide rapidly, thereby producing a large population in a short time.

The cytoplasm of some bacteria contains noteworthy internal membranes with attached photosynthetic pigments (Fig. 4-5). These are never enclosed into membrane-bound chloroplasts, and may either be scattered throughout the cytoplasm or associated together forming stacks of chlorophyllous membranes. The appearance of photosynthetic bacteria is often similar to that of blue-green algae, which will be studied in the next chapter.

Certain other granular cytoplasmic inclusions are noted in some bacterial species. Many of these granules have been shown to be composed of starch or starch-like substances and they probably function in food storage since they often disappear when the available food supplies become depleted. Some bacteria produce granules of sulfur or protein in their cytoplasm depending on environmental conditions. (Fig. 4-6).

Bacterial cells lack a membrane-bound nucleus which is typical of the cells of higher organisms. However, nuclear material is present in the bacterial cell. It often appears as a fibrous material occupying a relatively large volume near the center of the cell (Fig. 4-4). This material is composed of DNA and has been shown in many species to form a single circular chromosome, which is probably attached to the mesosome (Fig. 4-7). The DNA of this chromosome duplicates itself prior to cell division, and divides when the cell divides or just before it divides.

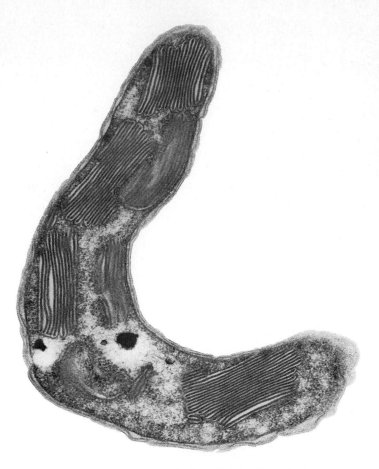

Figure 4-5. This electron micrograph of a cell of *Ectothiorhodospira* distinctly shows an internal membrane system. Photosynthetic pigments are attached to these internal membranes, and this bacterium is self-supporting. (Courtesy of S. F. Conti, University of Kentucky.)

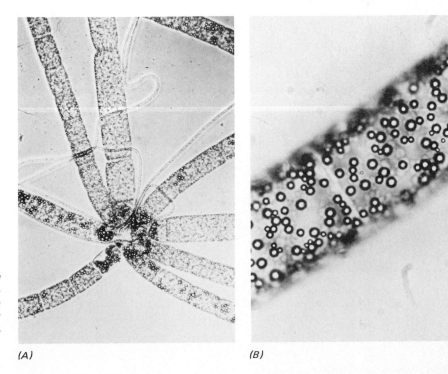

Figure 4-6. *Thiothrix* is a bacterium which has the capacity to form sulfur granules in its cytoplasm. A shows several filaments of *Thiothrix* and B is a closeup of a single filament showing the numerous sulfur granules in greater detail. (Courtesy of Sheril Burton, Brigham Young University.)

(A)

(B)

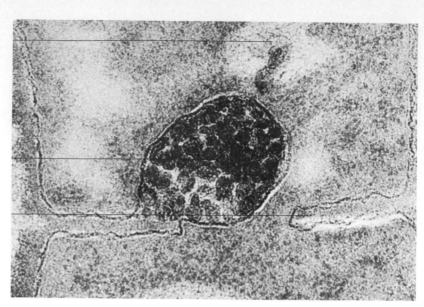

DNA

Mesosome

Cross wall

Figure 4-7. This electron micrograph of *Bacillus* shows a mesosome active in the formation of a new cross wall. It also shows a strand of DNA attached to the mesosome. (Courtesy of D. G. Lundgren, Syracuse University. Used by permission of the American Society for Microbiology.)

However, true mitosis does not occur. That is, the orderly step-by-step separation of chromatids characteristic of dividing chromosomes of higher plants and animals does not occur in bacteria.

One other cytoplasmic structure produced by some bacteria should be mentioned here. The *flagellum* is a thin cytoplasmic projection that apparently originates in the cell membrane and extends externally through the cell wall (Fig. 4-8). Flagella are usually longer than the cell and are occasionally several times as long. The construction of the bacterial flagellum seems simple when compared to the flagellum produced by higher plants and animals. It is usually composed of from three to six twisted chains of the unique protein flagellum. Flagella move rapidly by a rotational motion and thereby are responsible for the motility of flagellated bacteria. The number and placement of flagella on the cell are significant in some systems of bacterial classification.

Bacteria as well as all other organisms have certain nutritional requirements. They require certain substances from their external environment in order to survive. Some plants synthesize all of their own cellular components from water, carbon dioxide and inorganic elements, or ions. Such organisms are said to be *autotrophic*. These organisms manufacture food products either by photosynthesis, if they obtain energy for the process from light, or by *chemosynthesis*, if they obtain energy from inorganic substances such as iron or sulfur. By far the majority of autotrophic organisms on earth are photosynthetic since bacteria alone contain chemosynthetic forms. Bacterial photosynthesis differs significantly from that of other organisms and no oxygen is produced as a biproduct.

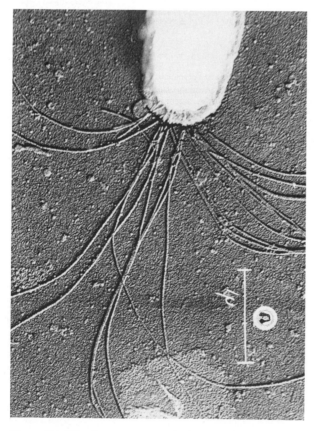

Figure 4-8. This electron micrograph of *Pseudomonas* has been prepared to demonstrate the numerous flagella at the end of the cell.

Organisms that must obtain at least some of their organic nutrients from other sources are *heterotrophic*. Heterotrophic organisms that obtain their necessary organic materials from dead cells or tissues, or from waste products are *saprophytic*. Conversely, organisms which obtain their organic nutrients from other living organisms are *parasitic*. Both saprophytic and parasitic bacteria are common. Many parasitic bacteria can be *pathogens* if they cause disease in their host organisms. Some normally saprophytic bacteria may also become pathogens under certain circumstances.

Most bacteria rely chiefly on cell division for reproduction. This is not equivalent to mitosis common in higher plants and animals. Bacterial cell division is often referred to as *binary fission*. Fission proceeds with the division of the nuclear material by *amitosis* which is merely division of the chromosome without passing through the typical steps of mitosis. As the nuclear material divides, or just after that time, the cell membrane invaginates from opposite edges of the cell toward the middle (Fig. 4-7). Thus, the cell is eventually separated into two regions. Wall material is then deposited between the membranes, again progressing from the edge of the cell toward the center. As the process is completed, the parts divide, and two daughter cells result, complete with their own walls, membranes and nuclear material (Fig. 4-9).

Cell division may occur extremely rapidly in many species of bacteria. It is often responsible for the development of huge populations in a short period of time. Thus, if a sufficient supply of nutrients and space is available, bacteria may grow exponentially, with the population doubling in a constant time interval. For example, under ideal conditions some bacterial cells may divide in 20 minutes or less. Starting with a single cell, then, that cell may divide to form two in 20 mins. Those two cells may then divide to form four cells after 40 mins, and eight cells after 60 mins. If this growth continued, 22,000,000,000,000,000,-000,000,000,000,000,000,000,000,000 bacterial cells would be produced after only 48 hr of growth! The startling fact concerning such growth is that this many bacterial cells would weigh about 24,000,000,000,000,000,000,-000,000 tons, which is several thousand times as much as the entire earth weighs.

Such growth may actually never occur in nature, or at best only occurs for short periods of time. Even so, is there any wonder that fresh milk may quickly spoil, or potato salad may become contaminated on an afternoon picnic, or that a human bacterial infection may cause serious illness or death within a matter of hours. The availability of nutrients and competition for space are two factors that often control the size of naturally-occurring bacterial populations.

Some bacterial cells form resting bodies when exposed to conditions unfavorable for continued growth. Such rest-

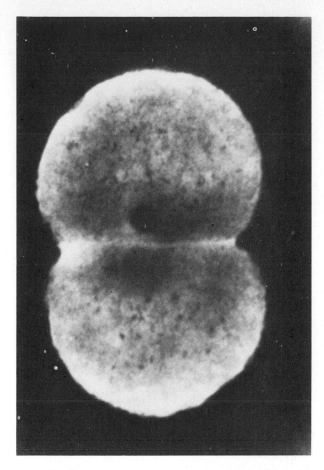

Figure 4-9. *Diplococcus pneumoniae* is a bacterium infamous for causing bacterial pneumonia. The cell in this electron micrograph has recently divided to form two daughter cells which have not separated. (Courtesy of the Center for Disease Control, Atlanta.)

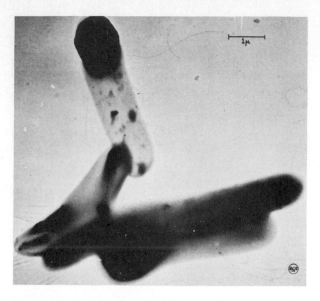

Figure 4-10. Many bacteria are able to produce resting bodies within their cells. These are known as endospores. An endospore is illustrated in this electron micrograph of *Clostridium tetani* as the dark body within the uppermost cell. This bacterium is the causal agent of tetanus. (Courtesy of Thomas F. Anderson, Institute for Cancer Research, Philadelphia. Used by permission of the American Medical Association.)

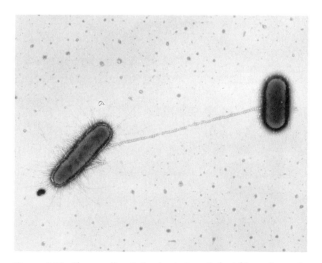

Figure 4-11. These cells of the bacterium *Escherichia coli* are in the process of conjugation. By this primitive sexual process, genetic material is transferred through the conjugation tube from one cell to the other. (Electron micrograph courtesy of Charles Brinton, Jr., University of Pittsburg.)

ing bodies are produced within the bacterial cell and are referred to as *endospores* (Fig. 4-10). These structures have thick walls, contain nuclear material, and are incredibly resistant to a wide variety of harsh environmental conditions. Endospores of many species, for instance, may remain alive after several hours of boiling in water which may cause considerable problems when trying to sterilize materials. Likewise, endospores of other bacteria may resist prolonged exposures to temperatures less than −250°C. All endospores are resistant to drying and low nutrient concentrations in their environment. Endospores swell on being exposed to conditions favorable for vegetative growth, and the cell wall fractures. A single complete bacterial cell emerges and begins to divide to form a bacterial population. Endospores thus function to carry the bacterium through periods of unfavorable environmental conditions.

Sexual reproduction has often been stated to be absent among bacteria. Indeed, a sexual process similar to that in higher organisms is absent. However, certain species do achieve some degree of genetic recombination as was first shown in the middle 1940's and has since been demonstrated many times. Such recombination occurs by three methods. *Transformation* involves the transfer of a strand of DNA from a dead donor cell to a receptor which then incorporates the DNA into its own genetic makeup. A rudimentary sexual process known as *conjugation* involves physical contact between mating pairs of cells with the subsequent transfer of genetic material from one cell to another (Fig. 4-11). Genetic information can also be transferred from cell to cell by bacterial viruses during a process known as *transduction*. The details of occurrence of these processes are currently under study in a number of laboratories throughout the world.

CLASSIFICATION

The classification of bacteria has proven to be difficult for a number of reasons. For instance, most bacteria are very small and often are difficult to study adequately with the ordinary light microscope. Other species grow in highly specialized environments in nature and are difficult to grow in the laboratory. Still others are pathogenic to humans and so must be handled with special care.

Perhaps the most difficult aspect of bacterial classification is in determining what constitutes a natural species. In higher plants and animals this is usually done on the basis of compatibiltiy in sexual reproduction and differences in shape, size, color, and so forth between comparable organisms. However, sexual reproduction in bacteria is apparently limited to a few species and even there is often infrequent. Likewise, detectable differences between comparable strains of bacteria are often negligible.

Because of these problems, bacteria are usually classified into highly arbitrary categories solely for convenience. Traditionally, classification has been based on size and shape of individual cells. Thus, three broad categories of bacteria have been long recognized. The simplest bacteria are spherical cells and are referred to as *cocci*. Rod-shaped bacteria are known as *bacilli* or *rods* (Fig. 4-12), and helically-twisted cells or *spirilla* (Fig. 4-13) comprise the third major type. In addition to single cells, cocci, for instance, may divide in several directions to form irregular clumps of cells known as *staphylocci* (Fig. 4-14). If division occurs to produce a chain or filament of cells, such a filament is known as a *streptococcus* (Fig. 4-15). Several other types of bacteria are similarly defined on the basis of their form.

During recent years some bacteriologists have used chemical means to differentiate bacterial groups. For instance, the products produced by bacteria from degrading a simple sugar can be easily isolated and identified. Since such products differ widely among different bacterial species, those that produce similar products are often placed in the same category of classification. Such a system is not satisfactory since obviously unrelated bacteria may produce the same breakdown products.

The most modern approaches to bacterial classification involve using biochemical and genetic tests. One of the most promising new methods involves isolating DNA molecules from the nuclei of a large number of cells from a single bacterial strain. This DNA is then incubated with the DNA of a second strain. If the molecules are similar, they will associate closely with each other. The degree of association of the DNA molecules can be accurately measured so that the closeness of relationship between the test organisms can be determined. This test is particularly promising since the DNA is always similar in construction among closely-related organisms and rather dissimilar among unrelated organisms. Such tests, known as *DNA hybridization tests,* may also prove valuable to help clear up problems of classification among certain higher organisms.

OCCURRENCE AND ECOLOGY

It is hard to imagine a group of organisms as widely distributed as bacteria. They occur in nearly all environments on earth. They are common and often present in tremendously large numbers both on and within soils, plants, animals, rotting logs, animal dung, and thousands of other substrata. In fact, there is scarcely any natural or artificial product known that is not a substrate for at least some kinds of bacteria.

Fresh-water and marine-aquatic as well as terrestrial species are known. Some species must have oxygen for their growth while others require its absence. Still others can live and grow either with or without oxygen. Hetero-

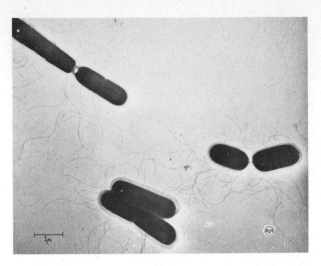

Figure 4-12. Bacteria have often been classified on the basis of individual cell shape. This electron micrograph shows the rod shape of *Clostridium tetani*. (Courtesy of Thomas F. Anderson, Institute for Cancer Research, Philadelphia. Used by permission of the American Medical Association.)

Figure 4-13. The human veneral disease syphilis is caused by the spirilla bacterium *Treponema pallidum*. (Courtesy of the Center for Disease Control, Atlanta.)

Figure 4-14. Irregular clumps of cocci are known as staphylococci. This electron micrography shows *Staphylococcus aureus* which is important to humans since it can cause food poisoning, hospital staph infections, and certain kidney diseases. (Courtesy of Thomas F. Anderson, Institute for Cancer Research, Philadelphia.)

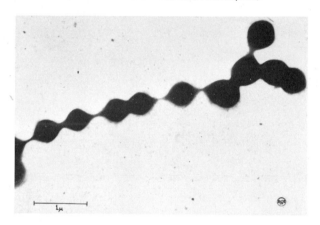

Figure 4-15. Cocci may divide to produce chains of cells similar to this electron micrograph of *Streptococcus pyogenes*. (Courtesy of David B. Lackman. Used by permission of the American Society for Microbiology.)

trophic bacteria require organic substances in their habitat while others need only a few basic mineral nutrients. Some bacteria can exist at low temperatures for extended periods of time, and others are able to live at extremely-elevated temperatures. Several bacteria are common in hot springs throughout the world where they exist at temperatures as high as 90°C. This means that they actually occur in boiling water at high elevations (Figs. 4-16 and 4-17).

The distribution of bacteria on the bodies of animals is a rather interesting phenomenon. For instance, different species of bacteria occur on different parts of the human body in relationship to moisture present, quantity and quality of nutrients available, etc. Bacteria are also common within the human body where their distribution is dependent on similar factors.

IMPORTANCE

It is fair to say that the earth as we know it could not exist without the tiny, simple organisms known as bacteria. These plants are involved in thousands of processes, some of which are harmful and some of which are helpful to human life. In fact, few groups of organisms are so critically and directly significant to the well-being of our entire planet and to all of its inhabitants.

Nutrients available to the living organisms on earth are present in finite, often rather limited amounts. These nutrients, including such significant elements for growth as nitrogen, phosphorous, and potassium, would have been mostly irrevocably "locked up" in plant and animal tissues millions of years ago if it were not for bacteria (as well as fungi which will be studied later) decomposing these tissues and releasing their nutrients to be used again by other organisms. Thus, nutrients have flowed through living tissues for millions of years, and the nutrients comprising

Figure 4-16. Bacteria are able to grow at extremely high temperatures in hot springs throughout the world. This spring located in Yellowstone National Park supports a bacterial population. Glass slides are suspended on string from the branches which extend over the spring. Bacteria colonize the slides and then can be further studied. (Courtesy of Thomas Brock, University of Wisconsin.)

our own bodies were once a part of other living organisms. It is exciting to think that the protoplasm coursing in human cells is comprised of substances that may have once been a part of a wolf or coyote or even a delicate prairie wildflower. Equally exciting is the concept that someday the elements making up our bodies will probably be used to form the protoplasm of other living organisms. The cycling of nutrients accomplished by the decomposing abilities of bacteria and fungi gives people in one real sense the immortality they have always sought.

Many bacteria aid in increasing soil fertility. This in part is related to their ability to decompose dead tissues of other organisms, thus releasing nutrients to the soil. It is also due to two other significant factors. First, a large storehouse of mineral nutrients is contained in the parent rocks and small particles that form a soil. These nutrients are released into the soil solution by the actions of the small organisms occurring there. Bacteria and fungi are again particularly significant in this regard.

The second factor critical in soil fertility and plant growth is the ability of some bacteria to fix nitrogen. This process of *nitrogen fixation* occurs when the bacterium uses atmospheric nitrogen (N_2) to manufacture its own protein molecules. Very few organisms are able to use atmospheric nitrogen directly in their own growth but instead require nitrogen in the form of nitrates or ammonia. Thus, when nitrogen-fixing bacteria die and decompose, the nitrogen contained within their cells, now in a form that can be used by other organisms, is released and used in the growth of higher plants.

Nitrogen-fixing bacteria may occur freely in the soil or they may be associated with higher plants. For instance, important crop plants of the family Leguminosae, such as beans, alfalfa, soy beans, red clover, and peas, develop nodules on their roots that are inhabited by nitrogen-fixing bacteria of the genus *Rhizobium* (Figs. 4-18, 4-19 and 4-20). The nitrogen fixed by these bacteria becomes available for the growth of the plant. It has been demonstrated that infected plants grow much more rapidly and luxuriantly than noninfected plants (Fig. 4-21). The relationship between the bacterium and the higher plant is mutually beneficial since the bacterium uses photosynthetic products produced in the leaves of the infected plant and transferred to the roots. This process of nitrogen fixation is responsible for increasing yields of some of the world's most important crop plants. Current research to develop more efficient nitrogen fixers and to determine the exact methods of nitrogen fixation is being conducted in many research labs throughout the world.

Also on a positive note, bacteria are responsible for the production of many significant industrial and consumer products. Several cheeses, for instance, depend on bacteria for their delicate flavors and textures (Fig. 4-22). Cottage cheese and buttermilk are also made by using bac-

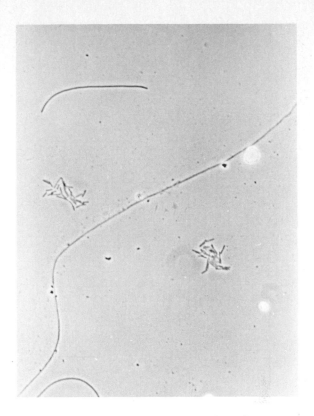

Figure 4-17. This photograph taken with the light microscope demonstrates two colonies of bacteria that became attached to the suspended glass slides shown in Fig. 4-16. (Courtesy of Thomas Brock, University of Wisconsin.)

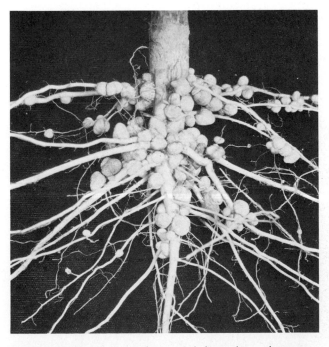

Figure 4-18. The nodules on the roots of this soybean plant are caused by the infection of the bacterium *Rhizobium*. This bacterium is a nitrogen fixer, and it has been demonstrated that soybeans and other legumes will grow much better when they are infected. (Courtesy of Nitragin Co., Milwaukee, Wisc.)

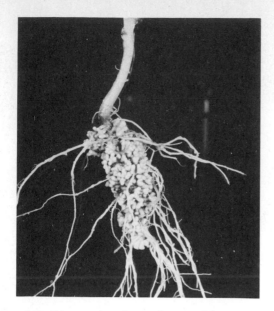

Figure 4-19. This pea plant shows a heavy nodulation caused by a nitrogen fixing bacterium. (Courtesy of the Nitragin Co., Milwaukee, Wisc.)

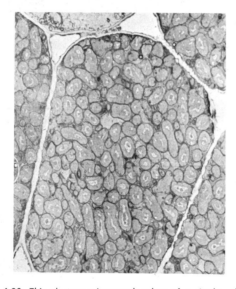

Figure 4-20. This electron micrograph taken of a single cell of a clover root is filled with numerous bacteria of the genus *Rhizobium*. (Courtesy of P. J. Dart.)

terial cultures. Several other products, including amino acids, enzymes, organic acids, steroids, and vitamins, are synthesized by various bacteria and then extracted industrially for commercial uses.

Bacteriologists specializing in the study of the relationships between bacteria and humans have come up with some very interesting findings within the past decade or so. For instance, they have shown that bacteria that normally inhabit the surface and openings of the human body are apparently significant for a variety of reasons. The presence of these bacteria often discourages the infection of other species that may be pathogenic. Bacteria in the vagina, for instance, are tolerant and actually create an acid environment (with a pH of around 4.5). Since most bacteria do not tolerate this degree of acidity, many vaginal infections are avoided thanks to the resident bacterial population.

Bacteriologists have also demonstrated that the bacteria resident in the human intestinal tract are highly beneficial. They provide at least some and probably all of certain vitamins required in our diet. They also protect against infection of the intestinal tract by pathogenic organisms. Even more startling are recent findings that indicate that the normal development of the intestinal tract is probably dependent on the resident bacterial population. Thus, in experimental animals, development of the intestinal tract is much different if the animal is raised in a bacteria-free environment (Fig. 4-23). More typical development can be obtained by inoculating bacteria-free individuals with normal resident bacteria. Development of certain other bodily organs also seems to be dependant on the resident bacterial population. Lymph nodes, for instance, tend to be less well developed in organisms lacking a resident bacterial flora. All of these studies are in early stages, and exciting information concerning the relationship between bacteria and human development will certainly be discovered during the next few years.

Several bacteria are agents of food poisoning. This poisoning ranges from mild intestinal discomfort perhaps accompanied by stomach ache and diarrhea to slow, pain-

Figure 4-21. The effects of nitrogen fixing bacteria on the growth of some plants is especially well documented in this figure. The plants on the left were inoculated with *Rhizobium* while those on the right were not. (Courtesy of the Nitragin Co., Milwaukee, Wisc.)

(A)

(B)

Figure 4-22. Several cheeses depend on bacteria for their delicate flavors and textures. These photographs show ripening blocks of cheese at A and an individual cheese block at B with a bacterial smear on its outer surface. (From N. F. Olson, *Ripened Semisoft Cheeses*. Reprinted by permission of Pfizer, Inc.)

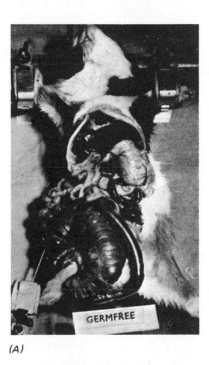

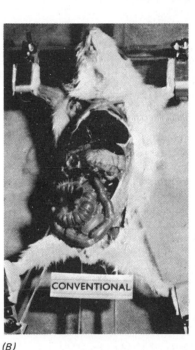

(A) GERMFREE

(B) CONVENTIONAL

Figure 4-23. The intestinal tract of two guinea pigs is shown in this figure. Note that the one that developed under bacterial free conditions is highly modified from the normal one. (Courtesy of the Walter Reed Army Institute of Research.)

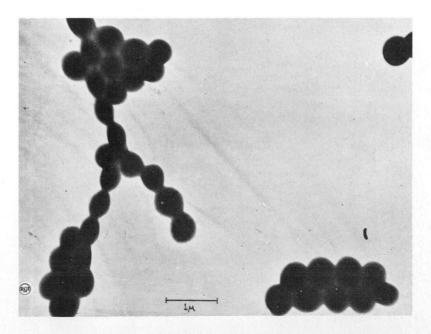

Figure 4-24. *Staphylococcus aureus* is probably the most common food poisoning agent. (Courtesy of Thomas F. Anderson, Institute for Cancer Research, Philadelphia. Used by permission of the American Medical Association.)

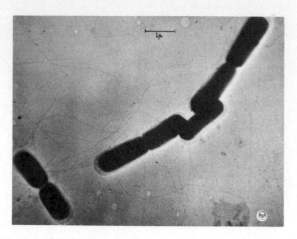

Figure 4-25. *Clostridium botulinum* produces toxins which are among the most potent known when it grows in anaerobic conditions. Adequate precautions must be used in food canning and preparation to insure against botulism poisoning. (Courtesy of Thomas F. Anderson, Institute for Cancer Research, Philadelphia. Used by permission of the American Medical Association.)

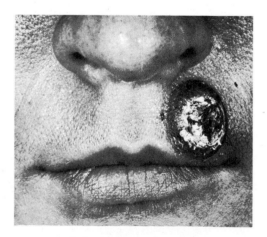

Figure 4-26. This photo shows a primary syphilitic lesion on the lip of a patient. (From S. Olansky and L. W. Shaffer, *Syphilis in Communicable and Infectious Diseases* (F. H. Top, ed.), 6th ed. The C. V. Mosby Co., St. Louis, 1968.)

ful death. *Staphylococcus aureus* is probably the most common food poisoning agent (Fig. 4-24). This bacterium produces a toxin that creates intestinal discomfort, nausea, diarrhea and general disability. The disability usually only lasts for a few hours and the death of a poisoned individual is rare. *Staphylococcus aureus* infects dairy products, bakery goods, potato salad, etc., and the adequate refrigeration of these products is a necessary preventative measure.

Clostridium botulinum (Fig. 4-25) is a common bacterium in many soils throughout the world. This bacterium is rarely troublesome, but when it grows in an oxygen-free environment, it produces toxins that are among the most potent known. The bacterium occasionally remains alive in canned vegetables or meat producing toxins. When the toxins are ingested they cause severe poisoning, and the mortality rate from botulism food poisoning is very high. It is imperative that adequate precautions be taken to eliminate this bacterium during food processing.

Each year millions of dollars are spent on controlling bacterial-caused diseases of plants and animals. Hundreds of thousands of lives and hundreds of tons of valuable food products are lost annually to such diseases and untold misery results. The list of human bacterial diseases reads like a horror story and includes the most dreaded of all diseases. Bubonic plague, cholera, diphtheria, gonorrhea, leprosy, spotted fevers, syphilis, tetanus, tuberculosis, typhoid fever, and whooping cough are all bacterial diseases affecting humans. The greatest strides in controlling such diseases have certainly been made during the past twenty years. Fortunately, many of these diseases are now largely controlled by immunization and improved sanitation. Even so, many are still dreaded and have the potential to form epidemics under certain circumstances. Venereal diseases, including gonorrhea and syphilis (Fig. 4-26), for instance, have reached epidemic proportions in many urban areas throughout the world. Remember the great strides people have taken in controlling bacterial diseases whenever anyone mentions that they long for "the good old days." Can the good old days have been so good when humans were constantly plagued by disease and its resultant death and disability? Such days weren't so long ago either since such critical antibiotics as penicillin have been available for scarcely a quarter century. Much research remains to be done in medical bacteriology, and new answers to age old disease problems are becoming realities with the advancement of this research.

FOSSIL BACTERIA

The fossil record of bacteria is poor since these organisms do not fossilize well. Thus, it is difficult to trace the geological history of bacteria on the basis of direct evidence alone. However, a knowledge of the earth's probable early

Figure 4-27. Tetanus is another bacterial disease dreaded by humans. This photograph of a painting of a soldier dying with tetanus gives graphic proof of the horrors of this disease. (Painting by Charles Bell. Used by permission of the President and Council of the Royal College of Surgeons, Edinburgh.)

conditions allows us to draw tentative conclusions bearing on the problem. The earliest atmosphere was probably very low in oxygen but rather high in such substances as nitrogen, carbon dioxide, and methane. Several experiments throughout the past decade have shown that if electrical sparks are passed through such an atmosphere, simple organic compounds may be produced. These discharges of electricity probably also occurred on the primitive earth so that simple organic compounds were probably found in ancient seas. Thus, the earliest organisms on earth most likely grew in an oxygen-free environment and used simple organic compounds as an energy source. Heterotrophic bacteria are certainly the most likely candidates as being such early organisms.

In addition to this theoretical evidence, recent discoveries of Precambrian microorganisms indicate that very probably bacteria were the first organisms to evolve on earth (Fig. 4-28). These tiny, simple plants probably diversified rapidly since even the earliest fossil bacteria discovered seem to be rather closely related to modern forms. Bacteria might have continued their evolutionary development to give rise to oxygen-producing algae. These algae evolved early from bacteria and were present in vast numbers through much of the Middle and Late Precambrian. They were responsible for the production of oxygen that was released to the developing atmosphere. Thus, the evolution of bacteria and subsequently of algae set the stage

Figure 4-28. This electron micrograph shows fossil rod-shaped bacteria approximately 2 billion years old. They were collected from the Gunflint chert in southern Ontario. (Courtesy of J. William Schopf, University of California at Los Angeles. From E. S. Barghorn and J. W. Schopf, *Science* 149:1365–1367. Copyright 1965 by the American Association for the Advancement of Science.)

for the evolution of all the higher forms of animal and plant life.

SUMMARY

Bacteria are simple unicellular or colonial organisms that demonstrate a cell wall and cell membrane surrounding the cytoplasm. The cytoplasm is generally simpler in construction than that of a higher plant or animal cell since it lacks well-defined membrane-bound organelles such as mitochondria, nuclei, and plastids. However, nuclear material is found in all bacterial cells and this material duplicates and divides at cell division, although the division is not mitotic.

Bacteria may be either heterotrophic or autotrophic. Heterotrophic species are saprophytic if they obtain their organic nutrients from dead tissues or waste products or parasitic if they obtain nutrients from other living organisms. Most autotrophic species are photosynthetic, although a few are chemosynthetic.

Most bacteria rely on cell division as their only or chief means of reproduction. Some species produce highly resistant endospores that may remain viable throughout long unfavorable periods to produce a new vegetative cell when favorable conditions return. A few bacteria demonstrate genetic recombination. This process is not equivalent to sexual reproduction among higher organisms since the union of gametes and subsequent meiosis do not occur.

Bacteria are critical in the earth's ecosystem and of direct consequence to humans for a variety of reasons. Several species fix atmospheric nitrogen which becomes available for use by other plants. Others are significant in breaking down parent soil material and thus increasing soil fertility. Most species are efficient decomposers which allows for recycling of the earth's nutrients. Several are important in industrial processes and are responsible for the formation of a large variety of commercial products. Still others are serious pathogens of humans as well as other animals and plants.

FURTHER READING

ALEXANDER, M., *Microbial Ecology,* John Wiley and Sons, New York, 1971.

BARGHOORN, ELSO, "The Oldest Fossils," *Scientific American,* 224:30–42, 1971.

BROCK, THOMAS D., *Biology of Microorganisms,* Prentice-Hall, Inc., Englewood Cliffs, New Jersey, 1974.

BROCK, THOMAS D. and KATHERINE M. BROCK, *Basic Microbiology with Applications,* Prentice-Hall, Inc., Englewood Cliffs, New Jersey, 1973.

DOBEL, C., *Anton van Leeuwenhoek and His "Little Animals,"* Dover Publications, New York, 1960.

HAYES, W., *The Genetics of Bacteria and their Viruses,* Second edition, John Wiley and Sons, New York, 1968.

SCHOPF, WILLIAM, "PreCambrian Micro-Organisms and Evolutionary Events Prior to the Origin of Vascular Plants," *Biological Review,* 45:319–352, 1970.

5 Introduction to the Algae: Division Cyanophyta

ERTS-1 satellite photograph of Utah Lake, Utah. The swirled patterns in the lake are blue-green algal blooms. (Imagery courtesy of NASA. Photography courtesy of National Environmental Satellite Service.)

INTRODUCTION TO THE ALGAE

Most people are aware that many ponds and streams contain green plants that are quite different from those occurring on land. Fishers often call these plants "moss" and their only concern with them is the irritation they feel when they dredge some up on a fish hook. Most often these plants are actually algae which are relatively simple, often unicellular or filamentous plants. Algae never produce true roots, stems, or leaves, and an algal plant is therefore a *thallus*.

Even though algae are often inconspicuous, they are the most prevalent green plants on earth. They are the predominant form of vegetation in all oceans, and in nearly any conceivable fresh water habitat.

Most algae are greenish to bright green in color because of the presence of chlorophyll in their cells. Several different types of chlorophyll occur in the algae of different divisions. These and other pigments are important in the classification of these plants. Likewise, simple sugars manufactured by photosynthesis are stored by algae in different ways and the nature of these storage products is important in algal classification. For instance, algae of the division Chlorophyta store starch while golden algae, or Chrysophyta, store oils.

Most algae demonstrate a true alternation of generations with the gametophyte predominating. The gametophyte produces gametes in cells known as *gametangia*. Gametangia are always unicellular in algae.

When two gametes unite, the resultant zygote is the first cell of the sporophyte generation. The zygote of an alga never produces an embryo as does the zygote of a higher plant. This combination of features broadly distinguishes the algae from all other groups of plants. Algae are separated into seven separate divisions based on differences in pigments, storage products, reproductive cycles, and morphology of the thallus. Each of these divisions will be discussed in the following chapters.

DIVISION CYANOPHYTA

GENERAL CHARACTERISTICS

The most primitive algal division is the Cyanophyta which contains the blue-green algae. Plants of this division are thought to be closely related to bacteria, and some botanists place them in the same division. However, blue-green algae differ from bacteria including photosynthetic forms in several important ways including pigmentation, type of photosynthesis, and morphology.

The number of different species of Cyanophyta is difficult to estimate. Even experts on the subject differ widely on the number of species they recognize in the separate genera. Much research is currently being done concerning the classification of these algae. Around 1,500 species of algae are usually placed in this division.

Blue-green algae contain a unique pigment complement consisting of chlorophyll *a, c*-phycocyanin, *c*-phycoerythrin, and various carotenes and xanthophylls. Chlorophyll *a* is the universal chlorophyll found in all green plants, while *c*-phycocyanin and *c*-phycoerythrin are blue and red pigments respectively that absorb light energy and thus aid in photosynthesis. The combination of these pigments often produces cells with a blue-green color which is responsible for the common name of the plants of this division. However, the influence of yellow carotenes and brownish xanthophylls may modify the coloration so that cells of almost any color may be found among various Cyanophyta.

Blue-green algae are very primitive plants exhibiting few of the general characteristics attributed to plants in earlier chapters. No well-defined nuclei occur in the cells of these algae. That is, nuclear membranes and organized chromosomes are both lacking in blue-green algal cells. However, a colorless granulate region near the center of the cell is often conspicuous. This region is variously known as the *central body,* the *nucleoplasm,* or the *centroplasm* (Fig. 5-1). It is the region of cellular control containing the genetic material of the cell.

The actual nature of the chromatin material of Cyanophyta is still in question. Some botanists have reported small rod-shaped particles in the central body of some species while others have reported a loose thread-like network. Future studies using the electron microscope will help to clarify the structure of the central body and the nature of the chromatin.

Since chromosomes either are lacking or organized differently than in a typical plant cell, true mitosis does not occur in blue-green algae. Mitotic phases have never been observed, and cell division appears to occur by the process of fission similar to that observed in bacteria. Each daughter cell receives approximately one half of the contents of the central body. Even though this sounds like a haphazard process, recent studies indicate that cell division in some Cyanophyta is more organized than previously believed.

True plastids including chloroplasts also are absent from cells of blue-green algae. Photosynthetic pigments occur on membranes dispersed in the cytoplasm surrounding the central body (Figs. 5-1 and 5-3). These dispersed pigments form a conspicuous, colored region of *chromoplasm* or *chloroplasm.* This region may be broad, nearly filling the cytoplasm of the cell, or it may be rather localized.

Gas vacuoles (Fig. 5-2) may be produced in the cells of many Cyanophyta. These small structures are filled with gasses and are bound by an unusual membrane. Their function is not definitely known, although they probably function to lend buoyancy to the cell and to protect the chloroplasm against strong sunlight.

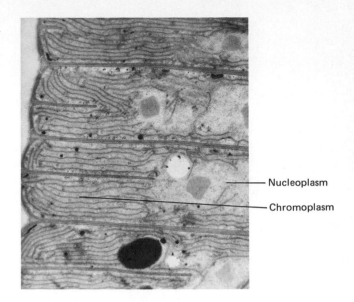

— Nucleoplasm

— Chromoplasm

Figure 5-1. This filament of cells of an unnamed blue-green alga shows central nucleoplasm and peripheral membranes with photosynthetic pigments (chromoplasm). (Electron micrograph courtesy of Norma Lang, University of California at Davis.)

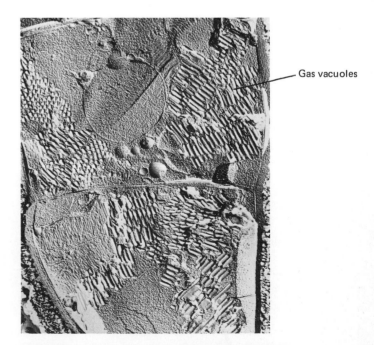

— Gas vacuoles

Figure 5-2. Gas vacuoles such as those illustrated in this electron micrograph are unusual features of Cyanophyta. They probably function in aiding the alga to rise and sink in the water. (Courtesy of Balzers Artiengesellschaft.)

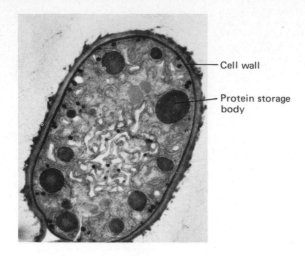

Cell wall

Protein storage body

Figure 5-3. Many Cyanophyta are able to produce thick-walled akinetes as resting spores. This electron micrograph of the akinete of *Anabaenopsis* shows the thick wall and internal structure. (Courtesy of Norma Lang, University of California at Davis.)

Figure 5-4. Several species of Cyanophyta were collected from the side of this greenhouse pot. Blue-green algae are common in nearly any moist terrestrial habitat.

Recent studies of the blue-green algal cell using the electron microscope indicate that storage products and other inclusions occur within the chromoplasm. Storage products are basically proteins (Figs. 5-1 and 5-3) and a unique carbohydrate, *cyanophycean starch*. Storage products are never stored in membrane-bound plastids as they are in typical plant cells.

The cell wall of blue-green algae is generally composed of several well-defined layers. The inner layers (or inner wall) are rigid. The outer wall (or sheath) contains large quantities of pectin and is often mucilaginous. The sheath may be highly colored, and it functions to some extent in allowing the algae to survive under adverse environmental conditions.

Reproduction of all blue-green algae is asexual. Even though fission is the basis for all cyanophyte reproduction (Fig. 5-1), a number of specialized systems for ensuring reproduction are known. Colonial blue-green algae often fragment, and each segment may grow into a new colony. Another specialized type of asexual reproduction common in many Cyanophyta occurs by the production of spores. A spore is any cell with the capacity to germinate to produce a new plant. Several kinds of spores are produced by different plants. Many Cyanophyta produce *akinetes* (Fig. 5-3). These are resistant, thick-walled, nonmotile spores formed directly when a vegetative cell encysts. Akinetes are capable of remaining dormant often for extended periods of time before germinating and hence are able to carry the species over periods unsuitable for vegetative growth. Other species produce nonmotile *endospores* that lack cell walls and are produced in groups within the cells of some species.

Sexual reproduction has traditionally been said to be absent among Cyanophyta. Indeed, this process does not appear to be common. However, recent studies indicate that genetic recombination apparently occurs in some species. Further study is needed to illuminate the method of this recombination and its frequency throughout the division.

No cells with flagella of any kind occur in Cyanophyta and the cells are therefore nonmotile. However, some species exhibit a peculiar gliding or oscillating type of motion. The mechanism of this motion is still in question, although it appears to be a function of protoplasmic streaming or cell membrane contraction and expansion.

OCCURRENCE AND ECOLOGY

Many blue-green algae are among the most widely distributed of all plants. Representative species grow in essentially all habitats from aquatic to extremely dry. The development of a population of Cyanophyta may be very rapid. It is common for several species to form large populations in short-lived habitats such as puddles formed from runoff that are left standing for only a week or two. Some Cyanophyta occur as dark green or brownish-black smears on the sides or roofs of buildings, on plant pots in a greenhouse (Fig. 5-4), or in the soil (Fig. 5-5), and in a tremendous variety of other habitats.

Most Cyanophyta occur basically in fresh water. Some occur in marine water, and a few species may occur in either marine or fresh water. Marine species often grow in the intertidal zone, especially of tropical waters. A few species float passively in the water, and are thus said to be *planktonic* or *plankters*. Plankton is comprised of both animal species, called *zooplankton,* and plants called *phytoplankton*. Planktonic organisms make up a very significant proportion of life on the earth.

Some blue-green algae exist at extremely high temperatures. Such algae are known as *thermophilic* and are common in hot springs throughout the world in water reaching as high as 75°C (Fig. 5-6). Much of the coloration of hot springs and thermal streams, including those in Yellowstone National Park, is caused by these thermophilic algae. It is fascinating to study one of these springs and to note the different colors in the course of its drainage stream. These colors often result from different species of blue-green algae with varying tolerances for high temperatures. Those able to withstand the highest temperatures are found nearest the hot water source. The more heat-sensitive species are found further down the course of the stream.

Other Cyanophyta are able to live under extremely cold conditions. Some species actually grow on the snow and ice of glaciers. Such algae are said to be *cryophylic*. The extent of occurrence of these cryophylic blue-green algae is not well known. However, under certain environmental circumstances in many parts of the world, a large population may develop in snow or ice coloring it green or blue-green.

Some species of Cyanophyta grow in extremely dry, hot environments (Fig. 5-7). Such algae are often found on deserts beneath or even inside of translucent stones, or within the soil or sand. They also occur in nearly any small permanent or temporary body of water on the desert. The importance of blue-greens in such habitats is now known to be greater than previously thought. In some cases they are extremely important in binding the soil, and thus in protecting against wind and water erosion.

Some blue-green algae occur on the surface of other plants, and are known as *epiphytes*. Others actually occur within other plants and are therefore *endophytes*. In addition, some Cyanophyta occur on or within certain animals, often in a relationship where both the algae and animal benefit (Fig. 5-8).

Figure 5-5. The blue-green algae *Nostoc* and *Oscillatoria* were collected from the layer of soil that was left in this pan.

Figure 5-6. Blue-green algae are common inhabitants of many hot springs and spring streams throughout the world. The stream illustrated in this photograph issues from a geyser in Yellowstone National Park. The dark colorations on either side of the stream are mats of Cyanophyta.

Figure 5-7. Blue-green algae are able to live in a tremendous variety of habitats. The sandstone wall illustrated in this figure was photographed in the southeastern Utah desert. A moist seam occurs along the wall as shown transversing the lower midregion of the photograph. This moist layer was completely covered with Cyanophyta which produced mucilage and in turn allowed the development of other algae.

IMPORTANCE

Cyanophyta are directly important to humans for many reasons. First, several species are able to use atmospheric nitrogen for their metabolism by the process of nitrogen fixation. Nitrogen fixed by these plants later becomes available for use by other plants when the algae die and decompose. This is a very significant process since all plants require nitrogen for their growth and development, and the sources of usable nitrogen for most plants are very limited. Few plants are able to use atmospheric nitrogen directly, and in some habitats blue-green algae supply a large proportion of the total nitrogen available.

The process of nitrogen fixation by Cyanophyta may be especially important in habitats such as deserts with an otherwise limited supply of nitrogen available. Likewise, certain blue-green algae are especially important as nitrogen fixers in cultivated rice fields (Fig. 5-9). The quantity of rice produced in the world would decrease sharply if it were not for these very important algae. Such a decrease would be catastrophic since rice is the most important world-wide food plant.

Blue-green algae are also important *primary producers.* A primary producer is any plant that manufactures simple sugars by photosynthesis. Primary producers are often eaten by *primary consumers,* which use the foods manufactured by the plant. A primary consumer may be subsequently consumed by a *secondary consumer* (Fig. 5-10), and a *food chain* or *energy chain* is thus established. All green plants are primary producers and are at the bottom of nearly all food chains. Thus, even though blue-green algae are not often eaten directly by humans, nonetheless they may be found at the base of several human food chains.

Cyanophyta, as well as all green plants, are critical in liberating oxygen as a biproduct of photosynthesis. This

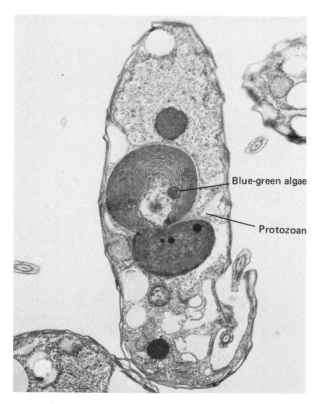

Blue-green algae

Protozoan

Figure 5-8. This electron micrograph shows a protozoan organism (*Cyanophora*) with blue-green algae growing symbiotically within it. (Courtesy of W. T. Hall, Electro-Nucleonics Laboratories.)

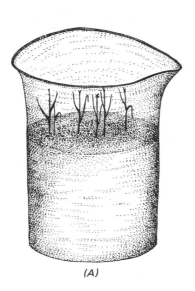

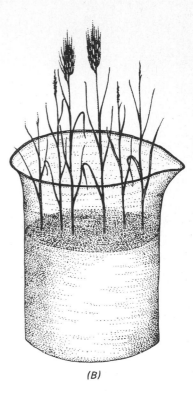

Figure 5-9. It has been known for some time that some Cyanophyta are critical in certain habitats since they have the ability to fix nitrogen. This is particularly the case in rice fields as demonstrated by this experiment. The rice seedlings in the beaker at A are growing without blue-green algae in the soil. The seedlings at B have these algae in their soil. The difference in growth is startling. (Redrawn from a photograph supplied by R. N. Singh, Banaras Hindu University.)

(A)

(B)

is essential for the maintenance of atmospheric oxygen which is used in respiration of nearly all living things. Also, since many blue-green algae are aquatic, they provide oxygen to their immediate environment which may then be used by other aquatic organisms and by the roots of higher plants growing in the same vicinity. For instance, blue-green algae are thought to be very important oxygen generators in rice fields where the oxygen is used by the roots of rice plants.

Many Cyanophyta figure prominently in water pollution. Under certain circumstances they may multiply very rapidly, sometimes completely choking a local water system. When this occurs, these algae are said to form a *bloom* (Fig. 5-11). Algal blooms often cause the water to have objectionable tastes and odors, and thus are a nuisance in municipal water supplies. Some bloom-forming Cyanophyta actually secrete toxins that may be harmful or fatal to other aquatic organisms and to terrestrial animals that may drink the water. Massive fish kills may result when such blooms develop. In addition, several cases have been reported where people have been adversely affected by drinking such water.

When the quantity of algae in a water supply increases to bloom proportions, bacteria and fungi also increase proportionately, acting to decompose the algae which die. These organisms are not photosynthetic and thus use oxygen from the water for their respiration. This is particularly critical at night when the rate of oxygen consumption is often greater than its rate of production by photosynthesis during the day. The result is an oxygen deficiency. Such a deficiency is often responsible for the extensive death of

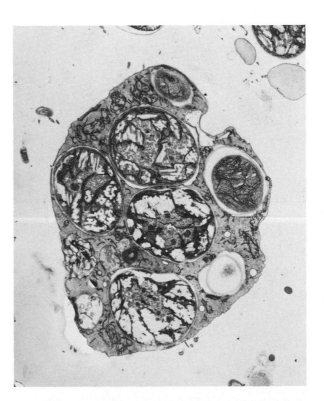

Figure 5-10. Many Cyanophyta are important in the food chains of other organisms. The amoeba shown in this electron micrograph contains several cells of *Microcystis,* a common lake-dwelling, blue-green alga. (Courtesy of B. A. Whitton, University of Durham.)

Figure 5-11. When environmental conditions are proper, an algal "bloom" such as the one illustrated on the surface of this small pond may result. The opener for this chapter shows a very large blue-green algal bloom photographed from the ERTS-1 satellite. These blooms are often critical in localized ecosystems since they may produce toxins and deplete oxygen to the extent that other aquatic organisms are killed.

many of the organisms of the aquatic community, including fish. A good deal of research is currently underway to develop methods for controlling these potentially dangerous blooms.

Many blue-green algae withstand relatively high levels of organic pollution, and some species are among the best pollution indicators. Because of this, Cyanophyta often are studied extensively by engineers and biologists in water-quality projects.

FOSSIL CYANOPHYTA

Fossil Cyanophyta (Figs. 5-12 and 5-13) have been reported from several different localities throughout the world. These algae were among the earliest living things on earth and have been collected as fossils from rocks as old as three billion years. It is interesting that many of the early fossil species are very much like present-day Cyanophyta. These plants have been very successful and have apparently changed little from the time of their first appearance.

Most botanists postulate that unicellular Cyanophyta are the most primitive forms. These simple forms apparently gave rise to filamentous species along one line, and to flat and three-dimensional colonial species along a second line.

REPRESENTATIVE CYANOPHYTA
UNICELLULAR OR SMALL
COLONIAL FORMS

Gloeocapsa

Gloeocapsa is a small, spherical blue-green alga which may be collected as a single cell, or more commonly as a loose

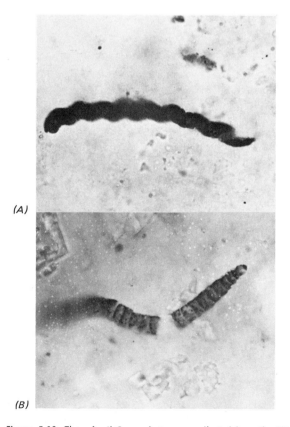

(A)

(B)

Figure 5-12. These fossil Cyanophyta were collected from the Bitter Springs Formation (Late Precambrian) of Australia. A is *Heliconema* and B is *Cephalophytarion*. Both of these closely resemble certain modern blue-green algae. (Courtesy of J. W. Schopf, University of California at Los Angeles.)

colony formed from a small group of cells (Fig. 5-14A). Each cell is surrounded by a gelatinous outer wall or sheath, and the entire colony is often surrounded by a common gelatinous sheath. The sheaths of the individual cells are often *lamellate* or layered, and may be highly colored.

Gloeocapsa occasionally occurs attached to submerged stones or pieces of wood, or as a true plankter. However, most species grow in moist terrestrial habitats where they often form a thin olive green film on rocks or soil. *Chroococcus* (Fig. 5-14C) is another spherical blue-green alga that is much like *Gloeocapsa*. This genus is a common plankter in many waters, although it occasionally forms films on moist soil or rocks.

AMORPHOUS AND PLATE-LIKE COLONIES

Microcystis

Microcystis (Fig. 5-15D) is a planktonic, often rather large, irregularly shaped colony. It is composed of many spherical cells ensheathed by a large amount of mucilage. This alga often forms blooms, especially in hard-water lakes where it occasionally causes fish kills by producing toxins and depleting oxygen. *Microcystis* also often is responsible for corroding concrete and metal walls of water storage reservoirs.

Merismopedia

Merismopedia forms rather small, square, or rectangular colonies one-cell thick (Figs. 5-15B and 5-16). The cells in this colony are arranged singly or in tight pairs, often forming regular vertical and horizontal rows. Such colonies vary in cell number from four to several hundred. Older colonies are generally larger with infolded and rolled edges. *Merismopedia* colonies vary in color from blue-green or gray to violet. *Merismopedia* is common in the plankton of many lakes and ponds throughout North America.

THREE DIMENSIONAL COLONIES

Although most species of *Microcystis* form irregular colonies, a few species are rather regular and spherical. *Chroococcus* also contains one species that forms small cubical colonies. In addition, a few other genera may contain species which form spherical colonies. However, for the most part, this is a rather rare shape for blue-green algae.

Eucapsis

Eucapsis forms colonies with the cells arranged into definite cubes (Fig. 5-15C). Each colony is composed of many cells although a separate sheath envelopes each group of eight or sixteen cells. This plant resembles *Merismopedia*

Figure 5-13. Many Precambrian and later fossil algal deposits such as these were formed by Cyanophyta. More and more information is accumulating to indicate that blue-green algae were among the earliest plants on earth and were responsible for changing the environment and allowing other plants and animals to develop.

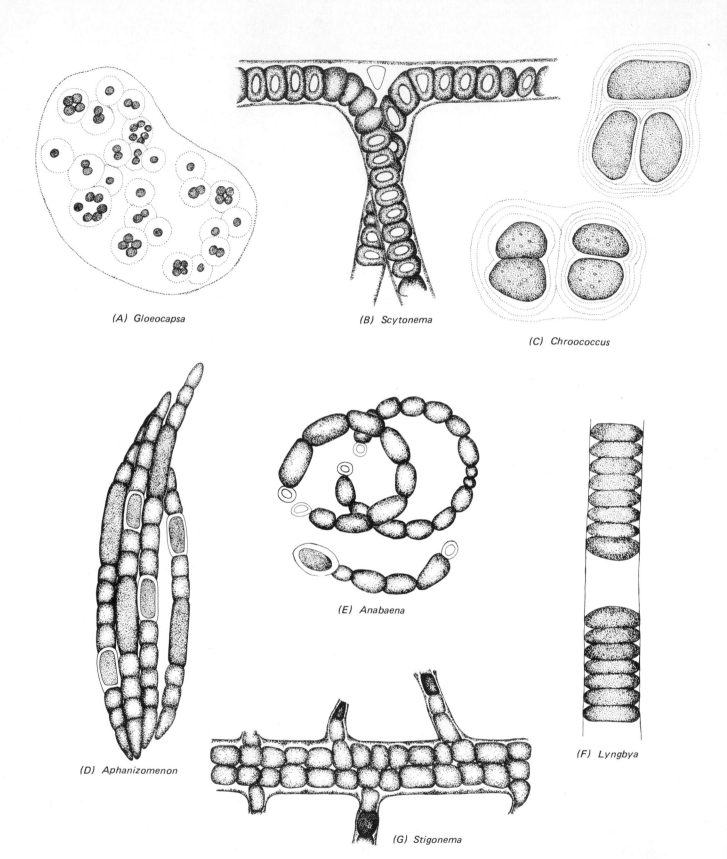

(A) Gloeocapsa

(B) Scytonema

(C) Chroococcus

(D) Aphanizomenon

(E) Anabaena

(F) Lyngbya

(G) Stigonema

Figure 5-14. Illustration of several Cyanophyta.

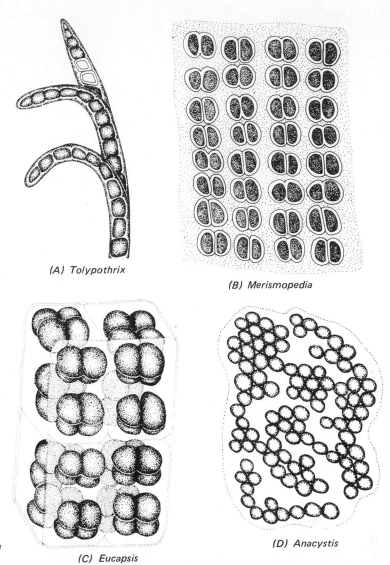

(A) Tolypothrix

(B) Merismopedia

(C) Eucapsis

(D) Anacystis

Figure 5-15. Four species of blue-green algae are shown here illustrating different types of colonies.

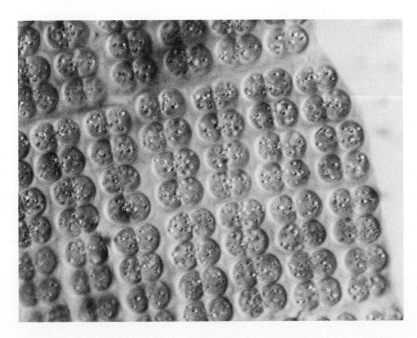

Figure 5-16. *Merismopedia* is a blue-green alga that produces characteristically flattened colonies 1 cell thick. The cells are often aligned into groups of 2 or 4.

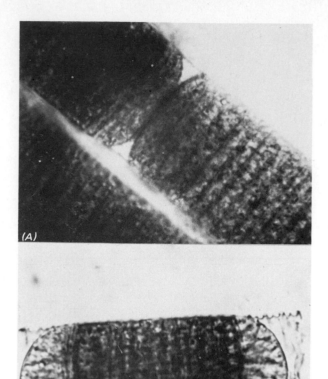

(A)

(B)

Figure 5-17. These photographs of *Oscillatoria princeps* show a separation disk at A and a short hormogonium at B.

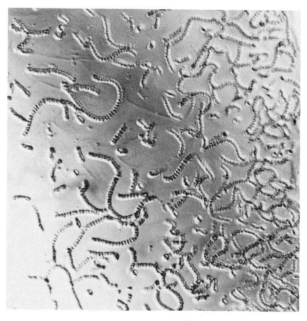

Figure 5-18. This photograph illustrates a colony of *Nostoc* filaments. Individual filaments of this alga secrete mucilage and form a rather rigid colonial matrix.

in face view since the cells are regularly arranged. However, cell division occurs in three directions rather than two, and therefore *Eucapsis* is cubical rather than planar. This alga is somewhat rare although it has been collected from the plankton of widely scattered North American lakes.

UNBRANCHED FILAMENTS

If a unicellular blue-green alga divided in a single direction, and the daughter cells did not separate, a *filament* of cells or *trichome* would result. This apparently occurred during the past, and filamentous blue-green algae are now the most common living types. Both branched and unbranched filamentous Cyanophyta are known. Branched Cyanophyta are further divided into true-branched and false-branched species.

Oscillatoria

Oscillatoria is a very widely distributed genus. Species occur in most marine and fresh water habitats as plankton or attached forms. Other species are common on or in the soil. Some species of *Oscillatoria* are often collected from ephemeral puddles or ponds even when they become very stagnant or putrid. *Oscillatoria* can be collected in nearly any ditch or gutter in the world. It is nearly impossible not to collect at least one or two species when making general algal collections from nearly any locality.

Oscillatoria forms cylindrical unbranched filaments lacking or with only a thin outer sheath (Fig. 5-17*A*). *Separation disks* or *necridia* are evident along the filaments. These are weak spots in the trichome formed from the death of a cell which cause the filament to break into short fragments known as *homogonia* (Fig. 5-17*B*). This method of reproduction is very successful as may be seen almost any summer when, in a matter of a week or two, *Oscillatoria* may turn a pond or lake olive green by its rapid reproduction.

Most species of *Oscillatoria* glide or oscillate from side to side. The exact mechanism of this movement and its reason for occurring are unknown, but this process is fascinating to observe with a microscope.

Nostoc

Nostoc is a colonial, blue-green alga composed of contorted filaments (Fig. 5-18). Each filament forms a copious sheath that fuses with that of other filaments to form a large, gelatinous, rounded colony. Occasionally these colonies grow so large that they are easily visible floating in the water. In fact, colonies occasionally become so large they are called *Nostoc* balls or *Nostoc* "apples" and are sometimes collected and eaten as a delicacy.

Nostoc cells are barrel-shaped or spherical, and a filament often appears as a string of beads. *Heterocysts* are produced along the filament. These enlarged, often empty-looking cells function in nitrogen fixation as well as cause the filament to fragment thus aiding in asexual reproduction. When the colony matures, akinetes are formed from some of the cells along the filament.

Nostoc is common on the surface of many moist, shady soils and occasionally is collected up to nearly 1 m (meter) deep in the soil. Most species are aquatic, where they occur as plankters, epiphytes or attached to the bottom.

Aphanizomenon

Filaments of *Aphanizomenon* are generally straight or only slightly curved (Fig. 5-14*D*). This alga is easily recognized because of the very large cylindrical akinetes along the trichome. Filaments characteristically are united into small or large bundles often visible to the unaided eye in lake plankton and give the water the appearance of containing an abundance of finely chopped grass or hay.

Aphanizomenon is occasionally a troublesome alga. It often forms toxin releasing blooms that cause fish kills and spoil water for agricultural and recreational uses. Fortunately, although this is a very widely distributed genus, troublesome blooms are not particularly common. However, they are increasing proportionately to the increase in organic pollution in our water supplies.

FALSE-BRANCHED FILAMENTS

False branches develop when the filament is separated into hormogonia which germinate without being liberated from the sheath. As these hormogonia begin to grow, they bend away from the direction of cell division of the original filament and thus form false branches. In other words, cell division in a falsely branched filament always occurs along only one axis just as in unbranched filaments. The appearance of branching arises merely from the bending of a filament.

Scytonema

Scytonema filaments are composed of large, barrel-shaped cells of uniform size imbedded in a wide conspicuous sheath (Fig. 5-14*B*). Heterocysts develop along the filament. False branches in *Scytonema* arise in pairs midway between two heterocysts (Fig. 5-19*A*).

Scytonema may be found in either aquatic or terrestrial habitats. It often occurs on soil moistened from the spray of a stream or waterfall, occasionally forming large, blue-green, felt-like masses. One species is common in hot springs.

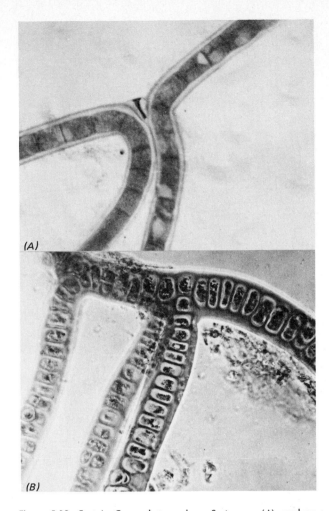

Figure 5-19. Certain Cyanophyta such as *Scytonema* (A) produce falsely branched filaments while others such as *Stigonema* (B) produce filaments with true branching.

Tolypothrix

Tolypothrix is similar to *Scytonema* in that it produces falsely branched filaments. However, it differs since the false branches occur singly rather than in pairs (Fig. 5-15*A*). These branches always arise immediately adjacent to a heterocyst rather than midway between two heterocysts.

Tolypothrix is generally aquatic although a few species grow on the soil. Most species occur as plankters or attached to submerged rocks or branches.

TRUE-BRANCHED FILAMENTS

True branches result from cell division in two directions forming a main axis with side or lateral branches. Branches are normally formed perpendicular to the main axis and may be frequent or rare among various species.

Stigonema

Stigonema is commonly collected from the surface of moist rocks. It is also occasionally found as a true plankter, or associated with other algae on the bottom of a slow moving stream or pond.

The main axis of *Stigonema* generally is composed of more than one row of cells (Fig. 5-14*G*). Lateral branches may be produced only on one side (Fig. 5-19*B*) or more commonly on all sides of this *multiseriate* axis depending upon the species. *Stigonema* filaments are imbedded in definite firm sheaths that may be clear or variously colored.

Hapalosiphon

Filaments of *Hapalosiphon* are similar in many respects to those of *Stigonema*. They differ chiefly in that the main axis of *Hapalosiphon* is often uniseriate rather than multiseriate, and the branches generally are produced from one side of the main axis only.

Halalosiphon is usually aquatic, although it is occasionally collected from the surface of moist soil.

SUMMARY

Algae are simple green plants that are common in most aquatic or moist environments. They differ widely in form and reproduction but none ever produces roots, stems, or leaves. Algae are classified on the basis of pigmentation, food storage products, morphology and reproduction. Sexually reproducing algae produce unicellular gametangia and the zygote never germinates to produce an embryo.

Cyanophyta are the simplest of all algae. They have no well-defined nuclei or plastids. Mitosis and meiosis never occur, and cell division is by fission. Asexual reproduction is common but sexual reproduction has never been definitely observed in any species. Because of this, blue-green algae do not conform to the generalized life cycle of a sexually reproducing plant and the terms sporophyte generation and gametophyte generation have no meaning among blue-green algae.

Several types of thalli occur in different Cyanophyta. Unicellular Cyanophyta are considered to be primitive forms, and the more complex species are thought to have been derived from them.

The division Cyanophyta is important for many reasons including nitrogen fixation, oxygen production, erosion protection, primary production and water pollution.

FURTHER READING

GENERAL REFERENCES ON ALGAE

CHAPMAN, V. J., *The Algae,* St. Martin's Press, Macmillan and Co., London, 1962.

HARDY, JOHN and HERBERT CURL JR., "The Candy-Colored Snow-Flaked Alpine Biome," *Natural History,* 71(9):74–78, 1972.

IDYLL, C. P., "The Anchovy Crisis," *Scientific American,* 228(6):22–29, 1973.

ISAACS, JOHN, "The Nature of Oceanic Life," *Scientific American,* 221(3):146–162, 1969.

JACKSON, DANIEL (Ed.), *Algae, Man and the Environment,* Syracuse Univ. Press, Syracuse, New York, 1968.

MORRIS, IAN, *An Introduction to the Algae,* Hutchinson and Co., Ltd., London, 1967.

PALMER, MERVIN, "Algae in Water Supplies," *United States Public Health Service Publication No. 657,* 1959.

PARKER, BRUCE, "Life in the Sky," *Natural History,* 75(1):56–62, 1970.

PRESCOTT, GERALD, *Algae of the Western Great Lakes Area,* Wm. C. Brown Co., Dubuque, Iowa, 1962.

PRESCOTT, GERALD, *How to Know the Fresh-Water Algae,* Second edition, Wm. C. Brown Co., Dubuque, Iowa, 1970.

SMITH, GILBERT, *The Fresh-Water Algae of the United States,* Second edition, McGraw-Hill Book Co., New York, 1950.

WEISS, FRANCIS, "The Useful Algae," *Scientific American,* 187(6):15–17, 1952.

ZAHL, PAUL, "Algae: The Life-Givers," *National Geographic,* 145(3):360–377, 1974.

SPECIFIC REFERENCES ON BLUE-GREEN ALGAE

BARGHOORN, ELSO, "The Oldest Fossils," *Scientific American,* 224(5):30–42, 1971.

CARR, N. G. and B. A. WHITTON (Eds.), *The Biology of Blue-Green Algae,* University of California Press, Berkeley, 1973.

CASTENHOLZ, R. W., "Thermophilic Blue-Green Algae and the Thermal Environment," *Bacteriological Review,* 33:476–504, 1969.

DROUET, FRANCIS, "Cyanophyta," In G. M. Smith, *Manual of Phycology*, 159–166. Ronald Press, New York, 1951.

ECHLIN, PATRICK, "The Blue-Green Algae," *Scientific American*, 214(6):74–80, 1966.

ECHLIN, PATRICK and I. MORRIS, "The Relationships Between Blue-Green Algae and Bacteria," *Biological Review*, 40:143–187, 1965.

FOGG, G. E., W. D. P. STEWART, P. FAY and A. E. WALSBY, *The Blue-Green Algae*, Academic Press, New York, 1973.

GANTT, ELIZABETH and S. CONTI, "Ultrastructure of Blue-Green Algae," *Journal of Bacteriology*, 97:1486–1493, 1969.

LANG, N., "The Fine Structure of Blue-Green Algae," *Annual Review of Microbiology*, 22:15–46, 1968.

SCHOPF, J. W. and J. B. BALCIC, "New Microorganisms from The Bitter Springs Formation (Late Precambrian) of the North-Central Amadeus Basin, Australia," *Journal of Paleontology*, 45:925–959, 1971.

WOLK, PETER, "Physiology and Cytological Chemistry of Blue-Green Algae," *Bacteriological Reviews*, 37(1):32–101, 1973.

Division Chlorophyta:
the Green Algae

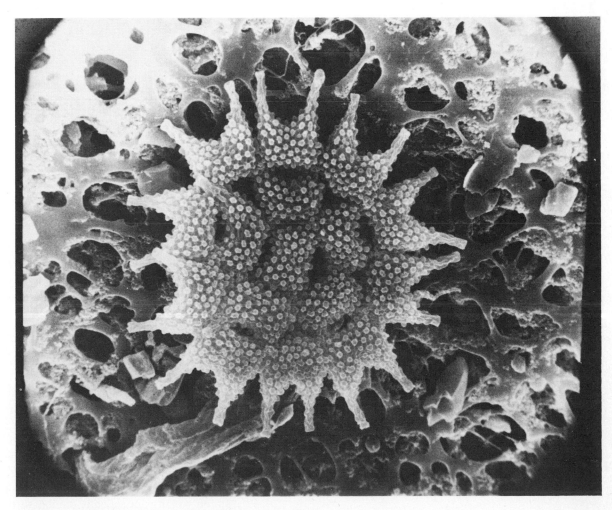

Scanning electron micrograph of the colonial green alga *Pediastrum*. (Courtesy of Jeremy Pickett-Heaps, University of Colorado.)

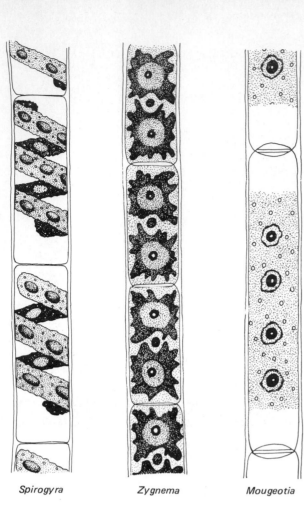

Spirogyra Zygnema Mougeotia Oedogonium

Figure 6-1. The chloroplast shape and size is often characteristic among green algae. These Chlorophyta demonstrate different chloroplast configurations.

GENERAL CHARACTERISTICS

Chlorophyta are common and prominent members of the earth's flora. As many as 20,000 species of these algae have been described, and undoubtedly more will be collected and named in the future.

These plants are known as green or grass-green algae because of their pigmentation. Chlorophylls *a* and *b* are the two prominent pigments generally present in a ratio that gives the plant a bright green color. Various carotene and xanthophyll pigments are also present occasionally coloring the cells yellow-green or orange. All higher plants have the same pigmentation as green algae. This is one evidence that Chlorophyta may be ancestral to higher plants.

Unlike Cyanophyta, the pigments of green algae are contained within well-defined, membrane-bound plastids. Chloroplasts are often very distinctive among Chlorophyta (Fig. 6-1) and are often used as one basis for classification. The number of chloroplasts in each cell varies from one to many according to species.

Photosynthetic products of Chlorophyta are stored as starch which is chemically similar to that stored by higher plants. This starch may be stored in amyloplasts or within the chloroplast itself. When it is stored within the chloro-

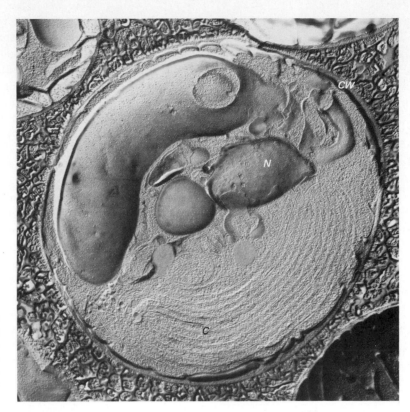

Figure 6-2. Even though many Chlorophyta are very simple organisms, they are much more complex than bacteria and blue-green algae. This freeze etch electron micrograph of the single-celled green alga *Cyanidium* shows a cell wall (CW), chloroplast (C), nucleus (N) and other organelles. (Courtesy of Andrew Staehelin, University of Colorado.)

plast, it is generally formed by a *pyrenoid* (Fig. 6-1). Each chloroplast of most Chlorophyta contains from one to several pyrenoids.

From one to several definite nuclei are present within each green algal cell. These nuclei are generally spherical or discoid (Fig. 6-2) and are bounded externally by a definite nuclear membrane. They contain well organized chromosomes, and thus true mitosis occurs.

The cell wall of green algae is commonly two layered, although the layering is usually not as conspicuous as in Cyanophyta. The inner cell wall is usually rigid and composed of cellulose, while the outer wall contains a large amount of pectin. This outer wall is slimy or mucilaginous in some species, and rigid and brittle in other species because of impregnation by calcium carbonate.

A wide variety of thallus types occurs among chlorophyta. Similar to Cyanophyta, those forms considered most primitive are unicellular. Both motile and nonmotile unicellular species are known. Most motile species exhibit flagella which undulate rapidly, causing the entire cell to move. The majority of Chlorophyta are filamentous. Both branched and nonbranched species are common. A few species form planar colonies and several are spherical. In some species, mature plants are composed of a constant, definite number of cells, and are termed *coenobic*. Many species are complex in structure, far more so than blue-green algae. Such complex species often have portions of the thallus specialized for certain functions. For instance, a *holdfast* (Fig. 6-3) which attaches the alga to

Figure 6-3. Often green algae demonstrate a thallus that has certain cells specialized for definite functions. Thus, this very young plant of *Oedogonium* sp. shows a holdfast that functions to attach the alga to the substrate.

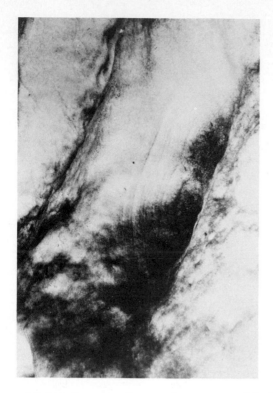

Figure 6-4. Several species of green algae are cryophilic. This snow bank is darkened near the center and right side by a species of *Chlamydomonas.*

some object in the water may be found at the base of the filament of many species. Likewise, certain cells or portions of the thallus are often specialized for reproduction. These usually appear much different from the rest of the thallus. Specialization of various parts of the thallus for separate functions is an important developmental step in plant evolution which represents a significant advancement over Cyanophyta.

Both sexual and asexual reproduction occur among Chlorophyta. Asexual reproduction is accomplised in a variety of ways. Some unicellular species merely divide by mitosis to form a new plant. Fragmentation is common among nearly all filamentous species.

Many Chlorophyta produce spores in asexual reproduction. Several kinds are produced. *Zoospores* are one common type. A zoospore is a flagellated cell able to germinate to form a new plant. In addition to zoospores, several types of nonmotile spores including akinetes are produced by Chlorophyta.

Sexual reproduction is common in most species, representing a tremendous advancement over Cyanophyta. The recombination of genetic material occurring by sexual reproduction coupled with natural selection is largely responsible for allowing plants to change and improve by evolution. Several variations in sexual reproduction occurring among Chlorophyta will be discussed on later pages.

Table 6-1 is a comparison of the characteristics of Cyanophyta and Chlorophyta.

TABLE 6-1. A Comparison of the Major Characteristics of Cyanophyta and Chlorophyta.

	Cyanophyta	*Chlorophyta*
Nuclei	Lacking; nuclear membrane and chromosomes absent.	Well defined; bounded by a nuclear membrane; chromosomes present.
Cell Division	Fission	Mitosis; Meiosis
Plastids	Lacking	Well defined; bounded by a plastid membrane.
Prominent Pigments	Chlorophyll *a* c-phycocyanin c-phycoerythrin	Chlorophyll *a* Chlorophyll *b*
Sexual Reproduction	Absent (or at least atypical and rare).	Frequent; occurs in most species.
Food Storage Product.	Cyanophycean starch; proteins	Starch

OCCURRENCE AND ECOLOGY

Green algae grow in almost any environment where light and moisture are present. In moist climates, it is not uncommon for the shingles of a roof to become green due to the growth of a green algae. Also, moist cement floors of diary barns or greenhouses, walls of buildings, moist soils, fallen trees, and many other suitable substrates are often covered with luxuriant growths of terrestrial Chlorophyta. Many species are epiphytic, and one genus, *Protococcus,* often forms the "moss" growing on the north side of trees in Northern Hemisphere forests.

About 90 percent of green algae are fresh water in occurrence and 10 percent are strictly marine. Marine Chlorophyta often occur attached in the intertidal zone. In certain localities they make up a prominent part of the local flora, particularly in polluted waters.

Fresh water green algae are widely distributed on earth, occurring in a great variety of habitats. Some species are very important plankters in streams and lakes. Others are equally significant as attached forms on the bottoms of the same waters.

Some species are able to withstand the harsh desert environment where they often occur in temporary pools, beneath translucent stones, or intermingled with mosses and lichens in the shade of other plants. Some Chlorophyta are cryophilic, growing on ice and snow (Fig. 6-4) and often turning it bright red or orange. These curious algae contain a red pigment in their plastids. Red snow is fairly common in mountainous regions of the world during spring months.

IMPORTANCE

Green algae are important primary producers, especially in fresh water lakes and streams. They are the base of many food chains in these habitats (Fig. 6-5). It is often

Figure 6-5. Algae are critical in the food chains of many animals. These tadpoles are grazing on green algae which make up a large part of their diet. (Courtesy of Paul A. Zahl, © 1973, National Geographic Society.)

Figure 6-6. Several species of Chlorophyta are currently under investigation as sources of food for livestock or humans. The algae being grown in these tanks will be fed to several species of fish in order to determine what species of fish can best use them. The fish can subsequently be harvested and the protein extracted and used directly by humans or as a supplement in the diet of livestock.

essential to understand these food chains and to know which algae are most important in them in order to properly manage important fisheries, recreational lakes, and irrigation reservoirs.

Green algae occasionally contaminate water supplies, although they are less important than Cyanophyta. Under certain circumstances some Chlorophyta form blooms with much the same results as blue-green algal blooms. Many green algae are very tolerant of polluted waters, and hence may produce large populations in such environments. For instance, *Cladophora* has become the most significant nuisance alga along polluted shores of the Great Lakes. It occasionally forms algal mats several feet thick covering many square miles. A few Chlorophyta cause unwanted tastes and odors in commercial water supplies.

Some Chlorophyta are eaten directly by humans, and certain species are an excellent source of nutrition. Research is presently being done on the feasibility of using these algae on a large scale as a source of protein for the world's population (Fig. 6-6). It is possible under experimental conditions to create a closed biological system in which algae grown in tanks are concentrated for food. Waste products of organisms using this food are recycled through the tanks where the algae use these nutrients for their own metabolism. Energy for such a system originates from light that allows photosynthesis to occur. Such systems will likely be used for long term space explorations. More important, modifications of such systems may be used in the future to provide the food necessary to feed large populations. For instance, tertiary treatment ponds of sewage treatment plants are natural sites for the rapid growth of algae since the temperature of these ponds is usually favorable and the ponds are very rich in nutrients. If algae could be grown on these ponds and harvested for use as food, two problems would be partially solved. First, they would provide a source of food and thus partially alleviate the protein shortage in the world; and second, since nutrients would be extracted by the algae from these treatment ponds, these nutrients would not be released into the environment to cause further pollution of our water systems. Preliminary studies have shown that as much as 90 percent of the mineral nutrients made available by sewage treatment could be used for the growth of algae in such a system.

FOSSIL CHLOROPHYTA

Green algae are rarely represented in the fossil record. However, some species precipitate from the water calcium carbonate that often impregnates their outer cell walls. These species become incrusted with lime deposits, and are more easily fossilized than other algae. Such species have left a comparatively good although incomplete fossil record, dating back at least to the Ordovician Period. A few green algae have been reported from as early as Pre-Cambrian times.

Chara is a peculiar green alga that has left a good fossil record. Species of *Chara* produce lime-encrusted gametangia that fossilize readily. These algae were apparently common during the Paleozoic and Mesozoic Eras. In many geological formations found in the western part of the United States, especially those of Jurassic Age, *Chara* and related genera have left extensive fossil material.

REPRESENTATIVE CHLOROPHYTA
ORDER VOLVOCALES

The division Chlorophyta is divided into many orders and families. Such orders often represent natural lines of evolution and development. One of the best examples of such an evolutionary line is found in the order Volvocales.

The order Volvocales contains several algae thought to be primitive Chlorophyta. This is evidenced by motility of the cells in the vegetative thallus. The most primitive genus of the order, *Chlamydomonas,* is unicellular. The most advanced is *Volvox,* which is a rather large spherical plant. A direct evolutionary development from *Chlamydomonas* through several other genera to *Volvox* probably occurred in this order.

Volvocales may often be collected from waters with high nitrogen content. Barnyard pools or polluted waters are good habitats for these algae.

Chlamydomonas

Chlamydomonas is a unicellular, biflagellate, single-celled alga. Each cell contains a single cup-shaped chloroplast with one large, evident pyrenoid. Cells are uninucleate, and contain the other organelles common in plant cells, although they are difficult to observe with the light microscope. A reddish or orange *eye spot* or *stigma* is often evident within the cell. This structure is composed of

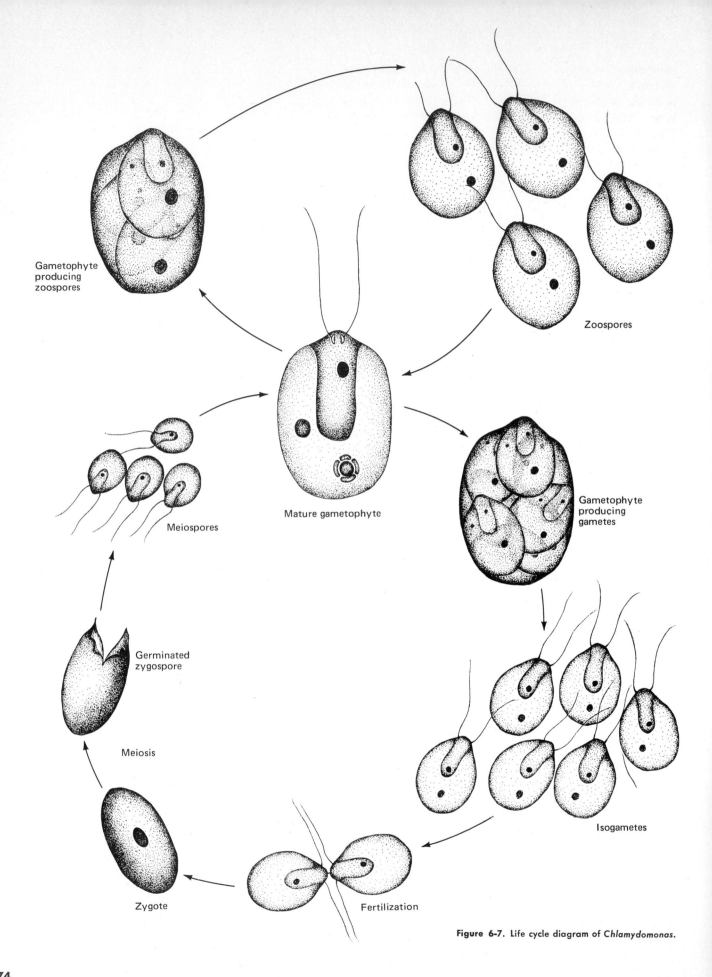

Gametophyte
producing
zoospores

Zoospores

Meiospores

Mature gametophyte

Gametophyte
producing
gametes

Germinated
zygospore

Meiosis

Zygote

Fertilization

Isogametes

Figure 6-7. Life cycle diagram of *Chlamydomonas*.

carotene pigments and is light sensitive. When the amount of light is too low for adequate photosynthesis, the organism is directed toward the light source. When too much light is present, *Chlamydomonas* will move away from the light. Thus, *Chlamydomonas* cells are often aggregated in regions of optimum light for photosynthesis.

Asexual reproduction in *Chlamydomonas* is by cell division or the production of zoospores (Fig. 6-7). Zoospore formation occurs when the protoplast of a cell divides to form several small daugther cells. The cell wall of the mother cell cracks open to release the zoospores which swim about rapidly and gradually enlarge to become the same size as the parent cell. If environmental conditions are optimum, this type of reproduction may produce large populations in a relatively short period of time.

When conditions become adverse for the vegetative growth of the organism, sexual reproduction is often initiated (Fig. 6-7). This process is similar in its initial stages to asexual reproduction. The protoplast divides to form several gametes within the cell wall of the mother cell. These gametes are released into the water where they swim actively for some time. The flagella of two gametes formed from separate plants entangle and evenutally the two cells unite to form a zygote. *Chlamydomonas* gametes are usually alike in shape and size and are termed *isogametes*. The union of isogametes is *isogamous* sexual reproduction.

The zygote that develops from *syngamy* (the fusion of gametes) often enters a resting stage, and is then termed a *zygospore*. This structure often functions to carry the alga through unfavorable circumstances such as cold or drought. The zygote or zygospore germinates by meiosis to produce four flagellated meiospores (or zoospores) that mature to become new *Chlamydomonas* plants.

Note that the gametophyte generation is the dominant, free-living generation. The sporophyte generation is highly reduced and is represented only by the zygote. Some botanists do not consider this to be a valid example of alternation of generations. However, the zygote fulfills both criteria for a true sporophyte generation since it is diploid and produces meiospores.

Gonium

Gonium (Fig. 6-8) is a colonial alga composed of from 4–32 cells, each of which is very similar to *Chlamydomonas*. These cells are arranged to form a flat rectangular plate one cell thick often with four or eight central cells surrounded by 12 or 24 peripheral cells. The entire colony is enclosed within a gelatinous envelope and is free swimming. Some species of *Gonium* are coenobic. For instance, *Gonium sociale* always has four cells in each colony.

Asexual and sexual reproduction in *Gonium* are similar to that of *Chlamydomonas*. The zygote of *Gonium* develops a thick wall and functions in carrying the alga through un-

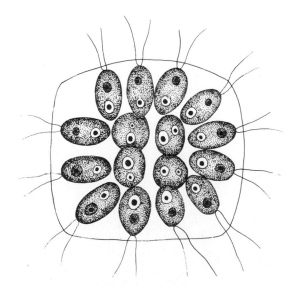

Figure 6-8. Many green algae produce plants of a characteristic shape and size. This figure shows *Gonium*, a common member of the Volvocales. It forms flat, angular colonies of *Chlamydomonas*-like cells.

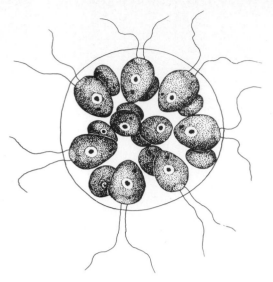

Figure 6-9. *Eudorina* is another common member of the order Volvocales. It forms small, spherical colonies.

favorable environmental circumstances. The zygospore germinates by meiosis producing four meiospores that usually remain together and divide to produce a mature colony.

Pleodorina

Pleodorina is closely related to *Gonium* although it is a spherical rather than a flat colony composed of from 32–126 *Chlamydomonas*-like cells (Fig. 3-3*B*). This alga differs from more primitive Volvocales since the colony contains cells of two distinct sizes. Some cells are small and never enter into reproduction. Reproductive cells are about twice as large as the smaller vegetative cells.

Sexual reproduction occurs when poor conditions for growth develop. Gametes are produced by the larger cells of the colony and are of two sizes. Some are relatively large and although motile are sluggish. Others are smaller and extremely active. A small gamete must unite with one of the larger gametes in fertilization. This type of reproduction is *anisogamous* and the gametes are *anisogametes*. The zygote formed from syngamy develops a thick wall and carries the alga through poor growing conditions.

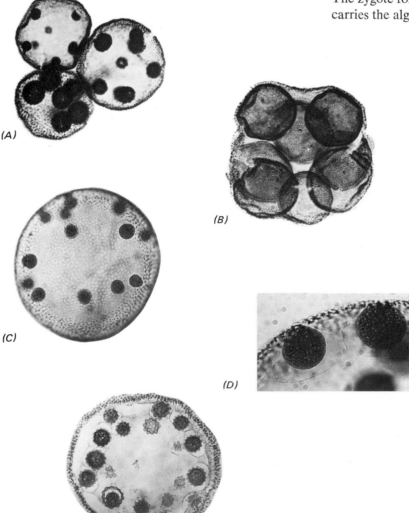

(A)

(B)

(C)

(D)

(E)

Figure 6-10. The most complex alga of the order Volvocales is *Volvox*. *A* shows three separate plants, each containing asexual several large daughter colonies. *C* shows a plant with several daughter colonies. *B* is an enlargement of a single plant with eggs and *D* is an enlargement of a portion of *C* showing two eggs. *E* shows a single plant with several zygospores.

The most complex genus in the order Volvocales is *Volvox*. This alga forms large, complex, spherical plants containing 500 to several thousand cells (Fig. 6-10). These cells are arranged at the periphery of the sphere, leaving a hollow center. Each cell is similar to *Chlamydomonas,* although they are interconnected by fine cytoplasmic strands.

In addition to the greater complexity of the vegetative plant, sexual reproduction is also more advanced in *Volvox*. Only a few cells of the entire colony function in sexual reproduction. Some of these reproductive cells enlarge to become rounded, nonmotile eggs (Figs. 6-10C and 6-10D). Others divide several times to form many small, highly motile, flagellate *sperm*. The egg is the female gamete, and the sperm are male gametes. This type of sexual reproduction is termed *oogamous*. Oogamy is the most advanced form of sexual reproduction and has developed from isogamy probably through anisogamy. The egg is not released from the parent colony, and the zygote develops in the center, hollow portion of the colony. It produces a thick, highly ornate wall to become a zygospore (Figs. 6-10E and 6-11) which overwinters the species. The zygote germinates by meiosis in the spring to produce four motile meiospores (Fig. 6-10). These meiospores divide to form the new colony. Thus, the *Volvox* gametophyte generation is dominant; the sporophyte generation is represented only by the zygote.

Asexual reproduction also occurs in *Volvox* and normally continues throughout the growing season. This process begins when a cell of the colony divides many times to produce a multicellular invagination into the mother colony. This small daughter colony (Figs. 6-10A, 6-10B, and 6-11) is released from the parent plant and divides rapidly to form a new *Volvox* colony.

Let us summarize the order Volvocales. *Chlamydomonas* is a small, unicellular biflagellate alga with a single cup-shaped chloroplast which reproduces sexually by isogamy. *Pleodorina* is more complex since it forms multicellular colonies and demonstrates anisogamous sexual reproduction. Each cell of the *Pleodorina* colony is very much like *Chlamydomonas,* and it is probable that *Pleodorina* evolved from *Chlamydomonas* or a similar plant. *Volvox* is the most complex genus of this order. It is a large, spherical colony with interconnected cells and oogamous sexual reproduction. Each cell is similar to *Chlamydomonas. Volvox* appears to have developed along a direct evolutionary line from *Chlamydomonas* or a related plant.

ORDER ULOTRICHALES

The order Ulotrichales contains many common green algae. Representatives of this order may be collected from

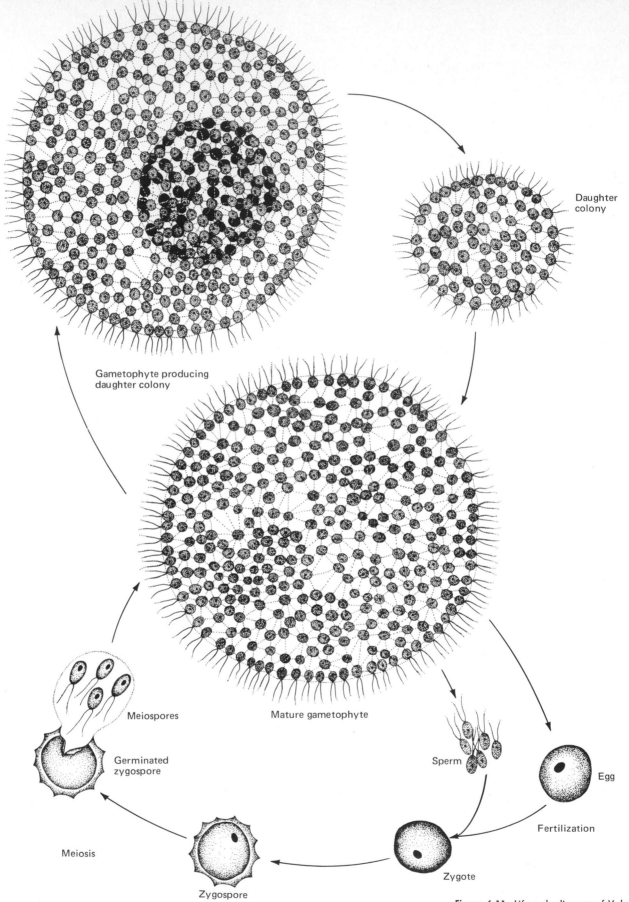

Gametophyte producing
daughter colony

Daughter
colony

Mature gametophyte

Meiospores

Germinated
zygospore

Meiosis

Sperm

Egg

Fertilization

Zygote

Zygospore

Figure 6-11. Life cycle diagram of *Volvox*.

many waters throughout the world. Some genera are especially common in rivers of temperate regions.

This order contains filamentous and either branched or unbranched species. Most botanists postulate that this order probably evolved from a unicellular alga similar to *Chlamydomonas*. Good evidence for this theory is that zoospores produced by various Ulotrichales are similar to *Chlamydomonas*. Sexual reproduction is well developed among Ulotrichales, varying from isogamy to oogamy. In addition, asexual reproduction by zoospores or fragmentation is frequent.

Ulothrix

Ulothrix is a widely-distributed alga growing in fresh and marine waters. Cool, well oxygenated streams often contain large populations of this alga. Some species grow on moist or wet soil. Most species of *Ulothrix* do not grow well in polluted waters, and so are good indicators of clean waters.

Ulothrix cells are barrel-shaped and aligned end to end forming long, uniseriate filaments (Fig. 6-12). The cells of the filament are alike except for the basal cell which is modified to form a holdfast.

Cells of *Ulothrix* each contain a characteristic, large, bracelet-shaped chloroplast near the cell wall (Fig. 6-13). This chloroplast generally contains several pyrenoids. The interior of the cell is occupied by a large central *vacuole*. This vacuole is penetrated by strands of cytoplasm containing the other organelles of a living cell.

Ulothrix produces zoospores throughout the spring and summer (Fig. 6-12). These spores are formed by cell division which produces from 2–32 small protoplasts each with four flagella. These are released from the mother cell and swim for a time (Fig. 6-13) before they become attached to the substrate and divide to form a new *Ulothrix* filament. When environmental conditions are proper, asexual reproduction may be rapid, forming a large population of plants in a short period of time.

When conditions for growth become poor, sexual reproduction is often instigated (Fig. 6-13). Early stages in this process are similar to zoospore formation. The protoplast of a cell divides several times to form many small gametes each with but two flagella. Every cell in a filament except the holdfast often divides simultaneously to produce gametes. These gametes are alike in size and shape and are released into water where they may swim for several hours. Gametes from two separate plants are necessary to produce a zygote. This important characteristic is demonstrated by most plants throughout the plant kingdom. Plants requiring two sexual strains for fertilization are usually said to be *heterothallic*. *Homothallic* plants are self-fertile.

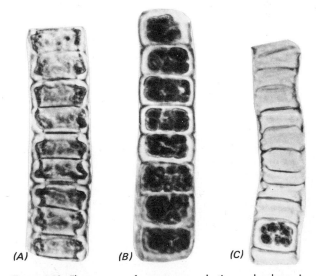

(A)　　　(B)　　　(C)

Figure 6-12. The process of zoospore production and release is documented in this figure. A shows a vegetative filament containing characteristic bracelet-shaped chloroplasts. The filament at B has produced zoospores, most of which have been released from the filament at C.

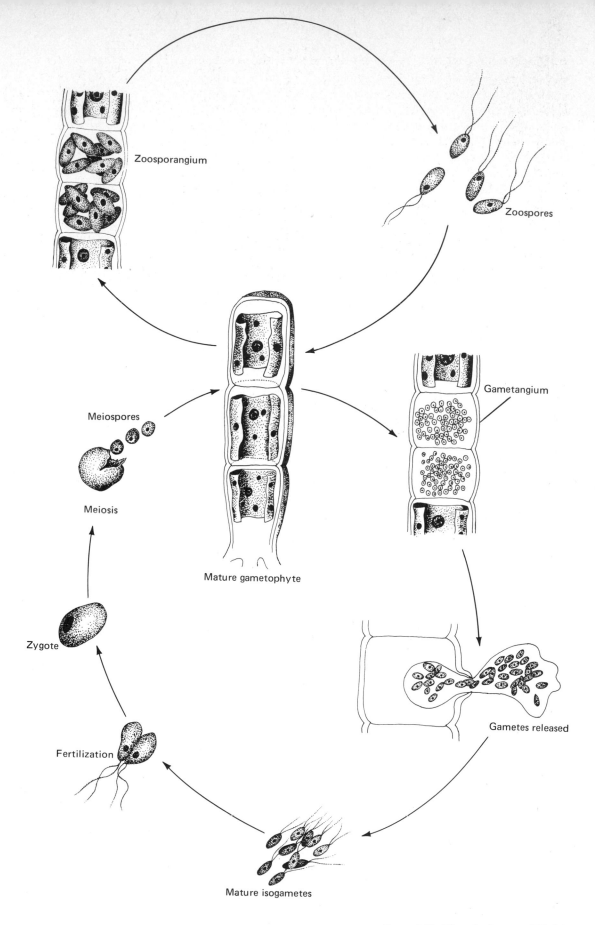

Zoosporangium

Zoospores

Gametangium

Meiospores

Mature gametophyte

Meiosis

Gametes released

Zygote

Fertilization

Mature isogametes

Figure 6-13. Life cycle diagram of *Ulothrix*.

The zygote resulting from isogamy develops a thick wall to become a resistant zygospore. The zygospore undergoes meiosis and germinates to produce four motile meiospores. The gametophyte generation of *Ulothrix* predominates, and the sporophyte generation is represented only by the zygote.

ORDER OEDOGONIALES

Oedogonium

Oedogonium is similar in many respects to *Ulothrix*. It grows well in flowing waters of high oxygen content, but may be collected from a wide variety of habitats such as ponds and lakes, moist soil, or as an epiphyte on aquatic plants (Fig. 6-14). Under certain environmental conditions, some waters become choked with troublesome *Oedogonium* mats.

Cells of *Oedogonium* are aligned end to end to form a long, unbranched, uniseriate filament similar to *Ulothrix*. These cells are basically alike except for the basal holdfast cell (Fig. 6-15). Each cell contains a single, net-shaped chloroplast at the periphery of the cell near the wall. This chloroplast contains many conspicuous pyrenoids that often become ensheathed by starch grains. Each cell contains a large central vacuole traversed by several strands of cytoplasm.

At the onset of sexual reproduction, certain cells along the filament become modified and appear different from the other cells. These cells become gametangia of two sizes. The larger produces the female gamete or egg and is known as an *oogonium* (Fig. 6-15). The smaller is the sperm-bearing gametangium or *antheridium* (Figs. 6-15 and 6-16). Sperm are large and motile and are generally produced in pairs in the antheridia. A single egg is produced in each oogonium. A fertilization pore develops in the oogonium when the egg reaches maturity. Sperm are attracted toward the oogonium when they are released from the antheridia. One or more sperm may enter the fertilization pore, but only one fertilizes the egg. The resulting zygote develops a thick wall and is resistant to unfavorable conditions. Later it germinates by meiosis to produce four motile meiospores, each of which may become a new *Oedogonium* plant. Since the zygote germinates directly by meiosis, it is the only diploid cell in the life history of this alga. Therefore *Oedogonium* exhibits a dominant gametophyte and a highly reduced sporophyte generation.

Asexual reproduction is common in *Oedogonium* throughout the spring and summer. This process occurs by the formation of zoospores (Figs. 6-15 and 6-17) or fragmentation of the thallus, and is often responsible for the development of very large populations.

Figure 6-14. Many plants grow as epiphytes on other plants. This photograph shows several *Oedogonium* filaments growing on another alga.

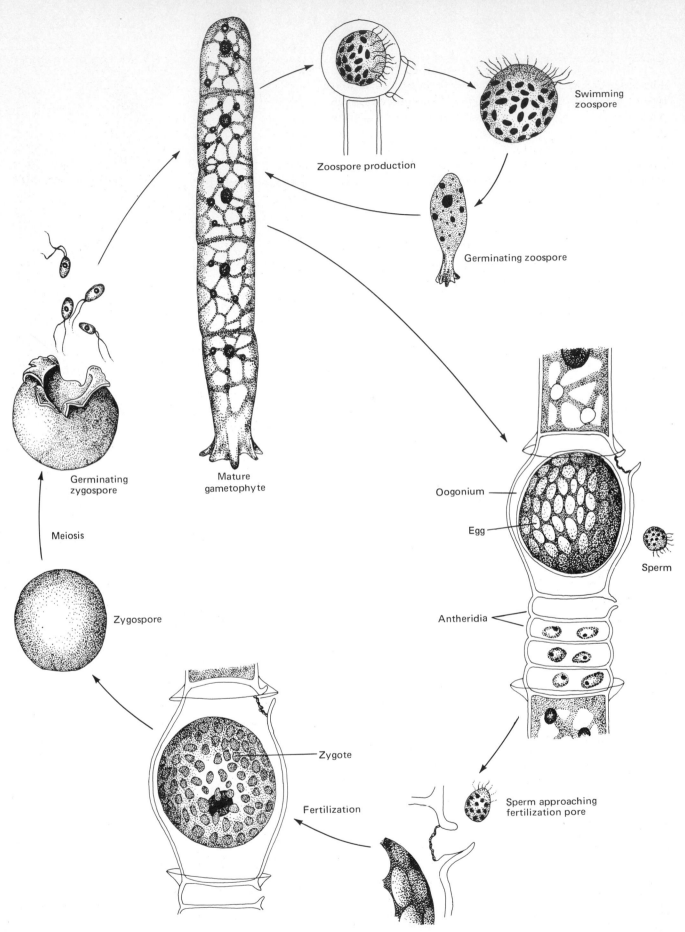

Swimming
zoospore

Zoospore production

Germinating zoospore

Mature
gametophyte

Germinating
zygospore

Meiosis

Zygospore

Zygote

Fertilization

Oogonium

Egg

Sperm

Antheridia

Sperm approaching
fertilization pore

Figure 6-15. Life cycle diagram of *Oedogonium*.

ORDER CONJUGALES

The order Conjugales contains Chlorophyta with a peculiar type of syngamy known as *conjugation*. This process occurs when the entire protoplast of a cell acts as a single gamete and fuses with the entire protoplast of another cell. Conjugales are prevalent throughout the world in both temperate and tropical regions. Almost any collection of algae contains one or more representatives of this order.

Spirogyra

Spirogyra is common in most fresh water ponds, puddles, lakes, and streams. It often forms luxuriant bright green "clouds" in these waters, and is occasionally a serious nuisance.

Cells of *Spirogyra* are aligned end to end to form uniseriate filaments of indefinite length. Each cell is cylindrical and contains one or more characteristic, helical chloroplast (Figs. 6-1 and 6-18). These chloroplasts are distinctive and make the identification of this genus easy. Each chloroplast normally contains several distinct pyrenoids, often with evident starch granules. Unlike many other filamentous green algae, most species of *Spirogyra* do not form basal holdfasts.

Asexual reproduction is common, occurring chiefly by fragmentation. Occasionally nonmotile spores or akinetes are formed by some cells of the filament.

Sexual reproduction occurs when the proper conditions for vegetative growth cease. Most species are self-sterile and the sexual process begins when two separate strands become closely aligned (Figs. 6-18 and 6-19*A*). Cells on each strand develop short protrusions which contact each other. Eventually the cell walls between them dissolve to form a complete *fertilization tube* connecting the two cells. After this occurs, the protoplast of each cell of one filament migrates through the tube and fuses with the protoplast of each cell of the other. Conjugation is a specialized case of isogamy, although some botanists consider it anisogamy since the protoplasts of only one sexual strain move through the fertilization tube.

The resultant zygotes develop thick walls and become resistant to adverse conditions. These germinate by meiosis generally in the early spring. Three of the resulting haploid nuclei disintegrate, and the fourth divides to produce a new *Spirogyra* filament. Again the zygote is the only cell of the sporophyte generation, whereas the gametophyte generation is prominent and well developed.

Zygnema and *Mougeotia* are also common genera of the order Conjugales (Fig. 6-1). *Zygnema* is characterized by two star-shaped chloroplasts contained in each cell. *Mougeotia* has very long cells with a single elongate chloroplast that often only partially fills the cell. The reproduction of these genera is similar to that of *Spirogyra* (Fig. 6-20).

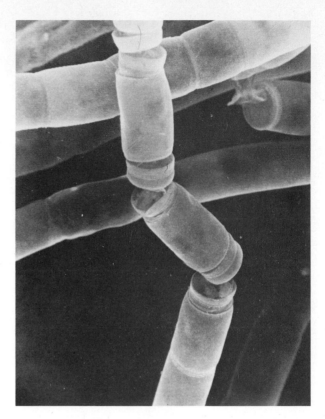

Figure 6-16. Antheridia of *Oedogonium* are disk-shaped and contain two sperm. This filament of cells has broken at the groups of antheridia. (Scanning electron micrograph courtesy of Jeremy Pickett-Heaps, University of Colorado.)

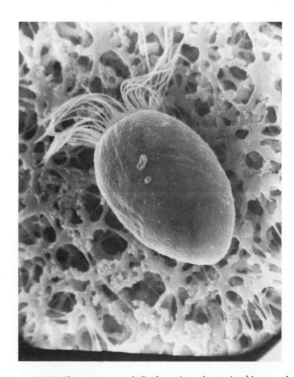

Figure 6-17. The zoospore of *Oedogonium* shown in this scanning electron micrograph is very distinctive because of its many flagella arranged in a ring. (Courtesy of Jeremy Pickett-Heaps, University of Colorado.)

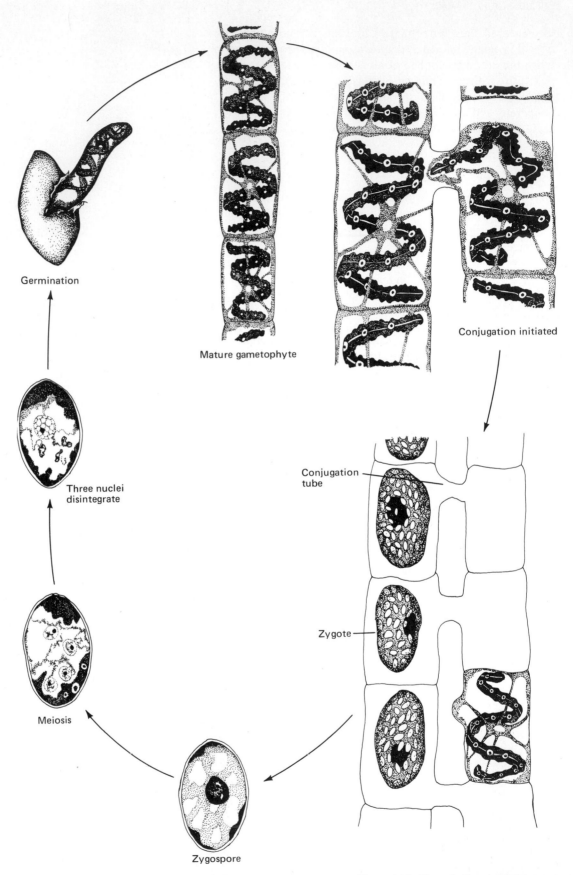

Germination

Mature gametophyte

Conjugation initiated

Three nuclei
disintegrate

Conjugation
tube

Zygote

Meiosis

Zygospore

Figure 6-18. Life cycle diagram of *Spirogyra*.

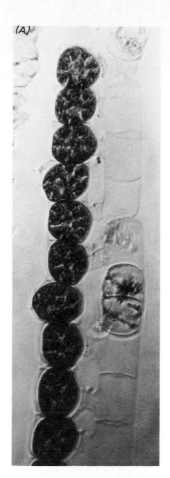

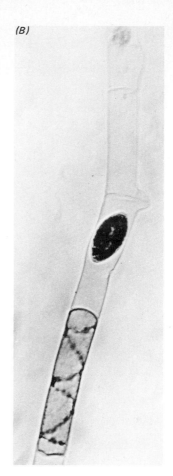

Figure 6-19. *Spirogyra* reproduces sexually by the process of conjugation. A shows two strands in close proximity which are connected by fertilization tubes. Gametes from the filament on the right have moved to the left through these tubes to form zygotes in the filament on the right side. B shows a single filament of a different species of *Spirogyra* which is self-fertile. A conjugation tube has formed between two cells on this filament and a gamete has migrated from the upper cell to the lower to form a zygote. The cell immediately below the one that contains the zygote has not entered into the sexual cycle.

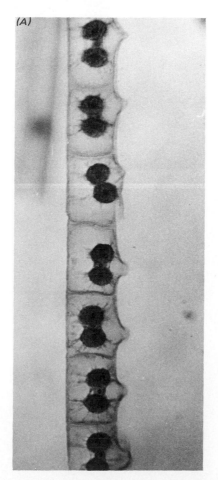

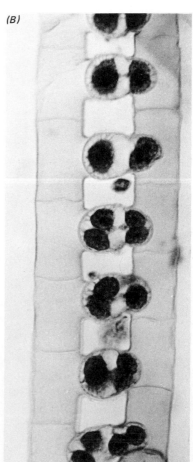

Figure 6-20. Green algae of the order Conjugales (Zygnematales) are characterized by distinctive chloroplast and cell shapes and especially by conjugation in sexual reproduction. These filaments of *Zygnema* show the process of conjugation. The filament at A is just beginning to produce fertilization tubes and those at B have conjugated. Zygotes are developing in the fertilization tubes.

(A)

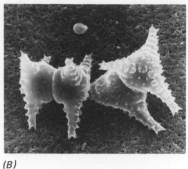
(B)

(C)

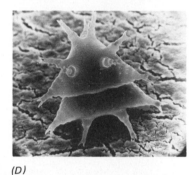
(D)

Figure 6-21. Desmids are Conjugales with characteristic, often beautiful shapes. These scanning electron micrographs show the desmids *Micrasterias* and *Staurastrum*. A and C show *Micrasterias*. The cell at C has recently undergone cell division. B and D are *Staurastrum*. The semicells are clear in all of these photographs. (Courtesy of Jeremy Pickett-Heaps, University of Colorado.)

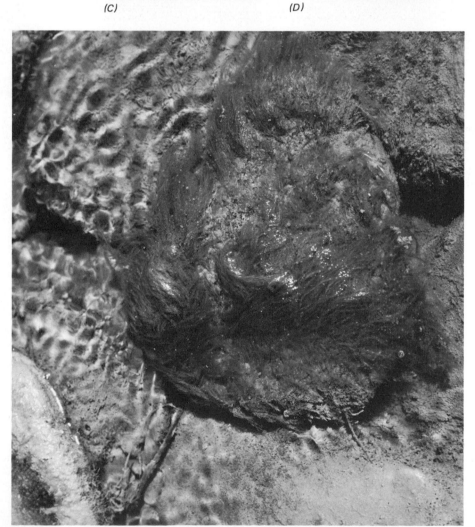

Figure 6-22. *Cladophora* is one alga that is in the upswing throughout much of the world. Apparently this is due to the ability of *Cladophora* to grow in waters that have been enriched by human activities. This rock covered with *Cladophora* was photographed in the spring in a high mountain stream in central Utah.

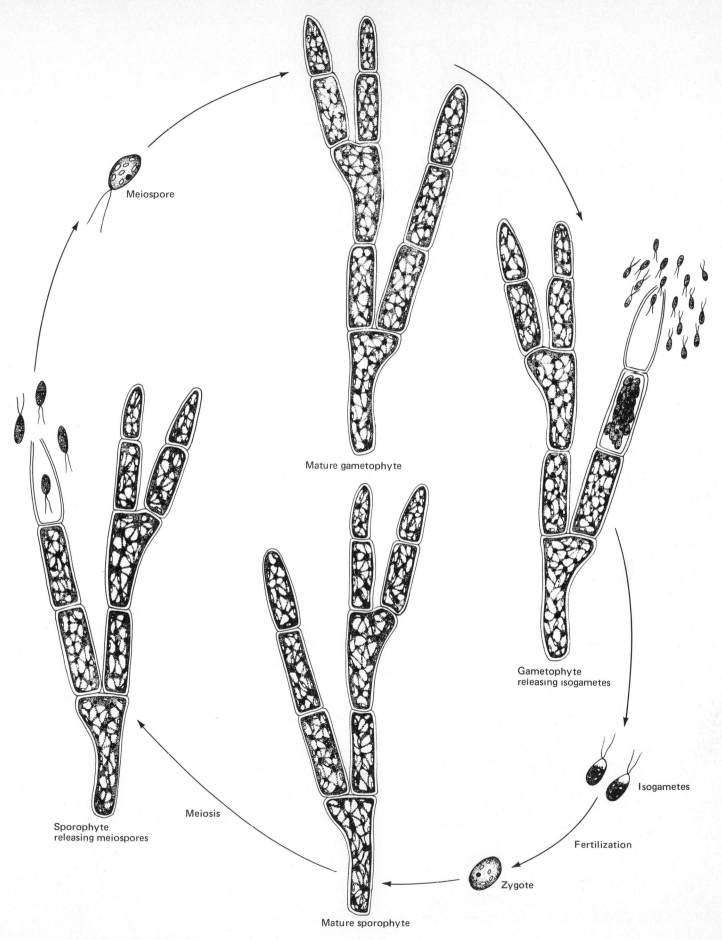

Meiospore

Mature gametophyte

Gametophyte
releasing isogametes

Sporophyte
releasing meiospores

Meiosis

Mature sporophyte

Isogametes

Fertilization

Zygote

Figure 6-23. Life cycle diagram of *Cladophora*.

(A)

(B)

Desmids are unicellular Chlorophyta which reproduce sexually by conjugation. These algae are common throughout the world, especially in nutrient-poor and acidic waters. They are among the most beautiful of all living organisms, since they are highly ornate and occur in many geometrical shapes and sizes (Fig. 6-21). Most desmids are constricted at the middle of the cell. The resulting narrowed region contains the nucleus and is known as the *isthmus*. Since the desmid is constricted, each cell is divided into two equal halves known as *semicells*. Each semicell contains one or more chloroplasts with one to several evident pyrenoids.

Asexual reproduction is common in desmids, occurring by cell division. When the cell divides, each semicell develops a new opposite semicell to produce two mature desmids.

ORDER: CLADOPHORALES

Algae of the order Cladophorales exhibit several characteristics not evident among algae studied previously. Cells of Cladophorales are multinucleate. These cells are aligned end to end to form uniseriate, repeatedly-branched filaments. Chloroplasts may be numerous and disk-shaped or more commonly single and net-shaped.

Cladophora

Cladophora is very common in most rivers and many lakes of temperate regions throughout the world (Fig. 6-22). It is generally attached to rocks or other submerged objects, and trails in the current downstream. Some species of *Cladophora* are marine, especially in intertidal zones.

Cells of *Cladophora* are cylindrical and attached end to end to form long filaments. These filaments are highly branched and almost feather-like, epecially near the apex of the plant. In its natural habitat, *Cladophora* resembles a tiny aquatic shrub. Each generally contains one net-shaped chloroplast near the cell wall. This plastid contains many pyrenoids that often obscure the rest of the contents of the cell. The cells are *multinucleate*.

Asexual reproduction of *Cladophora* occurs by fragmentation of the thallus or by zoospores (Fig. 6-23). This type of reproduction is often responsible for the development of large populations. *Cladophora* presently appears to be extending its range and occurring in larger populations. In many areas, such as the Great Lakes region of the United States and certain large temperate rivers, *Cladophora* has become an important nuisance form.

Sexual reproduction in *Cladophora* is isogamous (Fig. 6-23). Isogametes are produced near the ends of branches in gametangia that are similar to the other cells of the thallus. Two gametes fuse to produce a zygote that in some species germinates by mitosis to produce a diploid plant. This sporophyte plant is identical in appearance to the gametophyte plant. This type of alternation of generations in which the two generations appear exactly alike is known as *isomorphic* alternation. *Heteromorphic* alternation (the alternation of generations that appear different) occurs in most plants.

Certain cells of the sporophyte thallus produce motile meiospores. These meiospores swim for a time until they settle and germinate to produce a gametophyte thallus and thus begin the life cycle again.

SUMMARY

The Chlorophyta is a large division of plants that is widely distributed throughout the world in both marine and fresh waters. Green algae are important as primary producers, oxygen generators, water polluters, and pollution indicators, and as a possible source of food for humans and their domestic animals.

Asexual and sexual reproduction both occur in most green algae. Several types of spores, both motile and nonmotile, are produced by these plants. Most species are heteromorphic in alternation of generations, although some are isomorphic. Likewise, most species are heterothallic in sexual reproduction while others are homothallic.

Many different types of thalli are represented among various Chlorophyta. Some species are unicellular, others are colonial, and the majority are filamentous. Both branched and nonbranched filaments are common.

Green algae are more advanced than Cyanophyta. The former contain definite nuclei and plastids, divide by mitosis, and exhibit well-defined sexual reproduction. Cyanophyta and Chlorophyta are compared in Table 6-1.

FURTHER READING

IYENGAR, M. O. P., "Chlorophyta," In G. M. Smith, *Manual of Phycology*, 21–67, Ronald Press, New York, 1951.

KRAUSS, R. W., "Mass Culture of Algae for Food and Other Organic Compounds," *American Journal of Botany*, 49:524–435, 1962.

PARKER, BRUCE, "On the Evolution of Isogamy to Oogamy," In Bruce Parker and Malcolm Brown (Eds.), *Contributions to Phycology*, 47–51, Allen Press, Inc., Lawrence, Kansas, 1971.

WASSMAN, ROBERT and JOSEPH RAMUS, "Seaweed Invasion," *Natural History*, 82(10):24–36, 1973.

Also see the list of general references on algae at the end of Chapter 4.

(C)

(D)

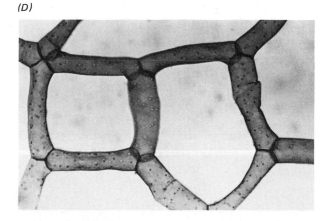

Figure 6-24. Many green algae are very distinctive in their form. Several Chlorophyta not discussed in the text are illustrated in this figure to point up the variation in this division. *Chara* (A) is a rather unusual oogamous green alga. This photograph shows the oogonium and a vesicle which contains several antheridia. These become surrounded by sterile cells making them appear to be multicellular. B shows the common marine alga *ulva* which is often known as sea lettuce. It is a flat colonial form. C shows *Codium* which is common in some marine environments. It is often known as dead man's fingers. D is *Hydrodictyon*. Cells of this green alga are large and multinucleate. They are united with several other cells to form a large net-shaped thallus.

7
Division Chrysophyta: the Golden Algae

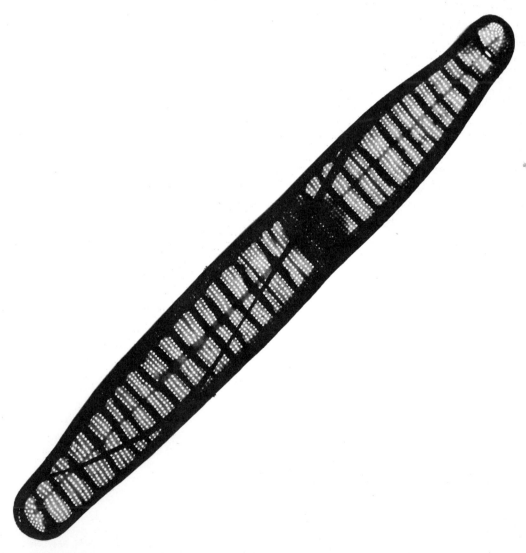

Electron micrograph of the diatom *Diatoma*.

The division Chrysophyta is a large division containing over 9,000 plant species. These algae are often divided into four classes. Two classes, the Xanthophyceae and Bacillariophyceae (diatoms), will be discussed.

CLASS XANTHOPHYCEAE

GENERAL CHARACTERISTICS

The class Xanthophyceae is unique for several reasons, and some botanists prefer to recognize it as the separate division Xanthophyta.

Xanthophyceae exhibit a wide diversity in thallus structure. Many species are unicellular and nonmotile. Others are unicellular and motile by amoeboid motion or by means of two unequal flagella. Some species are colonial and are similar in form to several Volvocales of the Chlorophyta division. Other species are typically filamentous, and a few are filamentous but lack cross walls. These latter forms are often long, containing hundreds of nuclei and other organelles, but they are not divided into separate cells. Plants with this type of thallus are said to be *coenocytic*.

True cell walls are present in most Xanthophyceae. This wall is composed of cellulose and pectic compounds. In addition, the cell wall of many species often becomes impregnated with silicon. The cell wall of some Xanthophyceae is composed of two equal overlapping halves. When species exhibiting this type of cell wall are examined in side view under the light microscope, the cell walls appear to be composed of two overlapping *H*-shaped pieces. This is one means of classifying certain species of Xanthophyceae.

Each cell contains one or more nucleus. These nuclei are small and difficult to examine in detail with a light microscope. Because of this, little information is available concerning their structure and genetic makeup.

Most Xanthophyceae are photosynthetic and autotrophic. However, a few species are colorless and apparently can ingest food particles or absorb nutrients directly through their cell membrane.

Photosynthetic species have rounded, disk-shaped chloroplasts that normally occur just inside the cell membrane lining the cell wall. The chloroplasts are yellow-green in color. This characteristic coloration is caused by the predominance of chlorophyll *a* and certain xanthophyll pigments. In addition to these prominent pigments, carotenes and a second unique chlorophyll, chlorophyll *e,* have been reported from plastids of Xanthophyceae. Some question still exists as to whether chlorophyll *e* actually occurs in these plastids, or whether it has been reported through experimental error.

Pyrenoids are generally absent from the chloroplasts of these algae, and starch is never stored. Most species store oils and the carbohydrate chrysolaminarin as the chief food reserves. It is interesting that chrysolaminarin is also stored in other classes of Chrysophyta and is very similar to a storage product in the brown algae (division Phaeophyta). This seems to some botanists to indicate that Chrysophyta may have given rise to Phaeophyta.

OCCURRENCE AND ECOLOGY

Most species of Xanthophyceae occur in fresh-water, commonly inhabiting streams and ponds. Other species occur on damp soil or stones, and, under certain conditions, may be very common within the soil. Epiphytic species are common, and some occur on damp walls or intermingled with mosses and other plants. Some species grow very well in stagnant waters, where they may form dense blooms.

A few species are marine, although the distribution and importance of these species are little known. It is probable that some are common plankters of the seas.

IMPORTANCE

The class Xanthophyceae is indirectly significant to humans for several reasons. These algae are important primary producers and are found in the food chains of many higher organisms. Of course they also are responsible for the production of oxygen by photosynthesis.

Some species cause water pollution. Dense, troublesome blooms are occasionally formed by some Xanthophyceae. These blooms may be responsible for fish kills and may produce undesirable taste and odor in water supplies.

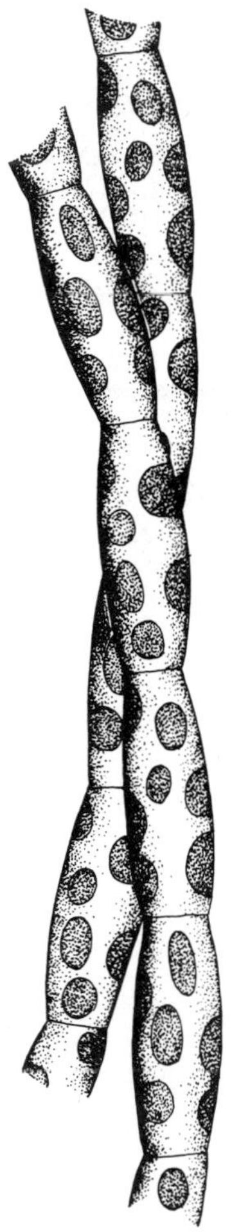

REPRESENTATIVE XANTHOPHYCEAE

Tribonema

Tribonema is a common alga of fresh-water streams and especially ponds or pools. This alga is composed of many barrel-shaped cells aligned end to end to form a uniseriate unbranched filament (Fig. 7-1). Each cell contains a single nucleus and several to many disk-shaped, yellow-green chloroplasts. Cell walls of *Tribonema* consist of two overlapping halves that often appear as *H*-shaped pieces when examined with the microscope. These halves separate easily when the alga is disturbed, causing the filament to break.

Asexual reproduction of *Tribonema* is common by fragmentation of the production of motile or nonmotile spores. Such reproduction may be very rapid under optimum conditions and may form dense blooms.

Sexual reproduction is isogamous, although one of the gametes loses its two flagella immediately prior to fertilization. The zygote is resistant and germinates by meiosis to produce new haploid plants. The gametophyte plant is the conspicuous, dominant generation, and the sporophyte plant is represented only by the zygote.

Figure 7-1. Chrysophyta vary widely in their size and form. Many, such as *Tribonema* illustrated here are similar to certain green algae. *Tribonema* is golden-green in color and contains several disk-shaped chloroplasts.

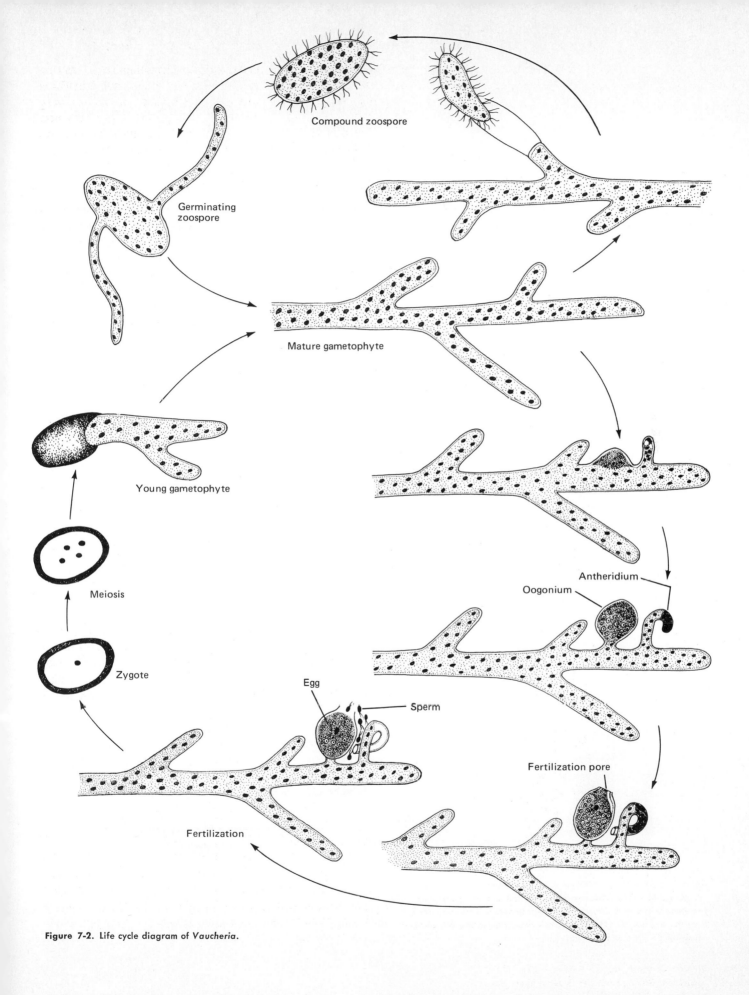

Figure 7-2. Life cycle diagram of *Vaucheria*.

Vaucheria is widespread in fresh and marine waters and on damp soils. It is common to find large felt-like masses of *Vaucheria* growing in the moist zone at the edge of a stream, under a bench in a greenhouse or submerged in shallow water at the edge of a lake.

The thallus of *Vaucheria* is a sparingly branched, co-enocytic filament (Fig. 7-2). Functionally, the entire thallus is composed of a single, giant, multinucleate cell. Nuclei are often difficult to discern because of the presence of many small, disk-shaped chloroplasts. These chloroplasts are yellow-green to green in color and are very prevalent in a layer of cytoplasm lining the walls. Small droplets of golden-colored oil often are evident in the filament as well.

Asexual reproduction is common and is by fragmentation or by the formation of large, multiflagellate zoospores which contain several nuclei. This zoospore is normally formed at the apex of a filament behind it. The protoplasm of the developing spore contracts somewhat from the wall and many pairs of flagella develop along its outer surface. Each pair of flagella is associated with a single nucleus. This spore is liberated when a portion of the cell wall dissolves to form an exit pore. The zoospore is active until it comes to rest and germinates to form a new filament.

Sexual reproduction also is common, especially when adverse growing conditions develop. This process is oogamous and is initiated by the formation of an oogonium and one or more adjacent antheridia (Figs. 7-2 and 7-3). A septum develops at the base of each gametangium separating it from the remainder of the filament. Many small, biflagellate sperm are produced within each antheridium. A single large uninucleate egg is produced within the oogonium. A fertilization pore develops in the oogonium to allow the entrance of sperm cells. These sperm are active for several minutes after they are released from the antheridia. Eventually one or more enters the fertilization pore where a single sperm fuses with the egg to produce the zygote. The zygote develops a thick wall and enters a resting stage that generally lasts throughout the winter. When proper conditions for vegetative growth reoccur, the zygote germinates by meiosis to produce a new haploid filament. As in nearly all of the algae we have studied, the gametophyte generation is thus dominant, and the sporophyte generation is represented only by the zygote.

CLASS BACILLARIOPHYCEAE: THE DIATOMS

Diatoms represent an extremely important and large group of algae. Botanists estimate that at least 6,000 species of diatoms occur on the earth at present, and several thousand fossil species are known as well.

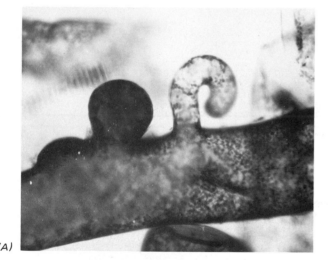

(A)

(B)

Figure 7-3. *Vaucheria* is a chrysophyte that demonstrates oogamous sexual reproduction. *A* shows a filament with immature gametangia. The gametangia at *B* are mature, and the sperm have been shed from the antheridium. Note the fertilization pore in the oogonium.

Diatoms have fascinated people since the development of the microscope. Many of the earliest microscopists studied these organisms because of their delicate shapes and beautiful symmetry and ornamentation. Because of this, diatoms are quite well known, although they are difficult to study in detail and to identify at the species level. Early diatomists published the results of their studies in many obscure books and journals. Because of limited communications among these scientists, many species that had already been named and described were renamed. Thus, a good deal of confusion exists concerning the classification of these organisms. Modern diatomists are clearing up this confusion with the use of advanced techniques including electron microscopy.

GENERAL CHARACTERISTICS

The thallus of a diatom is unicellular and either motile or nonmotile. A few species are aggregated into loose or, on occasion, rather tightly arranged colonies often of a characteristic shape and size. Motile species move in an unusual way which is still incompletely understood. Apparently this motility is caused by secretion of certain substances from the protoplast into a specialized groove on the surface of the diatom known as a *raphe* (Fig. 7-4). This material apparently acts as a piston pushing against the substrate and propelling the diatom in the same direction as the axis of the raphe. Only diatoms with a raphe are motile.

True cell walls are present in diatoms, although they generally contain little or no cellulose. These walls are composed largely of pectic compounds that become very heavily impregnated with silica. Silica may comprise 90 percent or more of the wall material in some species.

The walls are characteristic in another way. They are composed of two overlapping halves similar to a pill box or petri dish. Each half of a cell is termed a *valve*. The entire cell is often known as a *frustule*. The outer valve (which could correspond to the lid of the pill box) is known as the *epitheca* and the inner is the *hypotheca* (Fig 7-5). Diatom walls are generally highly ornate, particularly on the surface of the valve. Chambers, ridges, punctae, and striations of all shapes and sizes are common among different diatoms (Figs. 7-4, 7-5, and 7-6). Each species is sculptured slightly differently from all others. The ornamentation pattern of the cell wall is the most critical feature in identifying these organisms. It is difficult to observe the ornamentation on the wall when the cell is full of cytoplasm. Hence, special methods for "cleaning" or "clearing" diatoms have been developed. Generally this is done by boiling the diatoms in a strong acid to oxidize all organic material present. This removes the cytoplasm but does not alter the wall and thus allows unobstructed observation of the surface of the cells.

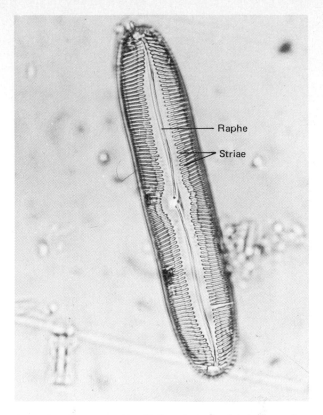

Raphe

Striae

Figure 7-4. Diatoms are identified on the basis of their shape, size, and especially their ornamentation. This photograph of the common diatom *Pinnularia* shows characteristic rib-like striations on the edge of the valve and a raphe at the middle.

(A)

(B)

Figure 7-5. Most diatoms are composed of overlapping cell wall halves known as valves. These scanning electron micrographs of the diatom *Achnanthes flexella* show the shape of the cell (A) and the way in which the two halves of the cell wall fit together (B). (Courtesy of James V. Allen, Brigham Young University.)

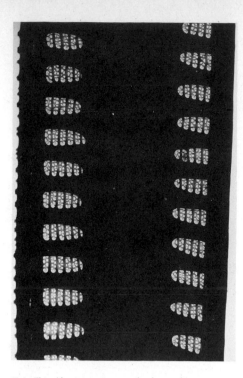

Figure 7-6. This electron micrograph shows the ornamentation on the cell wall of one species of *Synedra*.

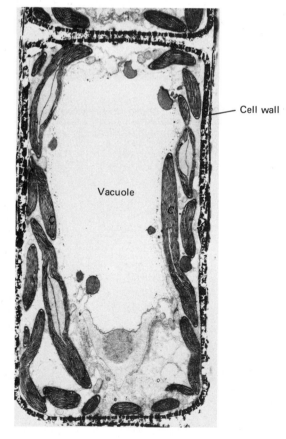

Figure 7-7. Thin cut sections of diatoms are difficult to obtain because of their glass cell walls. This excellent electron micrograph shows many marginal chloroplasts (C) surrounding a central vacuole. (Courtesy of Richard Crawford, University of Bristol.)

Most diatoms are chlorophyllous and autotrophic. Such species contain one or more typically golden or brownish-green colored chloroplasts (Fig. 7-7). These plastids are often characteristically shaped, and in the past botanists attempted to use these shapes for diatom identification. This method proved to be less satisfactory than a system based on ornamentation patterns and valve shape. Chloroplasts contain chlorophyll *a,* chlorophyll *c,* carotenes, and xanthophylls.

A few diatoms are non-green and heterotrophic. In addition, apparently many diatoms are able to live heterotrophically if they cannot manufacture their own food due to a lack of light. Pyrenoids are absent from diatom chloroplasts and oils and fats are the chief food reserves.

Asexual reproduction of diatoms is by cell division. When this process occurs, the nucleus divides by mitosis and the epitheca separates somewhat from the hypotheca. In many diatoms, each valve then develops a new hypotheca (Fig. 7-8). Thus, the hypotheca of the mother cell now functions as an epitheca of a daughter cell. The new cell arising from the hypotheca of the mother cell is smaller than the mother cell. When the smaller daughter cell redivides, its hypotheca again functions as an epitheca. Thus, it is apparent that some diatoms in the population continue to diminish in size. Because of this unique type of cell division, diatoms of the same species but of many different sizes may be found in any natural population.

The decrease in size cannot proceed indefinitely, and diatoms cease to divide when a lower size limit is reached. At this point, sexual reproduction is often initiated. The sexual process may be isogamous, anisogamous, or oogamous depending on the species. Two gametes unite to form a zygote that enlarges to two or three times the size of the gamete-producing cells and develops a silicon wall. This zygote is often termed an *auxospore.* An auxospore may have a smooth or ornamented wall depending on the species. It germinates by mitosis to become a diploid diatom plant of maximum size for that species. Diploid diatom cells produce gametes directly by meiosis without the formation of meiospores or a gametophyte generation. Therefore a true alternation of generations does not occur among diatoms.

Sexual reproduction apparently does not occur frequently among natural diatom populations. Most species appear to be homothallic and hence little genetic recombination occurs by sexual reproduction. This process appears to function primarily in producing cells of maximum size in a population so that asexual reproduction can continue.

OCCURRENCE AND ECOLOGY

Diatoms are very widely distributed over the earth in fresh water, marine water, and terrestrial habitats. Fresh water species are among the most common of all algae, and are present in high numbers in nearly all habitats including lakes and ponds and flowing waters (Fig. 7-9). They may be true plankters or may be attached to the bottom or to other aquatic plants. Diatoms form a thick brownish layer on the bottom of many rivers throughout the world. Likewise, they are often so abundant in lakes that the water becomes a brownish-green color. Fresh water diatoms are especially prevalent in temperate regions and most species grow best during the cooler months of the year.

Marine diatoms are prevalent in all oceans, especially those with cool or cold waters. These diatoms are the most abundant of all marine phytoplankton, and they often build up tremendously large populations.

Marine diatoms generally occur as plankters in the upper few feet of the ocean where light is optimum for photosynthesis. They may be found near shore and in the open sea as well. Several species occur as epiphytes on other algae in intertidal zones or attached to the surface of rocks or animals.

Many diatoms are common inhabitants of the soil (Fig. 7-10), and their importance in soil fertility may be much greater than previously believed. Other species occur as terrestrial epiphytes, on moist buildings or stones, or almost anywhere where light and moisture are present.

IMPORTANCE

Diatoms are unquestionably one of the most significant groups of plants on earth. They are very important primary producers, occurring both directly or indirectly in the food chains of many higher animals. These plants are responsible for the production of tremendous amounts of oxygen, much of which is released to the atmosphere where it becomes available for life processes of other living or-

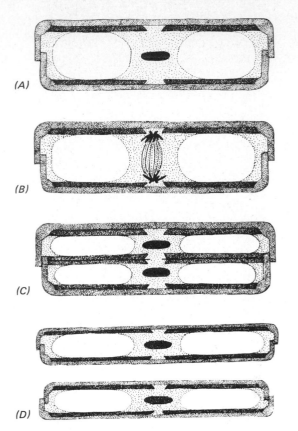

(A)

(B)

(C)

(D)

Figure 7-8. In the process of cell division of some diatoms, the epitheca separates from the hypotheca and each subsequently develops a new inner member. Thus one daughter cell is the same size as the mother cell but the other is slightly smaller. This process is illustrated in A–D.

Figure 7-9. Diatoms may be collected from most freshwater and marine habitats throughout the world. This researcher is collecting diatoms by scraping them from rocks on the bottom of a small desert stream.

Figure 7-10. Diatoms may play a much more important role in the soil than previously thought. This research worker is collecting diatoms from the soil beneath small shrubs in central Utah.

ganisms. In fact, it has been estimated that from 50–75 percent of all photosynthesis occurs in the oceans. Total photosynthesis occurring on earth has been estimated to be approximately 200 billion tons of carbon fixed each year. Thus, from 100–150 billion tons of carbon are fixed yearly in the sea, and an equal amount of oxygen is generated. Since much of this photosynthesis is accomplished by diatoms, the extreme importance of these organisms is evident.

Diatoms are also vital as primary producers in fresh water habitats. They make up an essential part of the diet of many consumer organisms and are responsible for the oxygenation and purification of many waters. Some species occur in highly polluted waters, while others are adversely affected even by small amounts of pollution. Thus, many diatoms are used as indicator organisms in water pollution

Figure 7-11. Diatoms are easily fossilized due to their glass cell walls and they may be found in large deposits. Such deposits are known as diatomaceous earth and they may be commercially mined as in this quarry owned by the Johns-Mansville Corporation. (Courtesy of the Johns-Mansville Corporation.)

Figure 7-12. These diatoms were collected from sediments in Utah Lake, Utah. They are fossil forms about 15,000 years old. *Navicula* (A), *Cocconeis* (B), *Diploneis* (C), and *Nitzschia* (D) are shown in this figure.

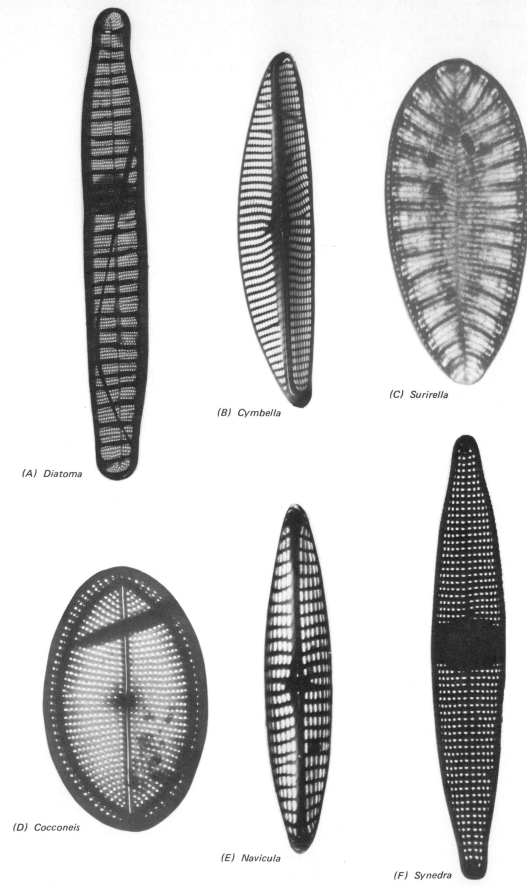

(A) Diatoma

(B) Cymbella

(C) Surirella

(D) Cocconeis

(E) Navicula

(F) Synedra

Figure 7-13. Several different diatoms are illustrated in this figure. Note their very different shapes and ornamentations.

studies. Species that tolerate pollution are often very prominent in those disturbed ecosystems.

In the geologic past, especially during the Tertiary Period, diatoms were apparently as prevalent as they are today. As these diatoms died and fell to the bottom of the sea, tremendous deposits of fossil diatom frustules developed in certain localities. This material, known as *diatomaceous earth,* is mined commercially (Fig. 7-11) and used as an abrasive, in paints, as an insulating material, and as a filtrant. The diatomaceous earth industry is economically significant in some parts of the world. Fossil diatoms are also important to geologists as indicator organisms or index fossils of some oil- and gas-bearing strata.

FOSSIL DIATOMS

The fossil record of diatoms is better than that of most algae since the silica cell wall is readily preservable. The oldest well-authenticated diatom fossils are from the Cretaceous Period. Diatoms were common and numerous by the Late Cretaceous. Most diatomaceous earth was deposited during the Tertiary Period, and many of the diatoms of these deposits are still represented in the modern flora (Fig. 7-12). A few reports of diatoms from deposits older than the Cretaceous Period have been made, and these organisms may have existed from as early as Late Paleozoic times.

REPRESENTATIVE DIATOMS

Diatoms are often divided into two orders based on their morphology. Diatoms which are round in face view are placed in the order Centrales. Elongate diatoms are placed in the order Pennales. As a whole, Centrales are more common in marine habitats, and Pennales are more prevalent in fresh water. However, both orders are well represented in both marine and freshwater habitats.

ORDER: CENTRALES

Cyclotella

Cyclotella is a common planktonic diatom containing both marine and freshwater species. This alga often builds up large populations especially in habitats rich in nutrients. *Cyclotella* blooms occasionally develop in ponds or lakes, coloring the water brownish-green. *Cyclotella* is perfectly round in face view and has large striations in a ring near the valve margin. The central portion of the valve lacks striations and is either perfectly smooth or contains fine punctae.

Melosira, Coccinodiscus, Stephanodiscus, and *Biddulphia* are other common diatoms of the order Centrales.

ORDER PENNALES

Navicula

Navicula is one of the most common freshwater diatom genera and contains many marine species as well. It is one of the largest genera of Pennales, containing several hundred species. These diatoms are symmetrically canoe-shaped in valve view with the middle region generally being wider than either end. (Fig. 7-13*E*). *Navicula* species are motile since a raphe is present on both the epitheca and the hypotheca. These diatoms live on rocks at the bottom of streams or lakes, or are epiphytic on other aquatic plants. *Caloneis, Diploneis, Gyrosigma,* and *Pinnularia* (Fig. 7-4) are all pennate diatoms closely related to *Navicula* and common in many freshwater habitats.

Fragilaria

Fragilaria species are pennate diatoms that lack a raphe. These diatoms are elongate, and may be shaped somewhat like *Navicula* or may be long and slender without the median swelling. *Fragilaria* species are often colonial with many individuals aligned side to side to form characteristic, long chains. The related genus *Synedra* (Fig. 7-13*F*) is very similar except that its cells are normally solitary rather than colonial. *Asterionella, Diatoma* (Fig. 7-13*A*), *Meridion,* and *Tabellaria* are common relatives of *Fragilaria*.

SUMMARY

The division Chrysophyta is composed of four distinct classes that differ widely from each other. The class Xanthophyceae is composed of algae that may be motile unicells, nonmotile unicells, colonial, filamentous, or coenocytic. The class Bacillariophyceae contains the diatoms. These algae are all unicellular and may be either motile or nonmotile.

Most algae of this division are chlorophyllous and autotrophic. Chlorophylls *a* and *e* have been reported from Xanthophyceae. Chlorophylls *a* and *c* are found in diatoms. Fats and oils and chrysolaminarin are the storage products of Chrysophyta. Pyrenoids are lacking in the chloroplasts of these algae, and starch is never stored.

Chrysophyta are important for a number of reasons. Diatoms in particular are extremely important primary producers and oxygen generators. Diatomaceous earth which is composed of fossil diatoms has many economic uses.

FURTHER READING

Conger, Paul, "Lesson of the Diatoms," *American Biology Teacher,* 18(6), 1956.

DRUM, R. W. and J. HOPKINS, "Diatom Locomotion: An Explanation," *Protoplasma,* 62:1–33, 1966.

LEWIN, JOYCE and ROBERT GUILLARD, "Diatoms," *Annual Review of Microbiology,* 17:373–414, 1963.

Also see the list of general references on algae at the end of Chapter 4.

Division Euglenophyta

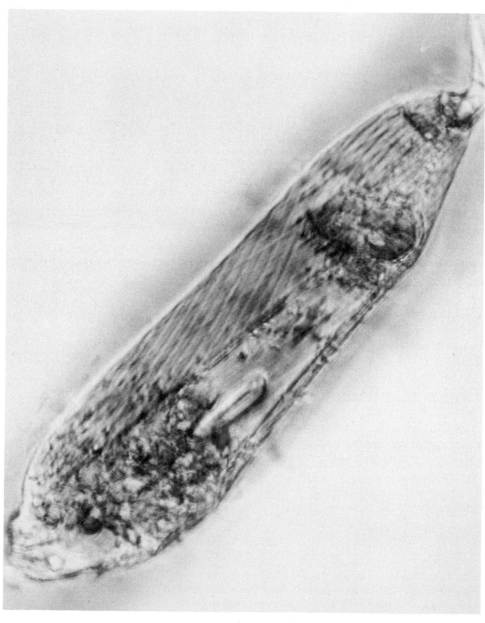

This light micrograph of *Euglena* shows internal details of the cell and the spiral striations in the periplast.

Most algae of the division Euglenophyta are unicellular and motile by means of one or two stout flagella. The majority are very active, unicellular swimmers, although a few move by amoeboid motion and some form small, nonmotile colonies. Approximately 500 species of algae are placed in this division.

Most Euglenophyta lack a true cell wall in the motile stage. However, the outer portion of the protoplast is rather firm and is known as a *periplast* (Fig. 8-1). The periplast is pliable so that many species change shape as they swim. Many Euglenophyta enter a resting stage when conditions become unfavorable for continued vegetative growth. Cells lose their flagella on entering this resting stage and often secrete a true cellulose cell wall. Resting cells are resistant to adverse conditions, and thus carry the alga through unfavorable periods.

Each cell exhibits a *gullet* (Fig. 8-1) which is an invagination of the cell apex. Flagella are attached at the side or base of the gullet. Food materials may be ingested into the gullet of some species where they become available to be used for nourishment.

Pigments within the chloroplasts of Euglenophyta include chlorophyll *a,* chlorophyll *b,* β carotene, and two unique xanthophylls. The chlorophylls predominate, and the cells of these algae are usually grass-green in color, similar to Chlorophyta. Several disk-shaped chloroplasts usually occur in each cell. Some species produce starshaped or ribbon-shaped chloroplasts. Pyrenoids are frequently present in these plastids.

The carbohydrate *paramylum* is the primary food reserve among Euglenophyta. This substance is similar to starch, and is accumulated on pyrenoids. Paramylum may be distinguished from starch since true starch stains a deep blue-black when treated with iodine solution, whereas paramylum does not. Paramylum often accumulates in grains of a specific shape and size. Thus, paramylum granules are often used in the classification of Euglenophyta.

Some euglenoid algae are not green, and therefore are heterotrophic. These heterotrophic species may either absorb dissolved food materials directly from the water, or ingest food particles into the gullet.

Most cells have a light-sensitive eyespot (Fig. 8-1) which directs the organism toward or away from light. This spot is often prominent in the cell. A few species appear bright red in color because of the presence of pigments in the cytoplasm. These pigments develop especially under high light intensity, probably protecting the chloroplasts from oxidation.

Asexual reproduction by cell division is common among Euglenophyta. This process may be very rapid under favorable circumstances, forming large populations. Sexual reproduction is not common, and probably does not occur

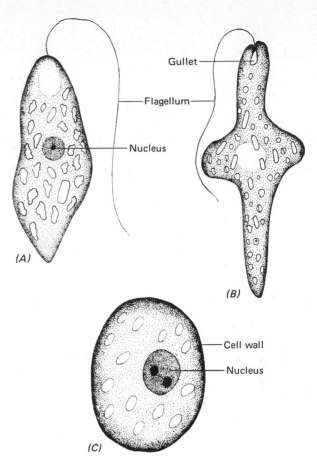

Figure 8-1. *Euglena* shows characteristics of both plants and animals. A and B show two motile cells and C is an encysted cell with the flagellum withdrawn and a cell wall present.

Figure 8-2. This photograph of the Bear River Bay Migratory Bird Refuge of Utah shows a dark coloration in the water. This was formed from a bloom of blood-red *Euglena* cells. (Courtesy of Lloyd Gunther, U. S. Department of Interior.)

in all species. It has been reported to occur by the fusion of isogametes or by conjugation of entire mature cells.

OCCURRENCE AND ECOLOGY

Euglenophyta are widely distributed. Most species are fresh water, although some are marine, and a few occur in water with very high concentrations of salt. Certain species are restricted to acidic waters such as moss bogs or industrial waste effluents. Fresh water species grow under most aquatic conditions and certain species grow very well in polluted waters. *Euglena viridis* is the most frequently mentioned pollution-tolerant alga in water pollution reports throughout the world.

When growth conditions are optimum, asexual reproduction occurs rapidly to build large populations of some Euglenophyta, especially *Euglena*. When this occurs, the water is often turned green. Likewise, some waters are occasionally turned red by *Euglena* blooms because of the red pigments in the cytoplasm of some species (Fig. 8-2).

IMPORTANCE

Euglenophyta are of little direct economic significance. However, these algae are primary producers, and under certain circumstances are quite important in this role. This is especially true of certain polluted environments where *Euglena* may be the most prevalent primary producer.

Euglenophyta are also employed as tools for scientific research (Fig. 8-3). Studies of the DNA and RNA of

Figure 8-3. *Euglena* is used for many scientific investigation including photosynthesis, respiration, and DNA and RNA studies. The flask illustrated here contains a high concentration of *Euglena* cells that well be harvested and used in research.

nuclei, mitochondria, and plastids are conducted on *Euglena* cells as are chloroplast development and enzyme system studies.

Euglenophyta are also important in understanding the evolutionary relationships between plants and animals since these algae exhibit characteristics of both plants and animals. Hence, from studying this division in detail, it may be possible to develop theories or draw conclusions concerning the evolution and development of other plant and animal divisions.

REPRESENTATIVE EUGLENOPHYTA

Euglena

Euglena (Fig. 8-4) may be collected from many fresh water localities including lakes, ditches, streams, and temporary pools. Barnyard pools rich in nitrates and phosphates are especially good localities for collecting specimens.

Euglena is a large genus containing several hundred species. Cells of *Euglena* are spherical to oval, generally with a central swelling (Fig. 8-2). These cells often change shape as the alga moves through the water. Usually many disk-shaped chloroplasts are present in each cell, and paramylum granules are common. A large, often conspicuous nucleus is present in each cell, and a red eyespot is normally evident near the gullet. These algae move very rapidly and are often difficult to observe with the microscope. Adding a drop of glycerine or concentrated sugar solution to the slide usually slows them down for better observation.

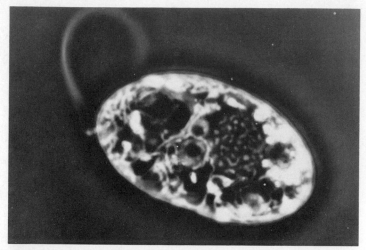

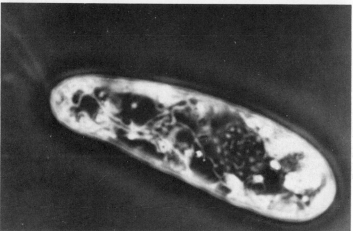

Figure 8-4. Many Euglenophyta change their shape as they move through the water as shown by these highly magnified specimens of living *Euglena gracilis* cells. (Courtesy of D. E. Buetow, University of Illinois.)

Asexual reproduction by cell division may occur very rapidly under favorable circumstances. Sexual reproduction has been reported in some species but has not been observed in most. The reason for this is not clear, and more life history studies of several species of *Euglena* are necessary before drawing conclusions.

Phacus

Phacus is similar in many respects to *Euglena.* It occurs in like habitats, and is similar in overall form and size. It differs from *Euglena* since the cells are flattened and often twisted, and the periplast is firmer and often rigid (Figs. 3-3*D* and 8-5). In addition, most species exhibit a posterior, tail-like extension on the cell.

SUMMARY

The division Euglenophyta contains mostly unicellular, flagellate algae. Some nonmotile species are known, and a few are colonial. These algae are mostly chlorophyllous, containing chlorophylls *a* and *b* as the predominant pigments. Paramylum is the storage product of Euglenophyta.

Some species lack chlorophyll and obtain their food by absorption directly through the cell membrane or by ingesting food particles into the gullet. Asexual reproduction is common and occurs by cell division. Sexual reproduction is less common and apparently absent in some species. When the sexual process occurs, it is by the fusion of isogametes or by conjugation of mature cells.

FURTHER READING

BUETOW, D. E. (Ed.), *The Biology of Euglena,* Vol. 1, *General Biology and Ultrastructure,* Academic Press, New York, 1968.

LEEDALE, GORDON, *Euglenoid Flagellates,* Prentice-Hall, Inc., Englewood Cliffs, New Jersey, 1967.

PALMER, JOHN, "*Euglena* and the tides," *Natural History,* 76(2):60–64, 1967.

WOLKEN, JEROME, *Euglena,* Rutgers University Press, New Brunswick, New Jersey, 1961.

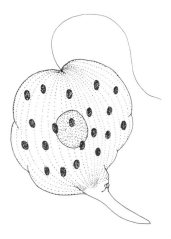

Figure 8-5. This figure shows *Phacus,* a close relative of *Euglena.* It differs by having an ornamented periplast, a posterior tail piece, and by being flattened.

Division Pyrrhophyta: the Dinoflagellates

Dinoflagellate blooms are common in both fresh and marine waters. This small storage reservoir supports such a bloom nearly every summer which results in undesirable taste and odor in the local culinary water.

(A)

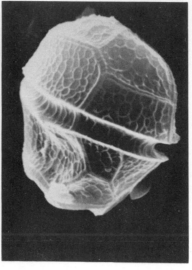
(B)

Figure 9-1. These scanning electron micrographs show the dinoflagellate *Peridinium*. A is *Peridinium leonis* and B is *Peridinium cinctum*. (Courtesy of John D. Dodge, Birkbeck College, University of London.)

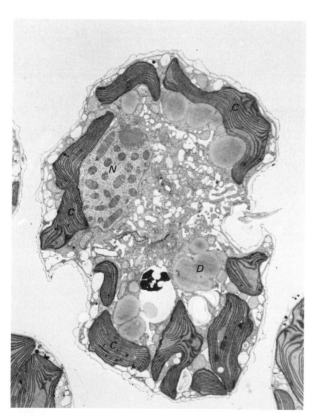

Figure 9-2. This figure shows an electron micrograph of a typical dinoflagellate cell. Note the organelles including the nucleus (N), chloroplasts (C) and oil droplets (D). (Courtesy of Brian Bibby and John Dodge, Birkbeck College, University of London. Used by permission of the British Phycological Association.)

Dinoflagellates are common plants of both fresh water and marine ecosystems. This division contains more than 1,000 species in about 125 genera.

GENERAL CHARACTERISTICS

Most species of *Pyrrhophyta* are unicellular and motile by means of two flagella. A few species form nonmotile colonies, but are closely related to motile species since they produce zoospores that resemble the vegetative cells of motile dinoflagellates.

Cells of most species have a true cellulose cell wall. In those lacking such a wall, the outer layer of cytoplasm is firm, similar to the periplast of *Euglena*. Dinoflagellates with cell walls are said to be armored. Some species produce smooth walls, although most produce walls composed of several overlapping cellulose plates that are cemented together (Fig. 9-1). These plates often form a definite, precise pattern. This is one characteristic used in classification of armored dinoflagellates. The cell shape and the wall pattern in many of these organisms are beautiful, making dinoflagellates among the prettiest of all plants.

Most Pyrrhophyta are autotrophic with chloroplasts containing chlorophyll *a*, chlorophyll *c*, carotenes, and certain unique xanthophylls. These pigments are present in a ratio that gives most dinoflagellates a greenish-brown or golden-brown color. Each cell may contain from one to several chloroplasts (Fig. 9-2) which may or may not have pyrenoids. Starch is the primary food storage product of dinoflagellates. Fats and oils are also stored by some species, occasionally in fairly large quantities, giving the cell a golden or reddish color.

Some species are heterotrophic and may absorb nutrients directly from the water or ingest food particles. A few species are reported to be parasites of marine animals.

Asexual reproduction by cell division occurs in all species. Each daughter cell commonly obtains one flagellum from the mother cell and rapidly develops a second flagellum as the daughter cell matures. Under certain favorable circumstances, asexual reproduction occurs very rapidly, creating large populations that often color the water. Many species have the ability to encyst to form a resistant cell that carries the organism through unfavorable environmental circumstances (Fig. 9-3).

Sexual reproduction has been reported for some species but has never been observed in most. This process is isogamous when it occurs but is apparently not common even among sexual species.

Many dinoflagellates are luminescent when their habitat is disturbed. Thus, the water in the wake of a ship or of rolling waves often glows brightly because of the presence of these bioluminescent algae. This phenomenon is responsible for a good deal of the folklore of the sea.

OCCURRENCE AND ECOLOGY

Dinoflagellates occur commonly in both fresh water and marine habitats. Marine species are generally more prevalent in warm seas both near shore and in open waters. They occur in largest numbers in the upper 3 ft of water, which is the region of maximum light penetration for photosynthesis. Second only to the diatoms, dinoflagellates are the most important group of phytoplankton in the world.

A few marine species are parasitic on fish or copepods, and some exist in a commensalism with certain marine animals where both organisms are benefited.

Fresh water species occur throughout the world both in cool and warm waters. Since most dinoflagellates are plankters, they are most common in lakes or other permanent standing waters. A few species are collected from rivers or streams.

IMPORTANCE

Indirectly, dinoflagellates are very important to humans. They are essential primary producers, particularly in warm marine waters. In such environments they make up a large proportion of the food present, and are responsible for the generation of large quantities of oxygen. A significant portion of the oxygen of the earth's atmosphere is generated by dinoflagellates.

Under certain conditions Pyrrhophyta are also very important in the formation of blooms. These blooms are frequently responsible for large fish kills both through oxygen depletion and toxin production. The infamous red tides that occur periodically throughout the world are caused by dinoflagellates (Fig. 9-4). Red tides are often devastating to aquatic life, occasionally causing the death of millions of fish and other organisms.

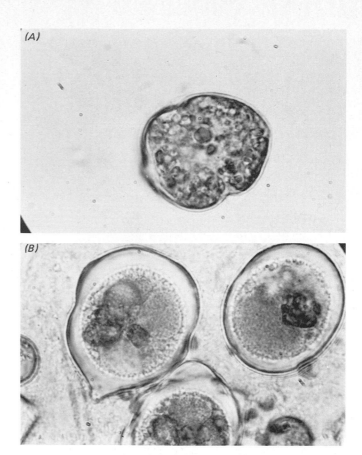

Figure 9-3. Under certain conditions many dinoflagellates are able to encyst. The photograph at A shows a motile dinoflagellate which has encysted at B. (Courtesy of Brian Bibby and John Dodge, Birkbeck College, University of London. Used by permission of the British Phycological Association.)

Figure 9-4. High altitude infrared photograph of a bloom of dino-flagellates. Such blooms are often known as *red tides*. This particular red tide has developed around a bed of brown algae or kelp. (Courtesy of Wheeler J. North, California Institute of Technology.)

When dinoflagellates occur in large populations in nearshore marine waters, filter-feeding mollusks such as clams and oysters use these cells for food. Toxins build up in the tissues of these mollusks, apparently with no adverse effects to the organism. However, when these shell fish are eaten by humans, they may cause paralysis or even death. Outbreaks of this Pyrrhophyta-caused poisoning occur frequently in some parts of the world.

Encysted dinoflagellate cells fossilize well, and many species are known from the fossil record. These structures exhibit characteristic shapes and sizes and are helpful as index fossils in oil exploration in some areas.

REPRESENTATIVE PYRRHOPHYTA

Classification of dinoflagellates depends upon several characteristics, including the presence or absence of a cell wall, cell shape, flagellar arrangement, and the arrangement of plates comprising the cell wall.

Gymnodinium

Gymnodinium is ovoid or pear-shaped and lacks a cell wall (Figs. 3-3*E* and 9-5). It is generally flattened and encircled by a groove or furrow. Two flagella are attached in this furrow, one trailing behind the organism and the other encircling the cell. Each cell contains several rod-shaped, greenish-golden plastids, giving the organism a golden or olive-golden color.

Most species of *Gymnodinium* are marine, although a few occur in fresh waters. This alga is easily identified because of its characteristic shape and coloration. It is often responsible for the red tides discussed earlier.

Ceratium

Ceratium is a widely-distributed dinoflagellate occurring in both fresh-water and marine environments. It is one of the most important fresh-water dinoflagellates, and often forms heavy growths that cause undesirable taste or odor in water supplies and occasionally cause fish kills. *Ceratium* is tolerant of widely diverse conditions and thus may inhabit any environment from clean cold waters to warm polluted waters.

Ceratium is characteristically shaped with prominent protruding horns (Fig. 9-6). One horn protrudes anteriorly and two to four protrude posteriorly. A cell wall composed of several overlapping cellulose plates is evident. A transverse furrow encircles the organism near its midregion, and two flagella are attached in this furrow. One flagellum lies entirely within the furrow encircling the organism and the other trails behind.

SUMMARY

The division Pyrrhophyta contains mostly unicellular, biflagellate algae commonly known as dinoflagellates. These algae are mostly photosynthetic, although some are heterotrophic, living as parasites or ingesting food particles.

Dinoflagellates occur in both marine and fresh waters, and affect humans in both positive and negative ways. They are critical primary oxygen producers, but they also form blooms which are both a nuisance and potentially dangerous to humans.

FURTHER READING

Dodge, John, "Fine Structure of the Pyrrhophyta," *Botanical Review*, 37(4):481–508, 1971.

Graham, H. W., "Pyrrhophyta," *In* G. M. Smith, *Manual of Phycology*, 105–118, Ronald Press Co., New York, 1951.

Zahl, Paul, "Sailing a Sea of Fire," *National Geographic*, 118:120–129, 1960.

Also see the list of general works on algae at the end of Chapter 4.

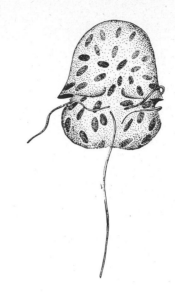

Figure 9-5. Some dinoflagellates such as *Gymnodinium*, illustrated here, lack a cell wall. They are known as *unarmored* dinoflagellates.

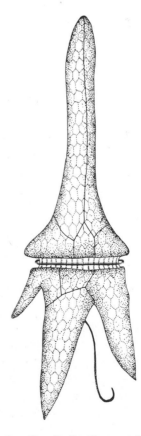

Figure 9-6. *Ceratium hirundinella*, illustrated here, is one of the most common freshwater dinoflagellates.

10 Division Phaeophyta: the Brown Algae

Postelsia is a common brown alga from the west coast of North America. (Courtesy of Jack Brotherson, Brigham Young University.)

The division Phaeophyta contains about 1,500 species of algae including the kelps and many common "seaweeds." These species are placed by most botanists into three classes based on life history and morphology.

GENERAL CHARACTERISTICS

The thallus of all brown algae is multicellular. Some of the less complex species are small, microscopic, branched filaments. On the other hand, the most complex species are large, specialized plants up to 150 ft long. These large conspicuous plants include the seaweeds and kelps common on many rocky marine shores throughout the world. Kelps are complex in structure with a basal holdfast, a stem-like *stipe,* and a flat, blade-like *lamina* that provides increased surface area for maximum photosynthesis (Figs. 10-1 and 10-2).

Cells of brown algae demonstrate characteristic, two-layered cell walls. The inner wall is rigid and composed of cellulose, while the outer wall is composed of pectic compounds which form the colloidal substance, *algin.* Algin is gelatinous and tends to give the plant a slimy texture when handled.

Cells of Phaeophyta all contain true nuclei. These nuclei are small and, although little is known of their structure, they appear to resemble the nuclei of other plants.

Each cell contains one or more chloroplast. These are located near the periphery of the cell and may be discoid, flattened and plate-like, or irregular in shape. These plastids are most often brown to brownish-green in color, imparting the characteristic brown coloration to the algae of this division. Chlorophylls *a* and *c,* along with carotenes and several prominent xanthophylls occur in these chloroplasts. The xanthophyll *fucoxanthin* is often present in large amounts and is responsible for the brown coloration characteristic of Phaeophyta. Chloroplasts of brown algae are similar in many respects to plastids of Chrysophyta and Pyrrhophyta and these three divisions appear to be closely related.

Pyrenoids have generally been reported to be absent from chloroplasts of brown algae. However, recent electron micrographs indicate that pyrenoid-like bodies may be associated with the plastids of some species. Regardless of this, these pyrenoids do not function in producing starch which is never stored by Phaeophyta. The carbohydrate *laminarin* is the chief storage product of Phaeophyta. In addition, fats, alcohols, and small amounts of other carbohydrates are reported to be stored by some species.

OCCURRENCE AND ECOLOGY

Essentially all brown algae are marine in occurrence (Fig. 10-3). These algae are most commonly attached in or just

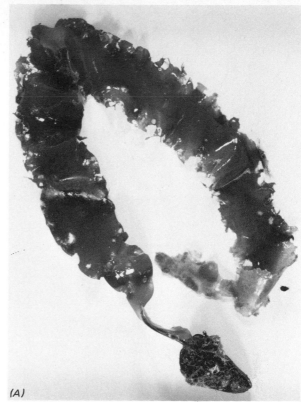
(A)

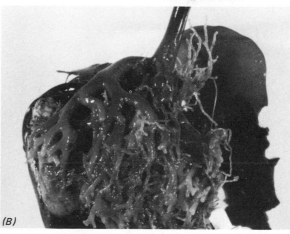
(B)

Figure 10-1. *Laminaria* is one of the common "seaweeds" along many rocky coasts. *A* shows an entire *Laminaria* plant, and *B* is a closeup of the holdfast attached to a mussel shell.

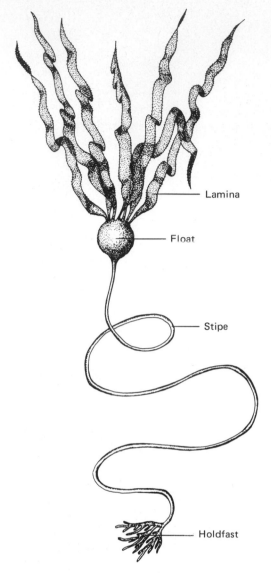

Lamina

Float

Stipe

Holdfast

Figure 10-2. The brown alga *Nereocystis* has a very long stipe and a group of photosynthetic laminae produced from a large float.

below the intertidal zone. Some occur in salt marshes and a few are free floating in the sea. *Sargassum,* for instance, develops in the intertidal zones of the southeastern United States and other nearby regions, but it becomes detached and swept to sea by storms. Dense, widespread mats of this alga are carried by oceanic currents in the North Atlantic Ocean to form the Sargasso Sea. This unusual habitat has existed for a long period of time, and the life of this region, including other algae, invertebrates, and fish, has become adapted to existing there.

Brown algae often form dense underwater "forests" in shallow, cold-water regions. The tips of these submerged plants may float on the surface of the seas forming mats. Such areas abound with life since they form habitats for many marine animals and other marine plants.

Brown algae are especially prevalent in cool and cold seas where the shoreline and shallow submerged zones are rocky. Some species occur in muddy habitats, but these areas rarely support large populations. Other species occur frequently in salt marshes, and a few may be found in relatively brackish water.

IMPORTANCE

Brown algae are important in the food chains of many marine animals. Some species are quite important in this role, especially in areas such as the Sargasso Sea and the nearshore kelp beds. However, dinoflagellates and diatoms undoubtedly are more important primary producers in most marine habitats than Phaeophyta.

Some Phaeophyta are directly useful to humans for several reasons. This is chiefly because of the presence of algin in the outer cell wall of many species, particularly the kelps. These algae are harvested commercially and the algin is extracted from the cell wall (Fig. 10-4). This substance is an efficient moisturizer and emulsifier and is used in many industries including ice-cream making, printing, pharmaceuticals, plastics, and synthetic rubber manufacturing.

Some brown algae are used directly as food by humans, particularly in Oriental cultures. For instance, Kombu, a common food in Japan, is made from several brown algae. It is used in soups or eaten with meat.

Simple fertilizers have been made from some Phaeophyta in the past. Because of their high concentration of phosphates, these plants were either burned and the ashes spread over the ground, or they were used directly by spreading **them** or plowing them into the ground. The use of brown algae as fertilizer has decreased during past years as a result of the more profitable manufacture of fertilizers in other ways. However, many coastal farmers still gather kelps in the fall and spread them over their fields or gardens. Somehow that has greater asthetic appeal to some than spreading a bag of commercial fertilizer.

(A)

Figure 10-3. Essentially all brown algae are marine in occurrence. The algal hummock illustrated in *A* was formed from detached brown algae washing ashore and entangling. *B* shows a tidal pool at low tide with several species of brown algae exposed. (Courtesy of Jack Brotherson, Brigham Young University.)

(B)

Figure 10-4. The *Kelmar* is a modern kelp harvester. The ship in this photograph is loaded with kelp and is returning to port. (Courtesy of the Kelco Company, San Diego.)

(A)

(B)

Figure 10-5. *Laminaria (A)* and *Egregia (B)* are two common brown algae found along American rocky coasts.

REPRESENTATIVE PHAEOPHYTA

Laminaria

Laminaria is a common seaweed found on many rocky shores around the world. Specimens may be collected from either coast of the United States where they are often found washed up on the shore. The thallus of *Laminaria* is composed of a basal holdfast (which is a series of root-like or peg-like protrusions), a tubular stipe, and an expanded, flattened lamina (Figs. 10-1 and 10-5). Different species of this genus vary from 1 to several feet long. Closely related genera contain species that may grow up to 150 feet or more long. It is interesting that the central tissue of the stipe and lamina of *Laminaria* contains elongate cells with swollen ends known as *trumpet hyphae,* which are thought to function in conducting food materials. These trumpet hyphae represent a primitive conduction system similar in some respects to the vascular tissue of higher plants which will be discussed in later chapters.

Oogamous sexual reproduction occurs in *Laminaria.* The sporophyte plant is dominant and consists of the thallus just discussed. Gametophyte plants are inconspicuous, few-celled microscopic filaments, each of which is unisexual. Male gametophyte plants produce antheridia and female gametophyte plants produce oogonia. Sperm are produced within the antheridia and are released into the sea when mature. They are attracted to the oogonium until one eventually fertilizes the egg. The resultant zygote grows by mitosis to produce a mature sporophyte. This sporophyte produces sporangia and subsequent meiospores that germinate to become gametophytes. Thus, the sporophyte of *Laminaria* predominates and the gametophyte is reduced in comparison.

Fucus

Fucus is frequently known as rock weed since it is one of the most common algae of rocky shores almost worldwide. The thallus is composed of a holdfast, stipe, and branched lamina. Air bladders (floats) are common and imbedded in the lamina. The tips of many branches of the lamina are inflated to form roughened *receptacles* (Fig. 10-7) in which gametangia are produced.

Reproduction of *Fucus* is chiefly sexual (Fig. 10-8). However, some species regenerate from a holdfast when the remainder of the plant is broken away. Gametes of *Fucus* are produced directly by meiosis by the sporophyte without the intervention of a gametophyte. Thus, no true alternation of generations occurs in *Fucus,* and the life history is similar to that of a higher animal. All species of *Fucus* are oogamous. Antheridia and oogonia are produced in chambers in the receptacles known as *conceptacles* (Fig. 10-9). According to the species, plants may be unisexual or bisexual. Bisexual plants may produce their gametangia in the same or in different conceptacles. Antheridia and oogonia are intermingled in the conceptacle with long, sterile, hair-like filaments or *paraphyses.* Sperm are released from antheridia as they mature and paraphyses aid in their discharge through an opening or *ostiole* in the apex of the conceptacle. Eggs are also released into the sea as they mature. Sperm swim in the ocean briefly and are attracted toward the eggs. One sperm fuses with each egg to produce a zygote. This zygote begins growth immediately to become a new sporophyte plant.

Sargassum

Some species of *Sargassum* are similar in appearance to *Fucus* which is closely related, although most species are more highly branched and complex in appearance than *Fucus* (Fig. 10-10). The thallus is differentiated into a holdfast, stipe, and lamina, but the lamina is often divided into small, feathery segments. Prominent stalked air bladders occur at branches along the lamina. These air bladders help to keep the plant floating upright off the bottom so that maximum surface is exposed to light for photosynthesis. *Sargassum* species are often very beautiful and delicate, often appearing as a miniature brownish shrub.

(A)

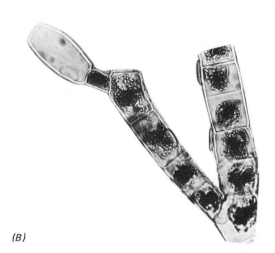

(B)

(C)

Figure 10-6. *Cladostephus verticillatus* (A) is a brown algae commonly found on some rocky shores of North America. *B* shows a sporophyte of *Ectocarpus.* This filamentous brown alga is isogamous and demonstrates an alternation of isomorphic generations. The sporangium at the upper left produces meiospores. Note the epiphytic diatoms along the thallus. C shows the tip of the lamina of the common brown alga *Dictyota.* The large cell at the apex divides repeatedly to produce the remaining cells. (A courtesy of Macmillan Science Company—Turtox/Cambosco.)

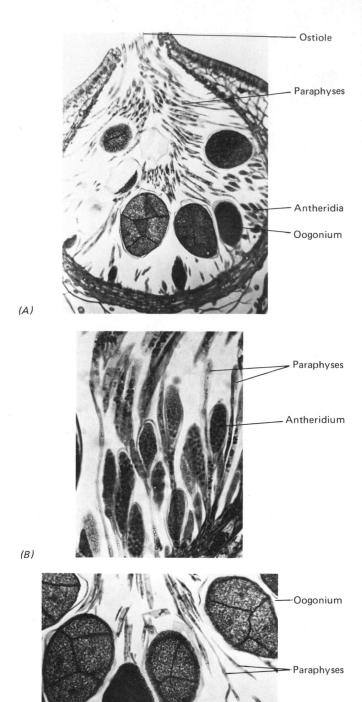

Ostiole

Paraphyses

Antheridia

Oogonium

(A)

Paraphyses

Antheridium

(B)

Oogonium

Paraphyses

(C)

Figure 10-8. The gametangia of *Fucus* are produced in conceptacles imbedded in receptacles at the tips of certain branches. A shows a conceptacle with both antheridia and oogonia. B is a closeup of several antheridia. C shows several oogonia with included eggs.

(A)

(B)

Figure 10-7. *Fucus* is one of the most common brown algae. It is so common along many rocky shores it is often called *rockweed*. A shows an entire thallus lacking the holdfast. B is an enlargement of a branch tip showing three receptacles. The light-colored spots on the receptacles are conceptacles.

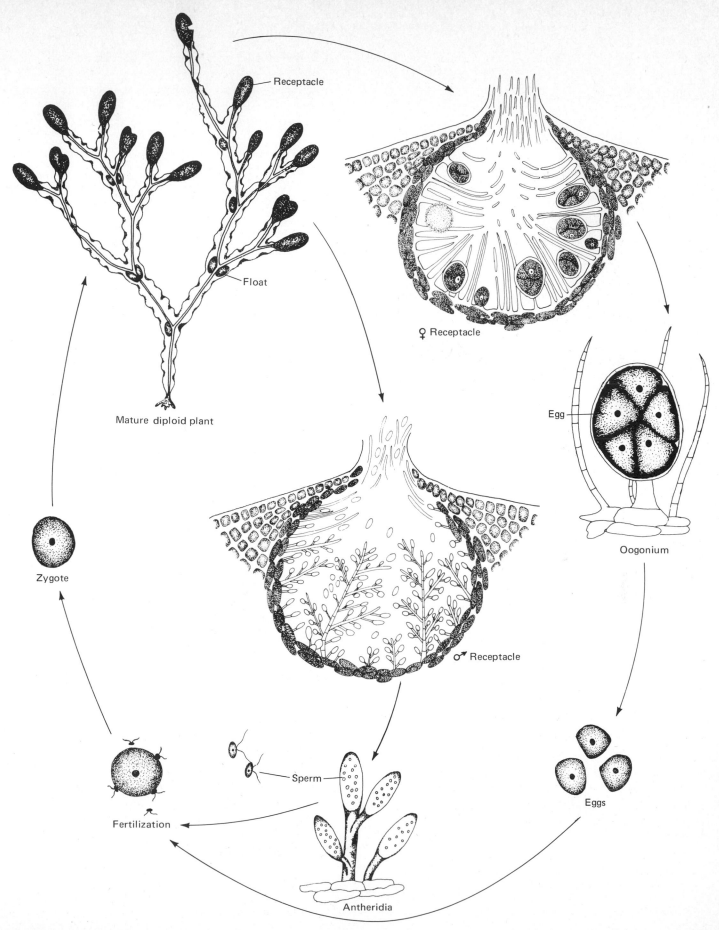

Receptacle

Float

Mature diploid plant

♀ Receptacle

Egg

Oogonium

Zygote

♂ Receptacle

Sperm

Eggs

Fertilization

Antheridia

Figure 10–9. Life cycle diagram of *Fucus*.

Figure 10-10. *Sargassum* is a brown alga closely related to *Fucus*. Note the floats along the thallus in this photograph.

Classification of many species of *Sargassum* is difficult since they are easily detached from the intertidal zones where they develop and are collected as fragments floating in the open sea. These can be identified only if sexual reproductive and vegetative portions of the plant are both present.

Sexual reproduction of *Sargassum* is oogamous and similar to that of *Fucus*.

SUMMARY

Phaeophyta are marine algae that occur especially in intertidal zones of cool to cold rocky shores. All species are multicellular and some are large and complex. The combination of chlorophylls *a* and *c* and fucoxanthin in the plastids of brown algae creates the brown-coloration characteristic of these plants. Laminarin and fats are the chief storage products. Starch is never stored.

Some species are directly important economically because of the presence of algin in their cell walls. These species are harvested and the algin is extracted for industrial use.

FURTHER READING

DAWSON, E. Y., *How to Know the Seaweeds,* Brown Publishing Co., Dubuque, Iowa, 1956.

DAWSON, E. Y., *Marine Botany,* Holt, Rinehart, and Winston, New York, 1966.

LEVRING, TORE, HEINZ HOPPE, and OTTO SCHMID, *Marine Algae,* Cram, DeGruyter and Co., Hamburg, 1969.

NORTH, W. J. (Ed.), *The Biology of Giant Kelp Beds (Macrocystis) in California,* Nova Hedwigia 32, 1971.

PAPENFUSS, F. G., "Phaeophyta," In G. M. Smith, *Manual of Phycology,* 119–158, Ronald Press Co., New York, 1951.

SMITH, G. M., *Marine Algae of the Montery Peninsula,* Second edition, Stanford University Press, Stanford, California, 1969.

Also see the list of general works on algae at the end of Chapter 4.

11 Division Rhodophyta: the Red Algae

Small encrusting colonies of red algae on a tide pool stone. These colonies are only a few millimeters in diameter and are bright red.

(A)

(B)

Figure 11-1. Marine red algae are among the most beautiful of all plants. These bright red seaweeds are *Porphyra* (A) and *Ceramium* (B).

Rhodophyta, or red algae, are common marine plants. About 4,000 species are known. These algae are placed in a single class with two subclasses based on complexity of the thallus and characteristics of the sexual life cycle.

GENERAL CHARACTERISTICS

Rhodophyta are distinguished from all other algae except Cyanophyta by the absence of motile cells of any kind. This is one strong evidence that has caused many botanists to suggest that Rhodophyta evolved from blue-green algal ancestors.

The thallus of most Rhodophyta is multicellular, although a few unicellular species are known. Most red algae are quite complex but lack the great size and vegetative development characteristic of the large Phaeophyta. Primitive red algae are often uniseriate, branched or unbranched filaments. Other species are flat sheets one cell thick and up to several inches in diameter (Fig. 11-1). The most complex Rhodophyta are basically filamentous, although the filaments are often fused to form rather large, three-dimensional thalli of various shapes and sizes. Several species are multiseriate in their main axis and repeatedly branched, sometimes resembling miniature, bright red shrubs (Fig. 11-1). Others are club-shaped or expanded and balloon-like (Fig. 11-2).

Cell walls of red algae are generally two layered. The rigid inner wall is composed of cellulose while the outer wall is composed of gelatinous pectic compounds. This layer in many species contain *agar* which makes these red algae economically valuable.

The outer wall layer of some Rhodophyta becomes impregnated with calcium carbonate (Fig. 11-2). These encrusted algae are fairly well known from the fossil record since they fossilize readily.

Rhodophyta contain true, well-defined nuclei (Fig. 11-3). The simplest species are uninucleate but many of the more complex species contain several to many nuclei in each cell. These nuclei are similar to those of other plants.

Chloroplasts occur singly or in large numbers in each cell (Fig. 11-3). These plastids are simpler in structure than those of other plants as demonstrated by studies with the electron microscope. Many botanists feel that this simple construction of the chloroplast is another factor which aligns the red algae with Cyanophyta. These chloroplasts are generally disk-shaped and located near the periphery of the cell.

Chlorophyll *a,* and chlorophyll *d, r*-phycoerythrin and *r*-phycocyanin as well as carotenes and xanthophylls are present in the plastids of red algae. Chlorophyll *d* has been reported only from Rhodophyta. Some botanists feel that this pigment is an artifact that develops from chlorophyll *a* when chlorophyll is extracted from the cells for chemical study. The pigments *r*-phycoerythrin and *r*-phycocyanin are often present in relatively large amounts. It is *r*-phycoerythrin that is responsible for the characteristic red coloration of Rhodophyta. These pigments are chemically similar to the *c* forms found among Cyanophyta.

Starch is the main storage product of Rhodophyta (Fig. 11-3), although it is somewhat chemically different from that of green algae and higher plants. This starch is known as *floridean starch* and stains red rather than blue-black when treated with iodine. In addition to floridean starch, certain sugars and alcohols are also stored by some species. Apparently the formation of different types of storage products is at least partly dependent on environmental conditions.

OCCURRENCE AND ECOLOGY

The large majority of red algae occur in marine habitats. They are widely distributed throughout the world, although most occur in tropical to warm temperate oceans. Most Rhodophyta are attached to the substrate in or just below the intertidal zone.

Some species grow well at surprisingly great depths in clear waters. Such species have been reported as deep as 600 feet in certain tropical localities. These plants are able to grow at such great depths under conditions of greatly restricted light because of *r*-phycocyanin and *r*-phycoerythrin in the plastids. These pigments efficiently trap the energy of blue light which penetrates to the greatest depth and transfer it to chlorophyll for photosynthesis. It is interesting to note that deep water Rhodophyta are often much redder in color than those of shallow waters as a result of the high concentration of *r*-phycoerythrin.

Some red algae occur in fresh water environments (Fig. 11-4). Such species are rarely red, but are usually green or violet-green. These algae are normally restricted to fast, well-oxygenated, cold streams. A few species are terrestrial where they occasionally form colored scums on the surface of damp soils.

IMPORTANCE

The outer cell wall of some Rhodophyta contains the complex colloid agar. This substance is commercially valuable and many red algae are harvested in diverse areas of the earth to obtain it. Agar is especially useful in the production of growth medium for the commercial and scientific cultivation of bacteria and other microorganisms. Agar is also used to make dental impressions, capsules for pharmaceutical use, as a mild laxative, and so forth.

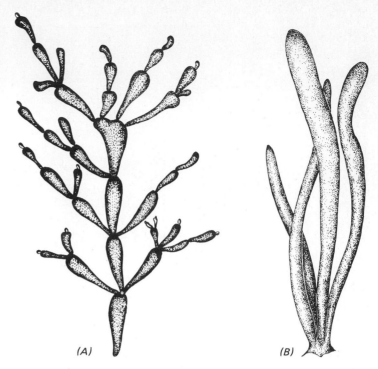

(A) (B)

Figure 11-2. The appearance of many red algae varies widely. For instance, *Corallopsis* (A) becomes heavily impregnated with calcium carbonate and so is heavy and bony to the touch. *Nemalion* (B) often resembles a red plastic filament or worm.

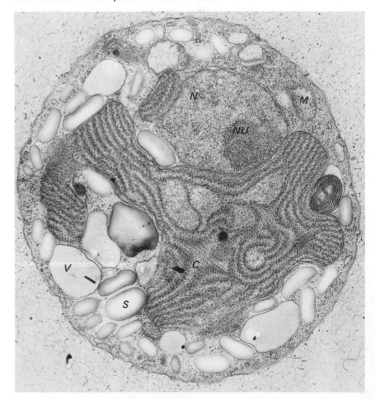

Figure 11-3. The organelles of red algae are similar to those in most other plant cells. This electron micrograph of the red alga *Porphyridium cruentum* clearly shows the nucleus (N) which contains a nucleolus (NU), a mitochondrion (M), chloroplast (C), starch grains (S), and vacuole (V). (Courtesy of Elizabeth Gantt, Smithsonian Institution.)

Figure 11-4. Some Rhodophyta are fresh water in occurrence. This research worker is collecting two species of red algae from a high mountain drainage in Idaho.

Other colloids including carrageenin are also obtained from the outer cell wall of some Rhodophyta (Fig. 11-5). Many of these absorb and hold water readily and retain flexibility when they dry. Because of this, they are valuable for use in water-base paints, hair dressings, laundry "starches," and several foods and cosmetics.

Many red algae are consumed directly as food by humans (Fig. 11-6). The commercial growth of edible species will undoubtedly increase in the future.

Some Rhodophyta play critical roles in the formation of reefs or atolls, especially in the South Pacific Ocean. These algae become encrusted with calcium carbonate to form a substrate for the attachment of other algae and animals, particularly corals. Such communities often develop into reefs. If the reef continues to increase in size it eventually may become an emergent atoll. Such small atolls may then develop further into an island complete with soil and a terrestrial flora and fauna. Thus, many of the so-called coral reefs and atolls of the South Pacific are actually basically algal structures on which corals became attached.

REPRESENTATIVE RHODOPHYTA

Bangia

Bangia is one of the least complex Rhodophyta (Fig. 11-7). It is composed of unbranched, usually uniseriate filaments. Each cell is uninucleate and contains a single star-shaped chloroplast, with a large central pyrenoid. Asexual reproduction is by spores produced singly in each cell of the filament. These spores are often amoeboid for a period of time before they become static and germinate to produce a new thallus. Sexual reproduction has not been adequately observed in *Bangia* and may or may not occur.

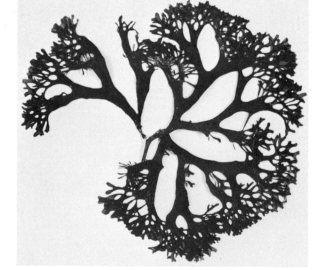

Figure 11-5. *Chondrus crispus* is an important red alga of Europe. This plant is collected and processed to yield the product carrageenin which has several industrial applications.

Figure 11-6. Rhodophyta are harvested commercially in the tidal waters of Japan. A shows a portion of a tidal flat "farm" with nets strung. Red algae become attached to these nets and are then harvested. B is a closer view showing the alga *Porphyra* ready for harvest. (Courtesy of Akio Miura, Tokyo University of Fisheries.)

Batrachospermum

Batrachospermum is one of the most common fresh water *Rhodophyta*. This alga may be collected from many regions of the world, especially from fast-flowing, well-oxygenated streams. *Batrachospermum* is often dark blue-green to grass green (or occasionally purplish) in color and forms characteristic tufts of vegetation on submerged stones or sticks.

The thallus of *Batrachospermum* contains a central multiseriate, filamentous stalk. Whorls of many branches occur at intervals along this stalk giving a characteristic "trimmed shrub" appearance (Fig. 11-8).

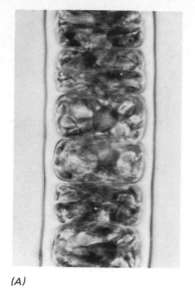

(A)

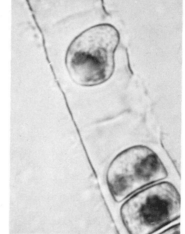

(B)

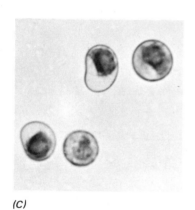

(C)

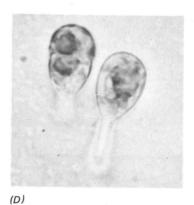

(D)

Figure 11-7. Many of the primitive red algae are rather similar to blue-green algae in appearance. Even so, red algae are more complex. These photographs show the primitive red alga *Bangia fuscopurpurea*. *A* shows a portion of a filament with a single stellate plastid and a central pyrenoid. *B* shows a filament releasing asexual spores. These may move by amoeboid motion for a period *(C)* prior to germinating *(D)*. (Courtesy of Milton Summerfield, Arizona State University.)

(A)

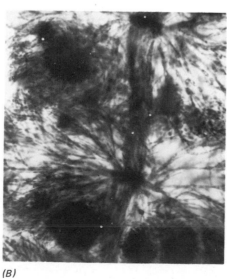

(B)

Figure 11-8. *Batrachospermum* shown in this figure is one common fresh water red alga. *A* is a branch of the thallus showing the characteristic tufted branching. *B* is a closeup of the thallus showing several very tight groups of small branches. These are the haploid carposporophytes.

The gametophyte plant predominates in the life cycle of *Batrachospermum*. This plant may be unisexual or bisexual. The male gametangium is the antheridium or *spermatogonium* which produces nonmotile male gametes known as *spermatia*. The female gametangium is the *carpogonium*. This structure has a swollen base containing the egg, and an elongate appendage or *trichogyne*. Spermatia are released from the male gametangium and are dispersed by water currents. A spermatium eventually contacts a carpogonium and its nucleus migrates into the trichogyne and down to the egg. It then fuses with the egg to form the zygote. The zygote subsequently divides by meiosis to produce a small cluster of haploid branches attached to the original gametophyte plant. These haploid filaments comprise the *carposporophyte* generation (Fig. 11-8). These carposporophyte branches produce haploid *carpospores* which germinate to form new gametophyte plants. *Batrachospermum* thus differs from all other algae previously studied since it produces an extra haploid generation, the carposporophyte.

Polysiphonia

The genus *Polysiphonia* contains several common marine algae that may be collected on the east and west coasts of North America. These algae are highly and intricately branched, forming delicate light red thalli (Fig. 11-9). The main axis of *Polysiphonia* is composed of a central filament of cells which is surrounded by four or more filaments. Thus, the *Polysiphonia* axis is round in cross section with a *central cell* surrounded by a ring of four or more *pericentral cells*. Each cell is uninucleate, and contains many small, reddish chloroplasts. These chloroplasts are especially prominent in the pericentral cells near the outer cell walls. Branches of the *Polysiphonia* thallus are similar in construction to the main axis.

Polysiphonia gametophyte plants are unisexual (Fig. 11-10). Gametangia are produced on small side branches. The male gametangium (antheridium or spermatogonium) produces spermatia that are released and carried about in water currents. Female gametangia (carpogonia) are exposed at the time the spermatia are released. One spermatium contacts the trichogyne (Fig. 11-11) and the nucleus of this spermatium passes into trichogyne and to the egg where fertilization occurs.

The resultant zygote divides by mitosis rather than by meiosis and produces a small *diploid* carposporophyte generation which remains attached to the female gametophyte (Fig. 11-11). This carposporophyte plant produces carpospores by mitosis which are therefore $2N$ in chromosome number. These carpospores are carried about in water currents for a period before they become attached and germinate to produce a second diploid plant, the *tetrasporophyte* plant. The tetrasporophyte represents the

Figure 11-9. This *Polysiphonia* plant shows the characteristic finely branched thallus.

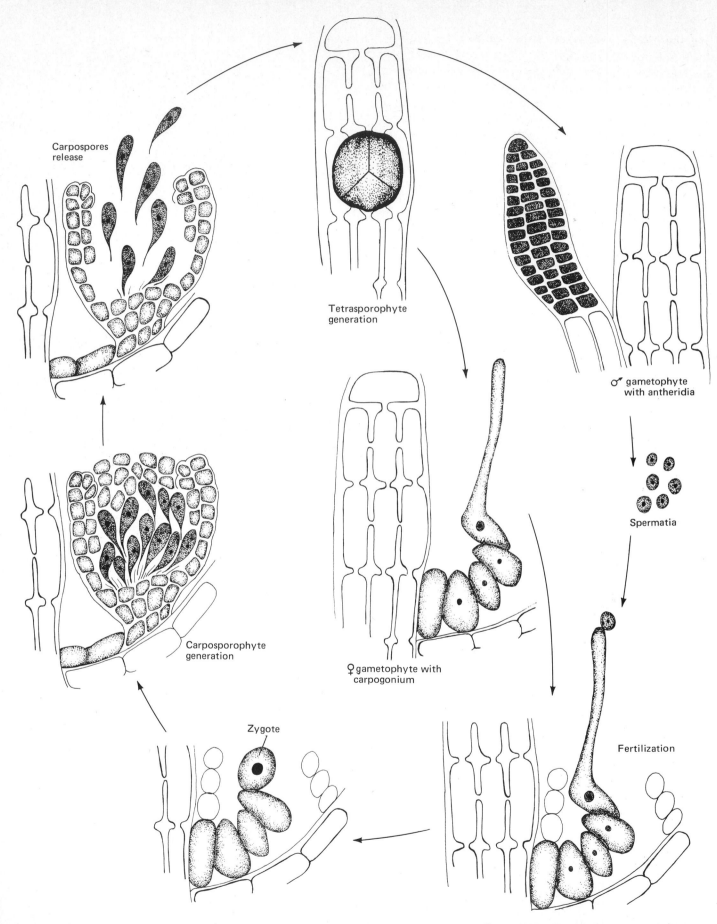

Carpospores
release

Tetrasporophyte
generation

♂ gametophyte
with antheridia

Spermatia

Carposporophyte
generation

♀ gametophyte with
carpogonium

Fertilization

Zygote

Figure 11-10. Life cycle diagram of *Polysiphonia*.

true sporophyte generation of *Polysiphonia* since it produces *tetraspores* (meiospores) by meiosis. Tetraspores are released from the tetrasporophyte and germinate to become new male and female gametophyte plants.

Polysiphonia differs from other algae previously studied since it produces a second diploid generation, the carposoporophyte. Hence, the carposporophyte plant is $1N$ in *Batrachospermum* and $2N$ in *Polysiphonia*. This unusual generation always develops directly from the zygote and is unique to *Rhodophyta*.

Life histories of many red algae differ distinctly from those of other algae. As a result of this and other evidence, scientists believe it is likely that these plants originated from blue-green algae and developed along a separate, distinct evolutionary pathway.

SUMMARY

Several similar characteristics indicate that Rhodophyta probably arose from Cyanophyta. These include the complete lack of motile cells in either division, the similar pigmentation of both divisions, and the primitive chloroplast in Rhodophyta which may have arisen from the chromoplasm of a blue-green alga.

Cells of red algae contain one or more nuclei and chloroplasts. The chloroplasts contain chlorophyll *a,* chlorophyll *d, r*-phycocyanin and *r*-phycoerythrin as well as carotenes and xanthophylls. Chlorophyll *d, r*-phycocynin and *r*-phycoerythrin are unique to Rhodophyta.

Red algae differ from most plants in their complex life cycles. For example, *Batrachospermum* exhibits an extra haploid generation and *Polysiphonia* exhibits and extra diploid generation. These unusual life cycles are unique and characteristic of *Rhodophyta*.

FURTHER READING

DAWSON, E. YALE, *Marine Botany,* Holt, Rinehart and Winston, New York, 1966.

DAWSON, E. YALE, *How to Know the Seaweeds,* Wm. C. Brown Publishing Co., Dubuque, Iowa, 1956.

DREW, K., "Rhodophyta," In G. M. Smith, *Manual of Phycology,* 167–191, Ronald Press Co., New York, 1951.

Also see the list of general works on algae at the end of Chapter 4.

(A)

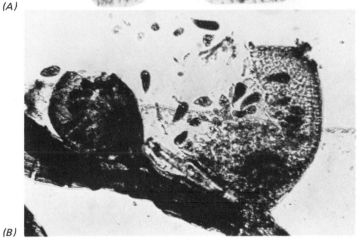

(B)

Figure 11-11. *Polysiphonia* differs from most plants by having an alternation of three generations. Mature cystocarps containing the carposporophyte are shown in A. B shows a ruptured cystocarp releasing diploid carpospores.

12

Introduction to the Fungi:
the Myxomycetes

Sporangia of the slime mold *Comatricha typhoides.* (Courtesy of
O'Neil Ray Collins, University of California at Berkeley.)

The fungi comprise a very large, diverse, and extremely important group of plants. Approximately 80,000 different fungi are presently known and new species are being discovered continuously. The study of fungi, called *mycology*, is a fascinating, ever expanding area of science that will undoubtedly expand during coming years.

GENERAL CHARACTERISTICS

The fungal thallus is highly varied among different groups. Some species are unicellular (yeasts), but most are either multicellular or coenocytic filaments. Fungi are similar in many respects to algae, and most botanists postulate that algae are probably ancestral to fungi.

Perhaps the most descriptive single characteristic of fungi is their lack of chlorophyll. Because of this, fungi must obtain an outside source of nourishment and are thus heterotrophic. Two basic types of fungi are recognized according to how they obtain their nutrients. Some fungi live on nutrients obtained from living plants or animals and are therefore parasitic. Others are nourished by decomposing dead plants or animals and are said to be saprophytic. Some can obtain nourishment from either living or dead organic matter and are thus *facultative* parasites or saprophytes.

Most fungi reproduce both sexually and asexually. Their methods of reproduction are often very similar to those of some algae. Fungal gametangia are always either unicellular or entirely absent. Sexual reproduction may be isogamous, anisogamous, or oogamous. Asexual reproduction occurs by fragmentation and by the production of several different types of spores.

OCCURRENCE AND ECOLOGY

Fungi are found in essentially all habitats, although they are present in low numbers in cold environments. Thus, very few ice or snow fungi have been discovered. Conversely, some species are able to live and even reproduce at temperatures as high as 60°C. These fungi may actually increase the temperature of their habitat by releasing heat from respiration. Such plants are occasionally responsible for the spontaneous combustion of stored agricultural products such as hay.

Fungi occur on the surface or in the upper few inches of the soil. They occur in fresh and marine waters, on and inside animals and plants, in association with algae to form lichens, and in many other habitats. Perhaps the only restriction to the multiplication of many fungi is their requirement for organic nutrients. Fungi are undoubtedly among the most widely distributed of all plant groups.

IMPORTANCE

In discussing the role of algae in the earth's ecosystem, it was mentioned that probably their most outstanding char-

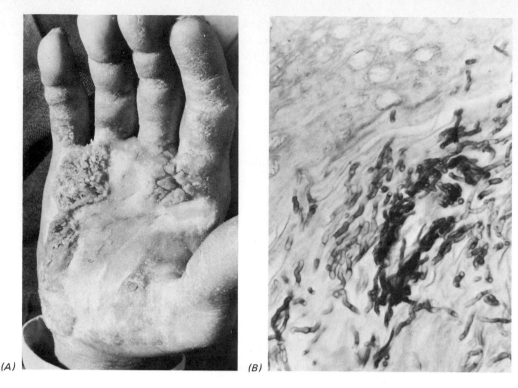

Figure 12-1. Many fungi are able to decompose organic matter with great efficiency. Living as well as dead organic materials are subject to their attack as illustrated in these photographs of a human hand (A and B) and the root of an oak tree (C). (A and B are courtesy of P. M. D. Martin and C is courtesy of the Dept. of Plant Pathology, Cornell University.)

acteristic is their ability to produce simple sugars by photosynthesis which then become available for use by the other organisms. Metabolic activities of plants (such as photosynthesis) will only occur as long as mineral nutrients are available for the synthesis of essential proteins and other compounds. Since these mineral nutrients are present in small, finite amounts in the soil or water of the earth, they rapidly would be used up if they were not returned to the habitat on the death of the organisms that used them for growth. Undoubtedly the process of organic decomposition is equally as important as organic synthesis. Bacteria and fungi are the decomposing agents of the earth. The significance of decomposers as nature's recycling organisms cannot be overemphasized. Some botanists have estimated that if the earth did not contain abundant fungi, it would soon be covered by a layer of undecomposed organic material several feet thick.

Fungi are able to decompose organic substances with great efficiency. Because of this, valuable products are often attacked and destroyed. Fungi attack such diverse products as jet fuels, leather, cloth, wood, plastics, rubber, and food. In addition, many fungi attack and destroy living organisms. Human beings, animals, and plants are all subject to their attack (Fig. 12-1). Billions of dollars are spent throughout the world in controlling fungal destruction and disease.

Several fungi are used directly and indirectly as foods. *Agaricus campestris,* the common, edible mushroom familiar as a food throughout the world, is a good example of an edible fungus. Truffles and morels are also highly prized, delicious fungi that command high prices in European and some domestic markets. Many other fungi are collected and valued as foods as well, but it is necessary to be extremely careful when identifying edible mushrooms since a few species are deadly poisonous.

Yeasts are used in food industries for brewing and baking. Other fungi are used to produce many organic compounds including important antibiotics.

CLASSIFICATION

Four major groups of fungi are often recognized, although the taxonomic level of these groups, and their limits, are subject to debate. The first group, containing the slime molds, is often placed in a separate subdivision of Mycophyta since it includes fungi with characteristics distinctly different from those of other fungi. These fungi are placed in the subdivision Myxomycotina, class Myxomycetes. Some mycologists believe these fungi are distinctive enough to warrant their separation from the remainder of fungi into a separate division, the Myxomycophyta.

Phycomycetes, or algal-like fungi, along with all remaining fungi, are placed into the subdivision Eumycotina. The class Phycomycetes contains fungi that are often

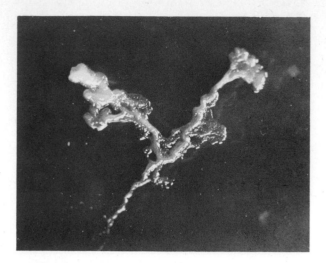

Figure 12-2. The vegetative plant of a slime mold is motile by amoeboid motion and is known as a plasmodium. The plasmodium of *Physarum* is illustrated here. This plasmodium is bright yellow in nature.

aquatic in habitat and are similar in appearance and life history to some filamentous green algae. The class Ascomycetes contains fungi with conspicuous, often cup- or disk-shaped fruiting bodies. The most advanced fungal class, the Basidiomycetes, contains nearly all commonly noticed fungi, including mushrooms or toadstools, earth stars, puffballs, shelf fungi, and others.

A fifth fungal group that is not a true class is also recognized. This group is the Fungi Imperfecti, or Deuteromycetes. Fungi placed in this group apparently have no sexual life cycle, or at least none has been discovered. This is not a natural taxonomic group of fungi since species from the other classes are often placed here if their sexual life cycle is unknown. Each of these classes will be discussed in the following chapters.

CLASS MYXOMYCETES

Myxomycetes, or slime molds, are widely distributed, although they are generally inconspicuous and seldom seen by the casual observer. Approximately 400 species are currently known.

GENERAL CHARACTERISTICS

The thallus of a slime mold is termed a *plasmodium* (Fig. 12-2). This structure is a multinucleate, often brightly colored mass of protoplasm which is motile by amoeboid movement. The plasmodium is coenocytic, and bounded on the outside by a flexible plasma membrane that permits the plasmodium to change shape readily.

The nuclei in this plasmodium are similar in organization to those of other plants. Many are present in each plasmodium (Fig. 12-3). Asexual reproduction occurs by fragmentation of the plasmodium as long as each fragment contains one or more nucleus.

Sexual reproduction occurs in most Myxomycetes (Fig. 12-4). This process is remarkably similar throughout most members of the class. Often it is initiated when conditions for vegetative growth become poor, such as during a drouth or following a frost. As reproduction proceeds, the plasmodium ceases to move, and gradually forms a sporangium. This structure is characteristically shaped according to the species. Meiospores are produced within the sporangium. These spores are also characteristically shaped and colored. They are released from the sporangium at maturity and blow about on air currents (Fig. 12-5). Ultimately, some of them land in a suitable, moist environment where each germinates to produce an amoeboid cell or a biflagellate, very motile cell. Two of these cells of different sexual strain ultimately fuse to produce a zygote. The zygote nucleus divides many times as

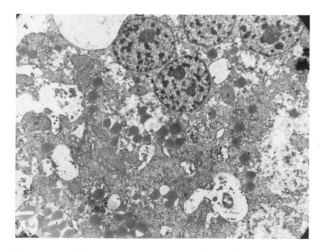

Figure 12-3. The slime mold plasmodium is not divided into individual separate cells. This electron micrograph through the plasmodium of *Physarum* shows four nuclei at the top right. (Courtesy of John Crawley, Clinical Research Centre, England.)

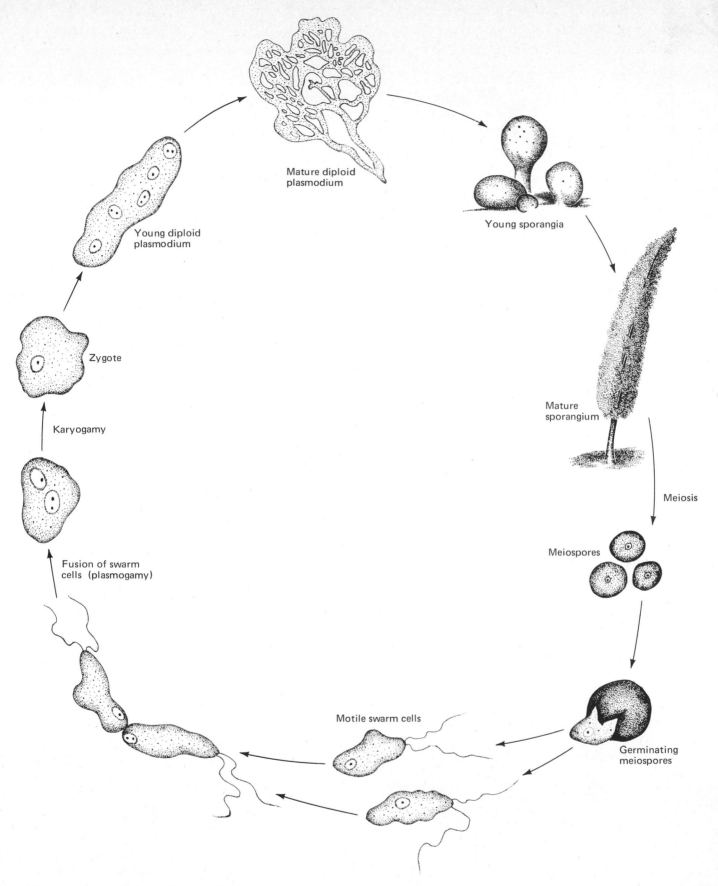

Mature diploid
plasmodium

Young diploid
plasmodium

Young sporangia

Zygote

Karyogamy

Mature
sporangium

Meiosis

Fusion of swarm
cells (plasmogamy)

Meiospores

Motile swarm cells

Germinating
meiospores

Figure 12-4. Life cycle diagram of a slime mold.

Figure 12-5. Meiospores of slime molds are produced in sporangia and are released to be disseminated by the wind. Stages in spore releasal from the slime molds *Hemitrichia* (A) and *Arcyria* (B) are illustrated here.

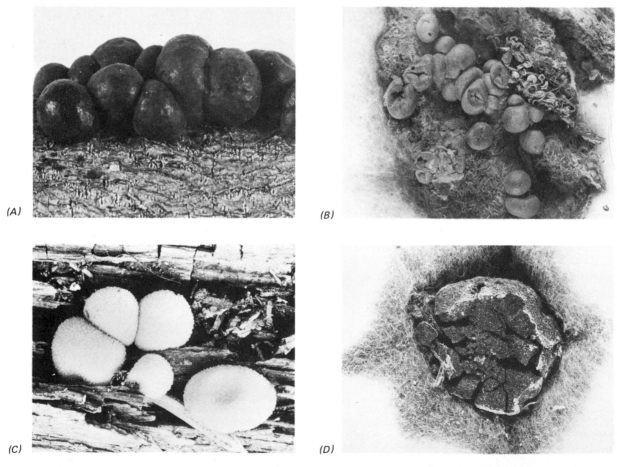

Figure 12-6. Slime mold sporangia vary widely in shape and size. The sporangia illustrated here are two species of *Fuligo* (A and D) and two species of *Lycogola* (B and C).

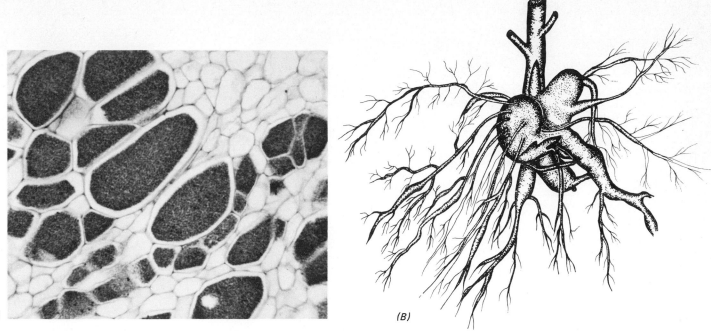

Figure 12-7. *Plasmodiophora brassicae* is a myxomycete responsible for the disease clubroot of cabbage. A is a cross section through a cabbage root showing enlarged cells containing spores. B is a drawing of a cabbage root infected with *Plasmodiophora*.

the cytoplasm increases in volume to form a mature plasmodium. Thus, the coenocytic plasmodium contains diploid muclei, and meiosis occurs in the sporangium at the time of meiospore production.

OCCURRENCE AND ECOLOGY

Slime molds are all terrestrial in occurrence and are widely distributed. They occur frequently in forests or woods under moist conditions, and they are nearly always associated with dead organic matter of some type. These fungi rarely are exposed to direct sunlight since they are very susceptible to heat and drying. Damp rotten logs or humus are favored habitats of many species. A few occur on cultivated shrubs or laws where they may form conspicuous plasmodia many inches in diameter. However, the plasmodium of most species is inconspicuous and only the sporangia are obvious.

IMPORTANCE

In recent years slime molds have become significant research tools for studies concerning the physical and chemical nature of protoplasm. Myxomycetes are also presently being studied for information on mitosis, sexual reproduction, and other problems.

Slime molds play a restricted role in the decomposition of organic materials since they generally feed by ingesting bacteria. A few Myxomycetes cause certain

(A)

(B)

Figure 12-8. Sporangia of the slime molds *Canatricha typhoides* (A) and *Trichia* (B) are illustrated in this figure. (Courtesy of O'Neil Ray Collins, University of California at Berkeley.)

significant plant diseases, such as club root of cabbage (Fig. 12-7).

REPRESENTATIVE MYXOMYCETES

Slime molds are classified largely on the basis of their sporangia which are highly variable (Fig. 12-8).

Physarum

Physarum is one of the most common slime molds. Species of *Physarum* are often gaily colored during their plasmodial stage, especially bright yellow. *Physarum* often may be collected within or on the under side of decaying logs. This fungus is formed from interconnecting net-like strands of protoplasm that form a fan-shaped plasmodium up to about 3 inches in diameter (Fig. 12-2). The heaviest concentration of protoplasm is generally at the leading edge of the plasmodium.

One plasmodium of *Physarum* produces several sporangia (Fig. 12-9) at sexual reproduction. These sporangia are generally quite small and tightly crowded. Each exhibits a definite outer layer known as the *peridium*, and an inner region of filaments or *capillitium*. Meiospores are produced among filaments of the capillitium.

Stemonitis

The plasmodium of *Stemonitis* is similar in many respects to that of *Physarum*, although it is not brightly colored. A few species of *Stemonitis* occur on lawns or on shrubs in cities where they appear as a bluish slime. If they are present in large amounts, they may do noticeable damage to the host, although often they are merely unsightly.

The sporangium of *Stemonitis* is also quite similar to that of *Physarum*. However, *Stemonitis* has a very thin, scarcely evident peridium. Its sporangium is more elongate and the capillitium is very lacy (Fig. 12-10). This sporangium contains a central sterile stalk, or *columella*, which is an extension of the stalk of the sporangium.

SUMMARY

Fungi are extremely important plants. They are devoid of chlorophyll, and are thus heterotrophic. Both parasitic and saprophytic species are known. The fungal thallus varies from a plasmodium to a highly developed cellular filament. When present gametangia are always unicellular.

Fungi are significant for many reasons. They are beneficial especially in decomposing large amounts of organic matter and thus recycling needed mineral nutrients. In this role fungi are both very necessary and serious pests. Some species cause serious plant and animal diseases, while others are used as foods, and still others produce antibiotics and other needed substances.

Slime molds are plasmodial and reproduce both asexually and sexually. These plants are inconspicuous and

Figure 12-9. At the onset of proper environmental conditions, the slime mold plasmodium comes to rest and produces one or more sporangia. This scanning electron micrograph of two sporangia of *Physarum polycephalum* shows this process. (Courtesy of Ilan Chet, Hebrew University of Jerusalem. From *Tissue and dell.* Used by permission of the Longman Group, Ltd.)

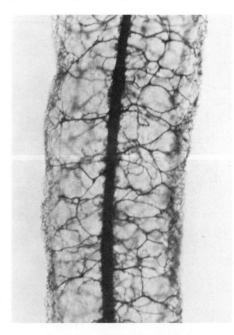

Figure 12-10. The tissues within a slime mold sporangium are illustrated in this longitudinal section through the sporangium of *Stemonitis*. The central columella surrounded by the lacy capillitium is clearly evident.

are often found in humid, dark habitats such as rotten logs and humus piles. A few cause significant plant diseases.

FURTHER READING

GENERAL WORKS ON FUNGI

AINSWORTH, G. C., "Historical Introduction to Mycology," In G. C. Ainsworth and A. S. Sussman (Eds.), *The Fungi,* 3–20. Academic Press, New York, 1965.

ALEXOPOULOS, C. J., *Introductory Mycology,* John Wiley and Sons, New York, 1962.

ALEXOPOULOS, C. J. and H. C. BOLD, *Algae and Fungi,* The Macmillan Co., New York, 1967.

CHRISTENSEN, C. M., *The Molds and Man,* McGraw Hill Book Co., New York, 1967.

HICKMAN, C. J., "Fungal Structure and Organization," In G. C. Ainsworth and A. S. Sussman (Eds.), *The Fungi,* 21–45, Academic Press, New York, 1965.

KAVALER, L., Mushrooms, *Molds and Miracles,* John Day Co., New York, 1967.

LARGE, E. C., *The Advance of the Fungi,* Dover Publishing Co., New York, 1962.

MARPLES, MARY, "Life on the Human Skin," *Scientific American,* 220(1):108–115, 1969.

MOORE-LANDECKER, ELIZABETH, *Fundamentals of the Fungi,* Prentice-Hall, Inc., Englewood Cliffs, New Jersey, 1972.

SPECIFIC WORKS ON MYXOMYCETES

BONNER, J. T., *The Cellular Slime Molds,* Princeton Univ. Press, Princeton, New Jersey, 1967.

CROWDER, W., "Marvels of Mycetozoa," *National Geographic,* 49:421–443, 1926.

GRAY, W. D. and C. J. ALEXOPOULOS, *Biology of the Myxomycetes,* Ronald Press, New York, 1968.

MARTIN, G. W. and C. J. ALEXOPOULOS, *The Myxomycetes,* Univ. of Iowa Press, Iowa, 1969.

13

Division Mycophyta, Class Phycomycetes

Phycomycetes are common in nearly any aquatic habitat including small freshwater streams such as the one pictured. (Courtesy of Sheril Burton, Brigham Young University.)

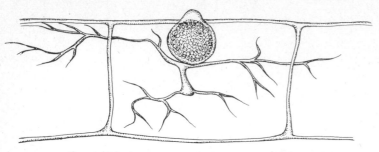

Figure 13-1. Some very simple fungi are unicellular and live entirely within the cell of a host. The fungus drawn here is unicellular but produces a network of rhizoids that aid in absorbing nutrients from the host cell.

Phycomycetes, or algal-like fungi, are common inhabitants of most aquatic and many terrestrial habitats. These fungi are distributed throughout the world, and are among the most effective of all decomposing organisms. More than 1,500 different species of Phycomycetes are presently known, and as these plants are studied more thoroughly during coming years, many new species probably will be discovered.

GENERAL CHARACTERISTICS

Phycomycetes are highly variable in their thallus, and the simplest of all fungi are placed into this class. These are unicellular, inconspicuous species, often occurring entirely within the cell of a host plant (Fig. 13-1). Other simple Phycomycetes are unicellular but have a basal tuft of root-like *rhizoids*. These attach the fungus to the substrate and absorb water and food materials. Rhizoids are not true roots since they are simple, often unicellular structures lacking vascular tissue. Most Phycomycetes are filamentous and often highly branched. A single fungal filament is termed a *hypha,* and a group of hyphae comprises a *mycelium.* Hyphae of Phycomycetes are generally coenocytic since septa are formed only at the base of reproductive structures or at the site of a wound. A few species produce false septae which are constrictions of the hypha rather than true cross walls.

Asexual reproduction is by fragmentation, and many species also produce asexual spores within a sporangium (Fig. 13-2). Often these are produced in large numbers and are released when the wall of the sporangium breaks open. Spores of Phycomycetes are common in the air. In fact, if a suitable nutrient source is left exposed to the air for a minute or two and then cultured for a few days, at least a few Phycomycetes will usually grow.

The way in which various fungal spores are produced is one essential characteristic used in the classification of fungi. Thus, bear in mind that asexual spores of many Phycomycetes are produced internally within a sporangium. Conversely, asexual spores of higher fungi (Ascomycetes and Basidiomycetes) are produced externally, and this is one distinction between these classes.

Sexual reproduction is also common among Phycomycetes. This process often begins when conditions adverse to vegetative growth develop. In fact, the zygote of many species forms a thick, resistant wall and functions in carrying the organism through these unfavorable periods. Sexual reproduction may be isogamous, anisogamous, or oogamous among the different species, and the wide variation in the exact method of sexual reproduction of these fungi is noteworthy.

Some disagreement exists among botanists concerning the classification of Phycomycetes. Several mycologists prefer to divide this class into four distinct classes depend-

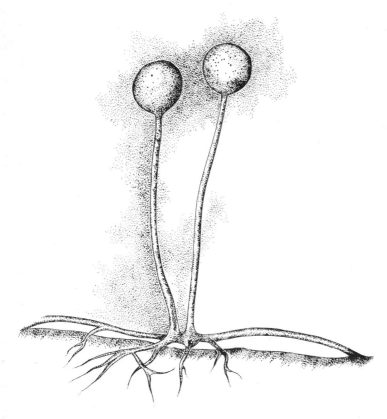

Figure 13-2. Asexual spores of most Phycomycetes are produced within a sporangium. The sporangia shown here are of *Rhizopus.* Spores are released when the wall of the sporangium ruptures.

ing on the type of sexual reproduction and especially on the number and type of flagella contained on motile cells. Other mycologists prefer to maintain the single class Phycomycetes and recognize several orders within this class based on these same criteria.

OCCURRENCE AND ECOLOGY

Phycomycetes undoubtedly are among the most widely distributed of all plants. These fungi are especially common in aquatic environments, although many species are terrestrial or parasitic. A few species are so widely distributed that they can be isolated from almost any region of the earth. For example, *Rhizopus,* the common bread mold, can be isolated from the upper inch of almost any soil throughout the world.

Many Phycomycetes are serious plant and animal parasites, occurring within or between the cells of their host. These parasitic fungi may occur wherever their host plant occurs. A few Phycomycetes are animal parasites and also are distributed as their host.

Most Phycomycetes are saprophytic and occur wherever moisture and nutrients are available (Fig. 13-3). Many of these plants can be isolated from almost any pond or stream by placing bits of apple, animal skin, cellophane, various seeds, and other organic substances in a wire basket and placing the basket in the water. Several species generally invade these substances and then may be cultured and studied in the laboratory.

IMPORTANCE

Phycomycetes are extremely beneficial to humans directly as well as to the entire ecosystem of the earth. In fact, most mycologists argue that these organisms, along with other fungi, are equally important as green plants because of their indispensable role as decomposers.

Phycomycetes are valuable especially as decomposers in the soil and in aquatic habitats. As such, they are responsible for recycling tremendous amounts of organic materials. These fungi are thus important in soil fertility since they make nutrients available to higher plants by decomposing dead plants and animals. Likewise, the fungus itself adds to these nutrients when it dies and is decomposed by other fungi or bacteria.

Because of the efficiency of Phycomycetes in decomposing organic materials, they are not only beneficial but are serious pests as well. Thus, some species are responsible for degrading manufactured structures such as wooden boats, docks, and pilings. A sizable amount of money is spent yearly to protect against such losses and to repair fungus-caused damage.

Some Phycomycetes cause plant diseases which are responsible for tremendous economic losses yearly (Figs. 13-4 and 13-5). It is difficult to understand the meaning

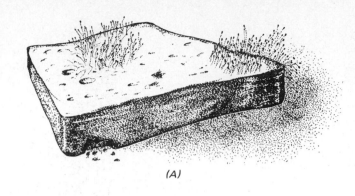

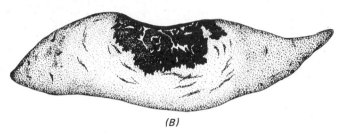

Figure 13-3. Many Phycomycetes infect stored and prepared foods causing great losses each year. *Rhizopus* illustrated here is one of the most important Phycomycetes to cause such infections. *A* shows an infected piece of bread, and *B* is a sweet potato in an advanced state of decay.

Figure 13-4. This figure shows a maple leaf infected with the phycomycete *Cylindrosporium* which causes a "shot hole" disease in which holes develop in the infected leaves. Fossil leaves have been collected that show the symptoms of similar shot hole diseases.

of this statement fully without visiting a farm with a crop infected by one of these pests. It is not uncommon for farmers to have their entire crop destroyed by a phycomycete. While this may represent a relatively small inconvenience to consumers in the form of a price increase, the result on the farmers is devastating. Their entire livelihoods may be eliminated by an attack of these serious plant pathogens.

Some Phycomycetes have played a significant role in world history. For instance, the great potato famine in Ireland in the mid-nineteenth century was caused by the phycomycete, *Phytophthora infestans* (Fig. 13-5). This organism causes potato blight which was responsible for the death by starvation of hundreds of thousands of Irish and for the immigration of large numbers of others to America. New York City still has a high population of Irish as a direct result of this potato famine. Many of these Irish im-

(A)

(B)

Figure 13-5. Potatoes that are resistant and susceptible to the pathogen *Phytophthora infestans*, the causal agent of Irish potato blight, are shown here. A shows susceptible and resistant plants side by side in the field. Note the extreme damage to the susceptible plants on the right. B shows potato tubers. Note the heavily infected, susceptible tuber at the bottom in comparison to the healthy ones. (Courtesy of U. S. Dept. of Agriculture.)

migrants worked on the first transcontinental railroad in America and were responsible for much of the construction of the Union Pacific Railroad. Several other examples of fungi affecting world affairs are cited by E. C. Large in his delightful book, *The Advance of the Fungi.*

REPRESENTATIVE PHYCOMYCETES

Several orders of Phycomycetes are recognized by mycologists. They will not be discussed here, but the supplementary reading list at the end of the chapter contains references to further information on these fungi. Three representative Phycomycetes will be discussed. The first, *Rhizopus,* is isogamous, *Allomyces* is anisogamous, and *Saprolegnia* is oogamous.

Rhizopus

The genus *Rhizopus* contains some of the most common Phycomycetes. Some species are called *bread molds* since they are common contaminants of bread (Fig. 13-3). *Rhizopus* is also prevalent in the soil and its spores are abundant in the air.

The thallus of *Rhizopus* is mycelial. The mycelium is visible without using the microscope and is generally white or grey in color. Filaments of this fungus grow along the surface of the substrate and penetrate it at various points with groups of rhizoids (Fig. 13-3). An upright-stalked sporangium often develops above the rhizoids. It consists of a sporangial stalk or *sporangiophore* which continues into the globular sporangium as the columella (Fig. 13-7). Many asexual sporangiospores are produced within this sporangium. These spores are released when the sporangial wall ruptures. These are buoyant and float readily on air currents until some of them come into agreeable habitats and germinate to produce new *Rhizopus* mycelia.

Sexual reproduction in *Rhizopus* is isogamous by conjugation (Fig. 13-8). This process is initiated when conditions for growth become poor. As this process begins, one hypha from each of two separate sexual strains grows up from the substrate for a short distance and contacts the other directly at the tip of each filament. A septum develops slightly behind the tip of the filament, and the two cells now in contact are gametangia. Eventually the wall between these gametangia breaks down and the protoplasts and nuclei of the two cells fuse. The resultant cell is a zygote, which subsequently develops a thick wall to become a zygospore (Fig. 13-7). This zygospore is resistant to fluctuating or harsh environmental conditions and may remain dormant for some time.

The zygote germinates under favorable growth conditions. A *germ sporangium* (Fig. 13-8) develops from the zygospore on germination. This structure appears similar to an asexual sporangium except that it arises from the

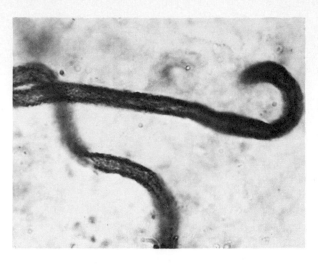

Figure 13-6. Phycomycetes are a very ancient group of fungi, as illustrated by this fossil Phycomycete, *Eomycetopsis robusta.* This fungus was collected from the Bitter Springs Formation (Late Precambrian) of Australia. (Courtesy of J. W. Schopf, University of California at Los Angeles.)

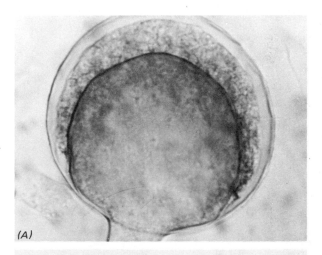

(A)

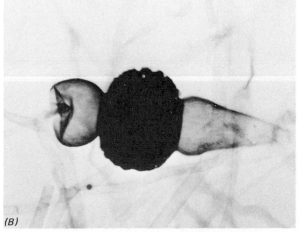

(B)

Figure 13-7. Stages in asexual and sexual reproduction of the common bread mold *Rhizopus* are illustrated here. A shows a sporangium and B shows conjugation of two hyphae to produce a zygospore.

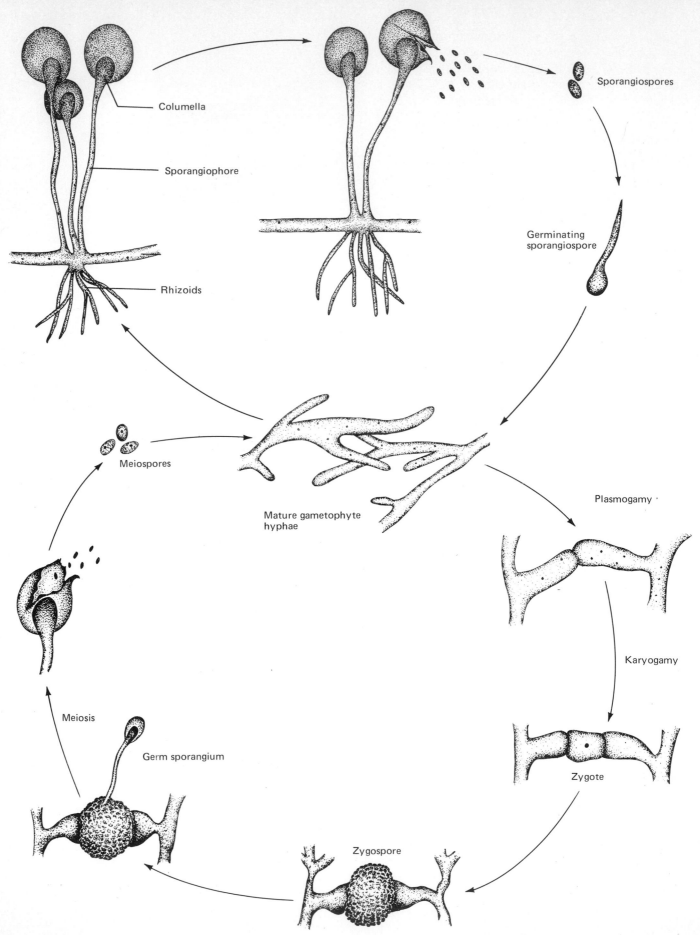

Columella

Sporangiophore

Rhizoids

Sporangiospores

Germinating
sporangiospore

Meiospores

Mature gametophyte
hyphae

Plasmogamy

Karyogamy

Meiosis

Germ sporangium

Zygote

Zygospore

Figure 13-8. Life cycle diagram of *Rhizopus*.

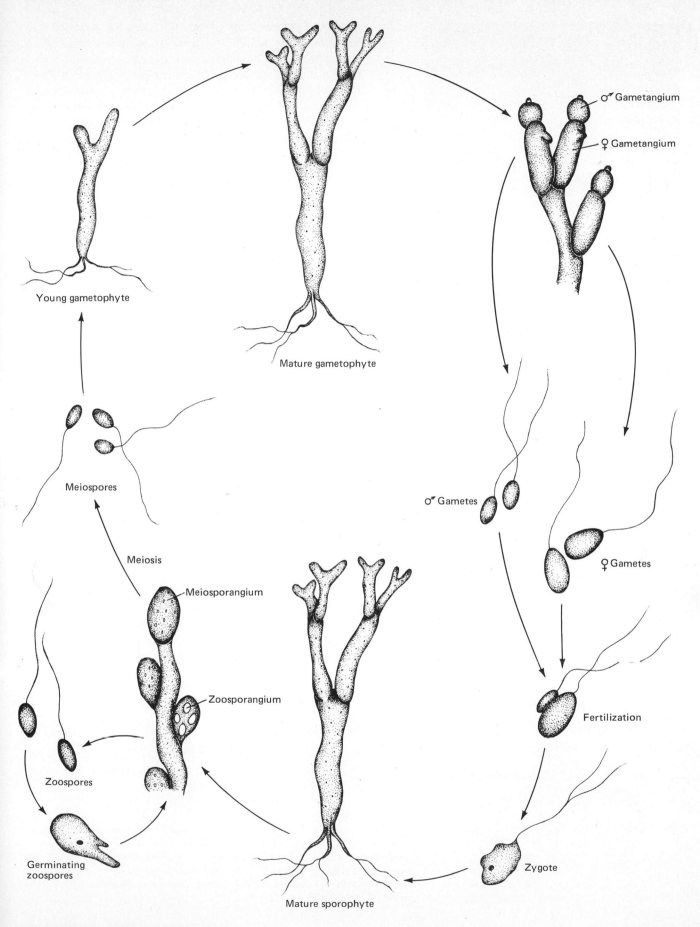

Young gametophyte

Mature gametophyte

♂ Gametangium

♀ Gametangium

Meiospores

Meiosis

Meiosporangium

Zoosporangium

♂ Gametes

♀ Gametes

Fertilization

Zoospores

Germinating
zoospores

Mature sporophyte

Zygote

Figure 13-9. Life cycle diagram of *Allomyces*.

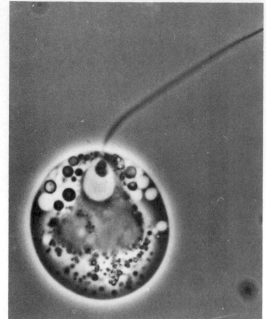

(A)

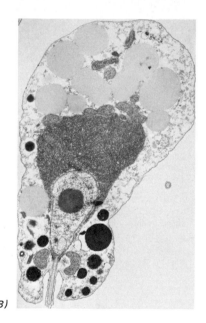

(B)

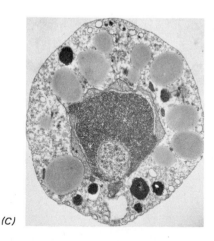

(C)

zygospore rather than from the vegetative mycelium. Meiosis occurs within the germ sporangium to produce meiospores. These spores are released and germinate under favorable circumstances to produce new vegetative, haploid *Rhizopus* plants. Thus, the gametophyte generation predominates and the zygote represents the only cell of the sporophyte generation. The life history of *Rhizopus* is thus heteromorphic, broadly resembling the life cycle of many algae.

Allomyces

Allomyces is a common aquatic fungus. It is often called a water mold since it forms a heavy white mycelium on seeds or insects that fall into the water.

The mycelium of *Allomyces* is composed of thick, trunk-like hyphae with prevalent basal rhizoids (Fig. 13-9). Hyphae are branched with most branches produced in pairs. Asexual sporangia are produced on the sporophyte thallus. These are zoosporangia which produce actively swimming zoospores (Fig. 13-10). Zoospores germinate to produce a new sporophyte plant similar to the parent plant.

Thick-walled, dark-colored sporangia are produced on the same sporophyte thallus. These are resistant to environmental change and may carry the fungus through periods unfavorable for vegetative growth. These resistant sporangia germinate by meiosis to produce motile meiospores (Fig. 13-10). Each meiospore may germinate to produce a gametophyte similar in appearance to the sporophyte. Thus, the life cycle of *Allomyces* demonstrates an alternation of isomorphic generations.

The gametophyte produces gametangia at the tip of some branches. The male gametangium is smaller and is produced immediately above the female gametangium (Fig. 13-11). Male gametes are released into the water and are small and motile. Female gametes are also released but are larger and more sluggish than the male gametes. A male and female gamete unite in the water to form a zygote which germinates immediately to become a new sporophyte plant. Sexual reproduction in *Allomyces* is thus anisogamous.

Saprolegnia

Saprolegnia is another common water mold (Fig. 13-12). This fungus is found in many aquatic habitats and often may be collected by placing a dead insect in the water. In a

Figure 13-10. Many of the Phycomycetes are aquatic in occurrence. One common "water mold" is *Allomyces*, which is illustrated here. A is a light micrograph of a zoospore. B and C are electron micrographs of meiospores. The meiospore at C has encysted just prior to germination. (A is courtesy of Melvin Fuller, University of Georgia. B and C are courtesy of Lauritz Olson, Universietets Genetiske Institut, Denmark.)

few days the insect becomes completely covered with a conspicuous mycelium of *Saprolegnia* or a closely related fungus.

Saprolegnia produces a haploid, whitish mycelium composed of branching hyphae. Asexual reproduction is common, occurring by fragmentation or zoospores. Zoosporangia are produced at the tips of certain hyphae and are slightly larger in diameter than the rest of the filament (Fig. 13-13). Zoospores are released at maturity from a pore which develops in the sporangium. These active spores swim for a time prior to settling and germinating to form a new gametophyte thallus.

In response to certain environmental stimuli, oogonia containing several eggs are produced along some hyphae (Figs. 13-13 and 13-14). Antheridia are produced along the same hyphae immediately beneath the oogonia. An antheridium contacts the side of an oogonium and a fertilization pore develops in the wall of the oogonium (Fig. 13-13). The eggs are fertilized by sperm that are deposited into the oogonium through the fertilization pore. The resultant zygotes develop thick walls to become overwintering zygospores. The following spring or when proper growth conditions return, these germinate by meiosis to form new gametophytes. Thus, *Saprolegnia* is oogamous and exhibits an alternation of heteromorphic generations.

SUMMARY

Phycomycetes are extremely important fungi. They are responsible for tremendous economic losses caused by decomposition of valuable materials and by causing many serious plant diseases. These fungi are also valuable as decomposers or "recyclers" in the ecosystem of the earth.

Phycomycetes are mostly filamentous and coenocytic, often developing a large mass of hyphae known as a mycelium. Septae are formed only at the base of reproductive structures and in response to wounding.

Asexual reproduction is by fragmentation or by the production of internal spores. Sexual reproduction is

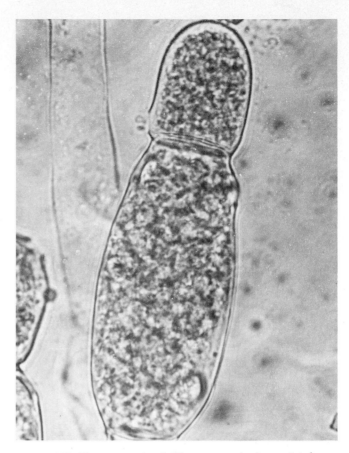

Figure 13-11. The gametangia of *Allomyces* are closely associated on the same plant. The male gametangium is smaller and directly above the female gametangium. Note the exit pore on the lower left of the male gametangium from which male gametes escape.

Figure 13-12. The brown trout illustrated here shows a severe infection of *Saprolegnia*, one of the common "water molds." (Courtesy of David White, Brigham Young University.)

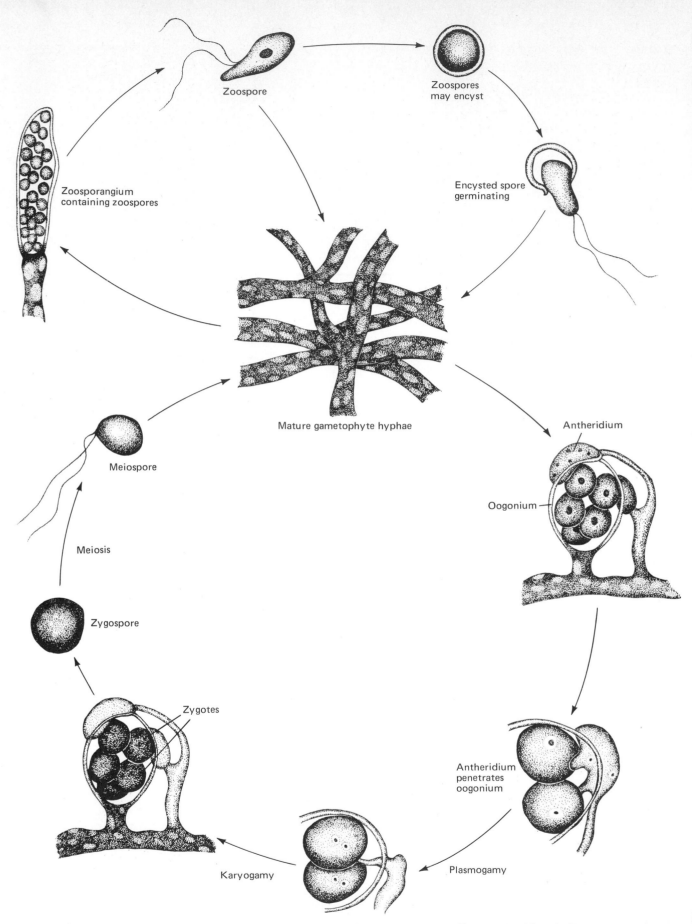

Zoospore

Zoospores
may encyst

Encysted spore
germinating

Zoosporangium
containing zoospores

Mature gametophyte hyphae

Antheridium

Oogonium

Meiospore

Meiosis

Zygospore

Zygotes

Karyogamy

Antheridium
penetrates
oogonium

Plasmogamy

Figure 13-13. Life cycle diagram of *Saprolegnia*.

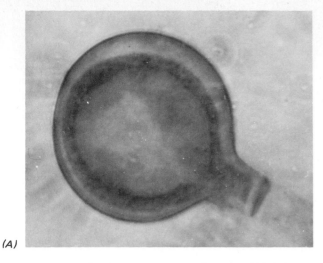

(A)

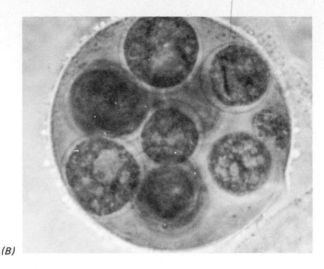

(B)

Figure 13-14. *Saprolegnia* is an oogamous "water mold." A is a young oogonium prior to the formation of eggs. The oogonium at B is mature and contains several eggs.

isogamous, anisogamous, or oogamous among different species. The alternation of generations of Phycomycetes may be isomorphic or heteromorphic.

FURTHER READING

FITZPATRICK, H. M., *The Lower Fungi—Phycomycetes,* McGraw Hill Book, Co., New York, 1930.

NIEDERHAUSER, JOHN and WILLIAM COBB, "The Late Blight of Potatoes," *Scientific American,* 200(5):100–112, 1959.

SMITH, G. M., *Cryptogamic Botany:* Vol. 1, *Algae and Fungi,* McGraw Hill Book Co., New York, 1955.

SPARROW, F. K., *Aquatic Phycomyetes,* Second edition, University of Michigan Press, Ann Arbor, 1960.

Also see the list of general works on fungi at the end of Chapter 12.

14 Division Mycophyta, Class Ascomycetes

Scanning electron micrograph of a conidiophore and conidia of *Aspergillus.* (Courtesy of Michiko and Junichi Tokunaga and K. Harada, Kyushu Dental College. Used by permission of Igaku Shoin, Ltd.)

Fungi of the classes Ascomycetes and Basidiomycetes are considerably more complex than Phycomycetes. Because of this, most mycologists refer to these two classes as the "higher fungi."

Approximately 12,000–15,000 species of Ascomycetes are presently known. This class contains may well-known fungi, including yeasts, morels, black molds, green molds, and powdery mildews. Many serious plant pathogens and a few serious animal pathogens (Fig. 14-1) are Ascomycetes.

GENERAL CHARACTERISTICS

Most Ascomycetes are filamentous, although yeasts are unicellular. Filamentous species are septate, a distinctive characteristic separating them from the coenocytic Phycomycetes. Each septum or cross wall has a simple pore

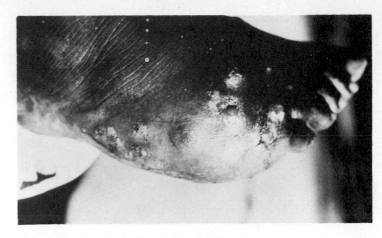

Figure 14-1. Many fungi are capable of causing infections in humans. This photograph shows a fungal infection (mycetoma) of a human foot. Mycetoma infections may be caused by Ascomycetes as well as other fungi. (Courtesy of F. Mariat, Institut de Pasteur, Paris.)

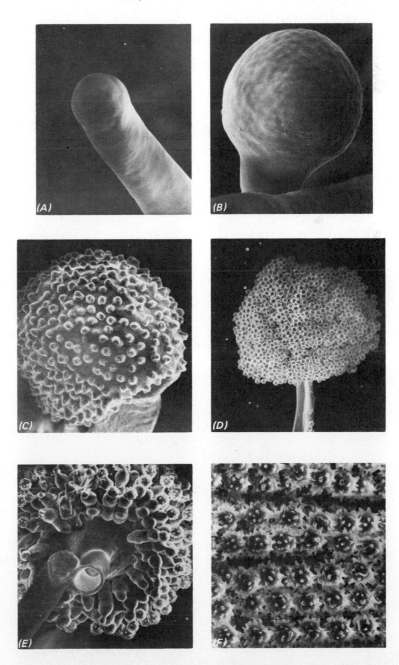

Figure 14-2. The asexual spores of Ascomycetes are known as conidia and are produced externally on conidiophores. These scanning electron micrographs show several stages in the production of conidia in *Aspergillus*. A shows the tip of a young conidiophore which has swollen at B. C shows the development of protrusions on which conidia will develop. D shows a mature conidiophore with attached conidia. E shows a view of the underside of a mature conidiophore. Note the broken hypha. F is a close view of conidia on the surface of the conidiophore. (Courtesy of Michiko and Junichi Tokunaga and K. Harada, Kyushu Dental College. Used by permission of Igaku Shoin, Ltd.)

(A)

(B)

Figure 14-3. The spores of many fungi are characteristically sculptured. These photographs of freeze-etch replicas of *Aspergillus* conidia show a distinct rodlet pattern on the spore wall. The photo shown at B is an enlargement of A. (Courtesy of W. M. Hess, Brigham Young University.)

in its center that is large enough for a nucleus to pass through.

Cell walls of Ascomycetes are firm and contain a high content of chitin and a comparatively low amount of cellulose. Each cell generally contains a single nucleus, although some species are multinucleate throughout their life, and the older cells of many Ascomycetes may become multinucleate.

Asexual reproduction occurs by fragmentation or the production of external spores known as *conidia* (Figs. 14-2 and 14-3). Conidia are usually produced on *conidiophores* which are specialized hyphae. In some species, conidia are produced in such great amounts that they color the substrate on which they grow. For instance, if an orange or other perishable food is left unrefrigerated, it will often "mold" to a greenish color. This coloration results from millions of conidia produced by the fungus which invades the orange. Conidia of Ascomycetes are common in the air and cause allergenic reactions in some people.

Ascomycetes do not produce motile cells of any kind. Even male gametes are nonmotile. In this respect these fungi are similar to red algae. On the basis of this and certain similarities in life histories, some botanists have postulated that these fungi have arisen from Rhodophyta. Others have proposed that Ascomycetes arose from the lower fungi. Evidence exists to support both theories, and it is possible that Ascomycetes actually arose from both ancestors.

Sexual reproduction is common for most Ascomycetes, but rare in some species. The sexual process is quite similar throughout the class. This process generally occurs within a fruiting structure known as an *ascocarp*. An ascocarp arises from a mycelium at the onset of sexual reproduction, and fertilization and meiosis occur within it. Three common types are prevalent. The *apothecium* is an open ascocarp that is often cup- or disk-shaped. A *perithecium* is a flask-shaped ascocarp that is enclosed except for an ostiole at its apex. The *cleistothecium* is a spherical, enclosed ascocarp that has no opening (Fig. 14-4).

Gametangia are usually located at the base of the ascocarp. The male gametangium is the antheridium, and the female gametangium is the *ascogonium*. The antheridium is always unicellular and is similar to the homologous structure in algae. The ascogonium is also unicellular and is especially similar to the carpogonium of some red algae. Often it has an elongate tip or trichogyne, and the female nuclei are in a swollen receptacle at its base. Antheridia are produced in close proximity to ascogonia (Fig. 14-5), although these gametangia normally are produced by opposite mating strains. Thus, most Ascomycetes are heterothallic.

The trichogyne is attracted toward the antheridium that it eventually contacts (Fig. 14-5). The walls between the trichogyne and the antheridium break down and male nuclei from the antheridium pass through the trichogyne into the base of the ascogonium where they pair with female nuclei. Then several hyphae begin to grow from the ascogonium toward the surface of the apthothecium or the cavity of the perithecium or cleistothecium. Each cell of these hyphae contains two nuclei, one from the male strain and one from the female strain (Fig. 14-5). Hyphae with this condition are said to be *dikaryotic*. Dikaryotic hyphae in Ascomycetes are termed *ascogenous hyphae*. These hyphae grow until they reach the surface or cavity of the ascocarp. When this occurs, each ascogenous hypha bends to form a hook. The next to the last cell of each ascogenous hypha then enlarges and elongates to become an *ascus*. The two nuclei in this ascus fuse to become the zygote. The zygote divides immediately by meiosis to produce four haploid nuclei. Each nucleus then normally divides by mitosis to produce eight haploid nuclei all within the ascus. Each of these nuclei becomes surrounded by a layer of cytoplasm and a wall to become an *ascospore* (Figs. 14-5 and 14-6). Thus, Ascomycetes are so named

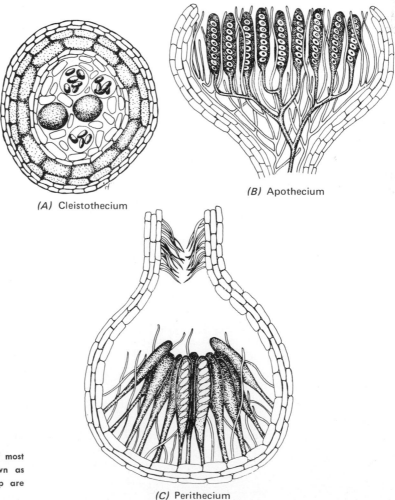

(A) Cleistothecium

(B) Apothecium

(C) Perithecium

Figure 14-4. The sexual reproductive process of most Ascomycetes occurs within fruiting bodies known as ascocarps. The three common types of ascocarp are illustrated in these longitudinal sections.

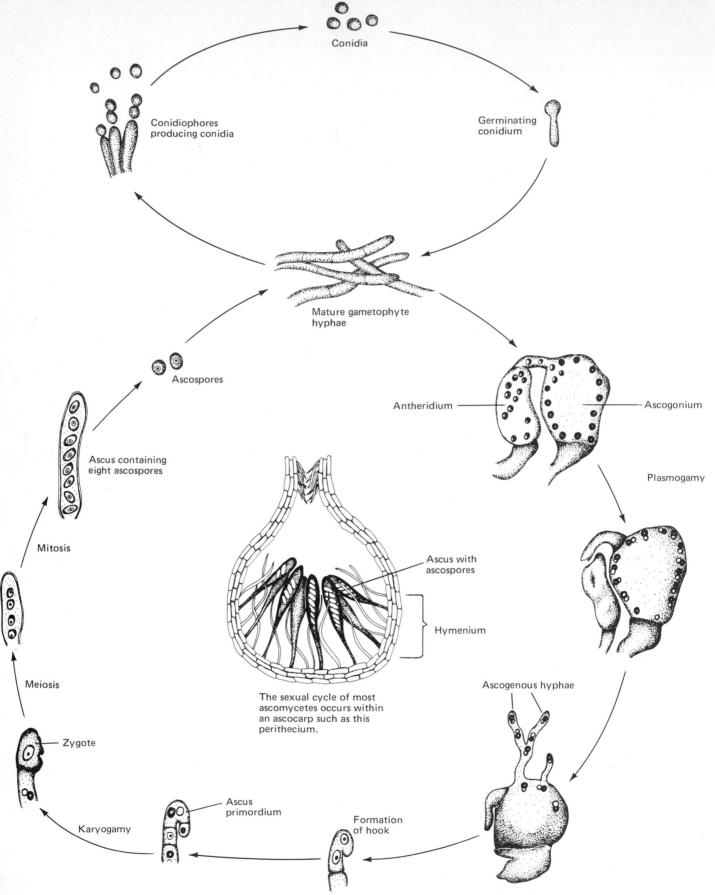

Conidia

Conidiophores producing conidia

Germinating conidium

Mature gametophyte hyphae

Antheridium

Ascogonium

Plasmogamy

Ascospores

Ascus containing eight ascospores

Ascus with ascospores

Hymenium

Mitosis

The sexual cycle of most ascomycetes occurs within an ascocarp such as this perithecium.

Ascogenous hyphae

Meiosis

Zygote

Ascus primordium

Formation of hook

Karyogamy

Figure 14-5. Life cycle diagram of an ascomycete.

because of the production of meiospores within a sac-like ascus. The production of these meiospores within the ascus is an especially significant characteristic that will be contrasted to meiospore formation in Basidiomycetes which occurs externally.

Asci are produced in a definite, recognizable layer in the apothecium and perithecium. This layer is termed the *hymenial layer* or *hymenium* (Fig. 14-5). Paraphyses are also produced in the hymenium and function in aiding spore dispersal. A cleistothecium does not produce a hymenium, and the asci are scattered throughout the interior of the ascocarp.

Ascospores are released or ejected from the asci at maturity (Fig. 14-7). The method of spore dispersal varies, but often a pore is formed at the apex of the ascus and the spores are "popped out" by the swelling of the paraphyses. They germinate under favorable conditions to produce a new male or female mycelium according to the genetic makeup of the spore.

OCCURRENCE AND ECOLOGY

Ascomycetes are widely distributed fungi. Both parasitic and saprophytic species are known. A few genera such as *Penicillium* and *Aspergillus* occur as either parasites or saprophytes on a vast number of different substrates. Some mycologists describe such fungi as being omnivorous.

Parasitic Ascomycetes mostly attack plants (Fig. 14-8), and to a lesser extent insects and higher animals. Several of these fungi are fascinating since they are extremely host specific. For instance, some Ascomycetes infect only one species of insect, only one sex of the insect, and only certain parts of the anatomy of that insect. Few other fungi are as specific in their host preference as are certain Ascomycetes.

Probably most Ascomycetes are saprophytic. These fungi are often collected in piles of humus, on rotting logs, on the forest floor, at the edge of marshes, or on any other moist, organic habitat. Many Ascomycetes commonly grow on animal dung. An interesting and amusing fungal garden may be grown by gathering dried horse or cow manure, putting it in a culture dish and keeping it moist for a week or two. Many fungi will appear on manure so prepared, and the beauty and diversity of these fungi are truly amazing.

A few Ascomycetes are aquatic in both marine and fresh water habitats. Some grow and complete their entire life cycle underground. Truffles (Fig. 14-9) are noteworthy examples. These fungi are highly prized as a gourmet food and often command very high prices in European markets. Thus, it is profitable to collect these fungi for sale to hungry rich people. Unfortunately, it is difficult to locate them since they grow underground. Happily, these fungi have a peculiar odor which dogs and pigs are adept at smelling and a trained "truffle sniffer" is valuable prop-

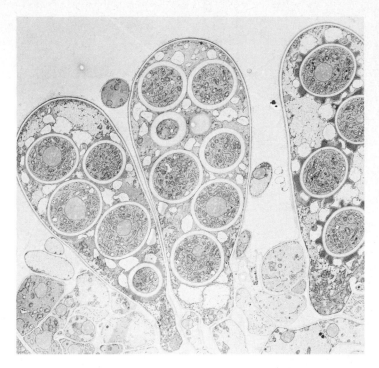

Figure 14-6. Karyogamy and meiosis in Ascomycetes occur within asci at the hymenial layer. This electron micrograph of a section of the hymenium of *Ascodesmis nigricans* shows three asci with ascospores at various stages of development. The ascus in the center is the youngest. (Courtesy of Charles Bracker, Purdue University.)

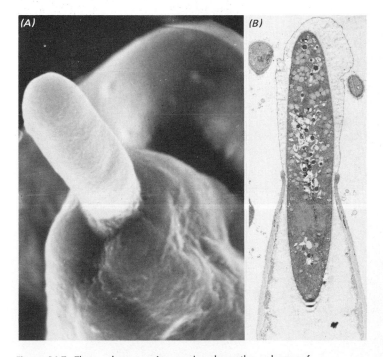

Figure 14-7. These electron micrographs show the release of ascospores from the ascus of *Lophodermella sulcigena*. A is a scanning micrograph showing surface details and B is a section through an ascus tip and ascopore. (Courtesy of Richard Campbell, University of Bristol.)

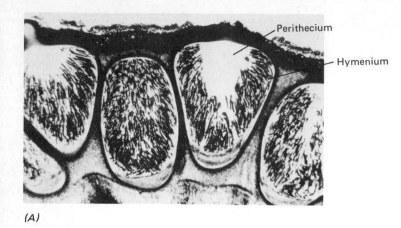

(A)

Perithecium

Hymenium

(B)

Figure 14-8. Several Ascomycetes are parasitic, causing both plant and animal diseases. These photographs show Ascomycetes parasitic on woody plants. *A* is a closeup of *Hypoxylon* which produces a cushion-shaped stroma with many perithecia imbedded within. *B* shows the thick, black stroma of *Dibotryon morbosum* on a branch of chokecherry.

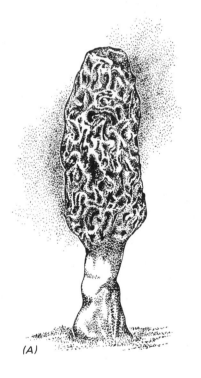

(A)

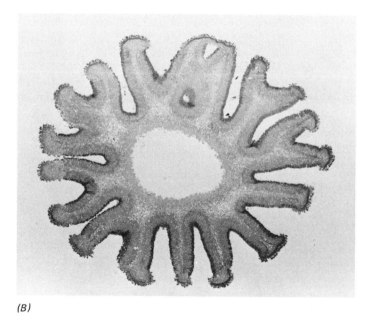

(B)

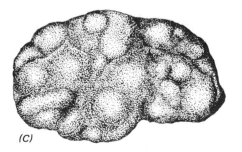

(C)

Figure 14-9. Several Ascomycetes are prized for their excellent flavor. *A* is an ascocarp of *Marchella esculenta*, the common delicious morel sought and prized as a gourmet food. *B* is a cross section through a *Morchella* ascocarp. This ascocarp is highly infolded and convoluted. The hymenial layer lines the ascocarp branches and depressions. *C* is an ascocarp of the truffle *Tuber*. Truffles grow and produce their ascocarps underground. Accordingly, they are difficult to find and are very expensive.

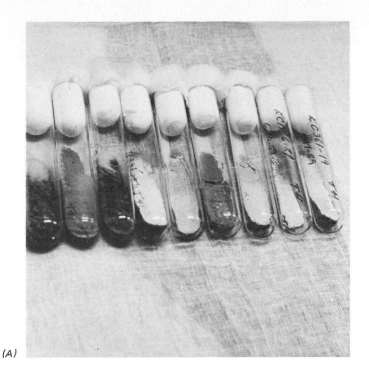

(A)

(B)

(C)

(D)

Figure 14-10. Eli Lilly and Company follows ten steps in its penicillin producing operation. The first is screening soil samples, then culture propagation, fermentation, filtration, isolation (to extract solvents), purification, sterile transfer of the dried crystalline antibiotic, testing, finishing, and distribution. These photographs depict the culture propagation process. Using sterile procedures, selected antibiotic strains are transferred to a series of agar slants (A). After several days incubation, the mold from the slants is transferred to a prepared nutrient flask (B). After continuous shaking, the growth in the flask has reached the optimum stage for transfer to larger vessels. C and D depict the isolation stage of pencillin production. With fast revolving machines (C), the antibiotic-laden broth from the filter is extracted with a solvent. Absorption techniques (passing a solution of the antibiotic through columns) are also used (D). (Courtesy of Eli Lilly and Company.)

(A)

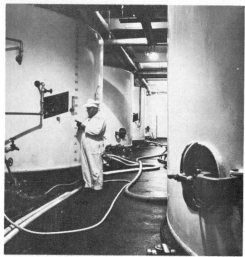

(B)

Figure 14-11. Yeasts are important *Ascomycetes* in the brewing and baking industries. These photographs show the preparation of yeast cultures *(A)* and fermentation tanks *(B)* used in brewing. (Courtesy of Joseph Schlitz Brewing Co.)

Figure 14-12. Yeasts are commercially important Ascomycetes because of their uses in several industries. This figure shows yeast culturing vats used in the commercial production of yeast. (Courtesy of Fleischmann Laboratories, Standard Brands, Inc.)

erty. In fact, it is not uncommon to see a prized truffle-sniffing pig fed table scraps and carted to and from the truffle grounds in a wheelbarrow. I suppose this is small consolation though, since the pig or dog is never allowed to eat the truffles which it is so adept at finding.

IMPORTANCE

Ascomycetes are among the most important plants on earth. These fungi are responsible both for tremendous direct economic losses and benefits. In addition, they are significant in the ecosystem, decomposing large amounts of organic matter.

Ascomycetes include some very destructive plant pathogens. Each year large sums of money are spent controlling plant diseases induced by Ascomycetes. In addition, some species are pathogenic to man both directly as parasites, and indirectly by producing toxins in foods. *Penicillium,* for instance, occasionally infects the skin or respiratory tract of humans, and can cause serious diseases even resulting in death.

Penicillium is also beneficial in antibiotic production which undoubtedly has allowed some of the readers of this book to do so. Secretions from this plant yield penicillin which is one of the world's foremost antibiotic agents (Fig. 14-10). This drug has been responsible for saving thousands of lives and for reducing the suffering from many bacterial diseases. Related Ascomycetes produce several other antibiotics, and many are used for synthesizing organic acids and alcohols.

Yeasts are Ascomycetes that are critical to baking and brewing industries (Fig. 14-11). These industries are economically very important since the majority of Americans and Europeans both like to eat bread and drink beer. In addition, cultivating yeasts commercially for brewing and baking has become a large industry since heavy usage requires large amounts of yeast (Fig. 14-12). Many people consume yeasts directly as a high-protein, high-vitamin diet supplement. Very likely, the direct use of yeasts for food will become more important in the future.

A few Ascomycetes such as truffles and morels are prized for their delicate flavors. The morels (*Morchella*) (Fig. 14-9) are among the most sought after fungi and are collected in the spring when the "oak trees are in mouse-ear stage." These fungi have not yet been successfully cultivated, and so must be gathered. To those who do collect *Morchella,* the flavor and satisfaction is well worth the effort. Again, care must be used in eating any wild mushroom because of the few deadly poisonous fungi that may be confused with edible species. Some Ascomycetes are indirectly responsible for fine flavors as a result of their use in the cheese industry. *Penicillium roqueforti,* for instance is responsible for imparting the superb flavor to roquefort cheese.

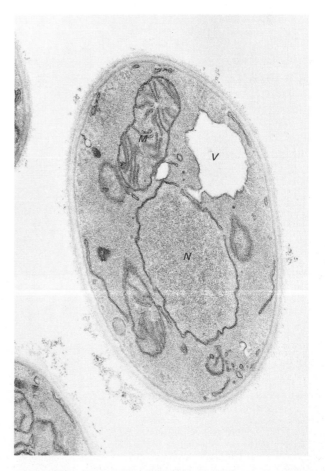

Figure 14-13. This electron micrograph through a yeast cell shows the nucleus *(N)*, mitochondria *(M)*, vacuole *(V)* and other organelles. (Courtesy of Masako Osumi, Japan Women's University.)

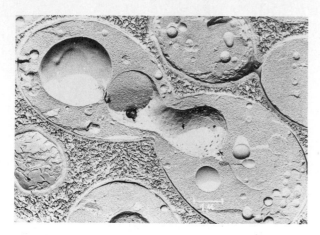

Figure 14-14. Yeast cells reproduce asexually by the process of budding as illustrated in this freeze etch replica of the yeast *Saccharomyces cerevisiae*. The dividing nucleus is clearly evident at the center. (Courtesy of Balzers Artiengesellschaft.)

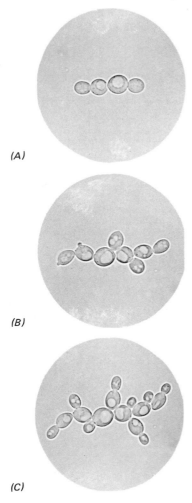

(A)

(B)

(C)

Figure 14-15. Yeast cells will occasionally bud very rapidly to produce pseudofilaments. This figure shows stages in the development of a yeast pseudofilament. A shows a short, simple pseudofilament. B shows early stages in branching, and C is a large, highly branched pseudofilament. (Courtesy of Fleischmann Laboratories, Standard Brands, Inc.)

In their role as decomposers, Ascomycetes are especially significant in reducing certain organic compounds which are not decomposed well by bacteria or other fungi.

REPRESENTATIVE ASCOMYCETES

Yeasts

Several genera of Ascomycetes are commonly referred to as yeasts. These fungi occur throughout the world and grow especially well on substances high in sugar content. Decaying fruits and flower nectaries are good places to collect yeasts. Others occur in the soil, in animal dung, in milk, and on plants and animals. A few yeasts occasionally attack humans and may cause serious diseases.

The different species of yeasts share many common characteristics. These include a unicellular habit, a more or less spherical shape, and a uninucleate condition (Fig. 14-13). Asexual reproduction also is similar among yeasts. This process, which is termed *budding,* proceeds by the formation of a small hemispherical protrusion on the side of a mature cell. Mitosis then occurs and one nucleus migrates into the protrusion while the other remains in the mother cell (Fig. 14-14). The protrusion or *bud* enlarges and eventually "pinches off" from the mother cell to form a new yeast plant. Under favorable circumstances budding occasionally proceeds so rapidly that buds form secondary buds before they separate from the mother cell, and a false filament may result (Fig. 14-15).

Sexual reproduction is also similar among most yeasts, although a few species apparently do not reproduce sexually. This process occurs when two haploid yeast cells fuse to form a diploid cell that acts as the zygote and becomes the ascus directly (Fig. 14-16). The diploid nucleus thus formed divides by meiosis to produce four haploid nuclei. These may remain undivided or may divide mitotically to form eight nuclei. The cytoplasm in the ascus divides about the nuclei to form ascospores that are released from the ascus to become new haploid yeast plants (Fig. 14-16).

Some variations in this life cycle according to the relative longevity of the diploid and haploid generations are noteworthy. In some species the diploid cell is produced early in the life cycle and the majority of vegetative cells are diploid. In others, the $2N$ cell divides by meiosis immediately after its formation, forming haploid cells that represent the majority of the vegetative cells. In still others the longevity of the two generations is approximately equivalent. Thus, even though the sexual process is basically alike in all yeasts, some species exhibit a predominant sporophyte generation, others a predominant gametophyte generation, and some an equally long-lived sporophyte and gametophyte generation.

As mentioned earlier, yeasts are used in brewing and baking industries. They are also important in other less

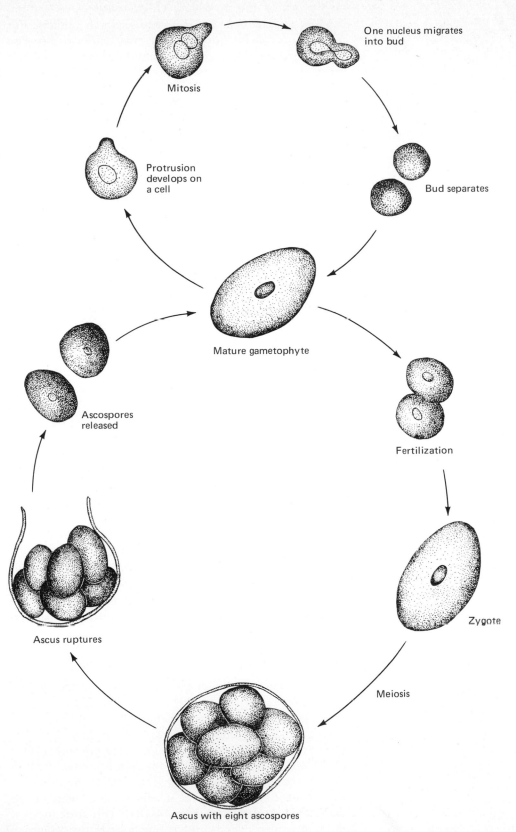

One nucleus migrates
into bud

Mitosis

Protrusion
develops on
a cell

Bud separates

Mature gametophyte

Ascospores
released

Fertilization

Ascus ruptures

Zygote

Meiosis

Ascus with eight ascospores

Figure 14-16. Life cycle diagram of a yeast.

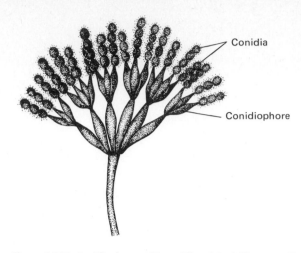

Figure 14-17. Conidiophores with conidia of *Penicillium* are illustrated here. This widely distributed fungus is economically important for a number of reasons including antibiotic production.

well-known roles such as decomposing organic materials and products rich in sugars. Other yeasts cause uncomfortable and, on occasion, serious diseases in humans. For instance, *Cryptococcus* is a yeast that causes the disease cryptococcosis. This organism may infect the pulmonary tract or skin, or more seriously, the spinal column and brain. Cryptococcosis is generally a rather mild, chronic infection, but at times it is very dangerous, causing damage to the central nervous system that results in mental disorders.

Penicillium

Penicillium (Fig. 14-17) is economically significant for several reasons. Of primary concern to humans is the production of antibiotic agents by certain species. These antibiotics were discovered by accident when it was noticed that bacteria cultured on nutrient agar plates did not grow well when close to *Penicillium* colonies. This important ascomycete is now grown commercially for antibiotics that are extracted and used medicinally.

Other species of *Penicillium* are useful for a variety of reasons. *Penicillium roqueforti* is responsible for the flavoring and coloration of roquefort cheese. However, the same organism may be responsible for undesirable flavors imparted to certain other cheeses. Many species of *Penicillium* and the closely related genus *Aspergillus* (Figs. 3-3*N*, 14-2, and 14-3) comprise the destructive blue and green molds. These molds attack almost any organic matter, often causing considerable damage to valuable products.

Penicillium may cause potentially dangerous diseases in man. Conidia of this fungus occasionally germinate on the skin or in the lungs to cause penicilliosis. This disease is generally not serious, although when located in the pulmonary tract it may be quite troublesome. The reason this fungus in generally innocuous but occasionally becomes pathogenic on humans is not fully understood. In fact, the entire science of medical mycology is in an early stage, and much future research is necessary here.

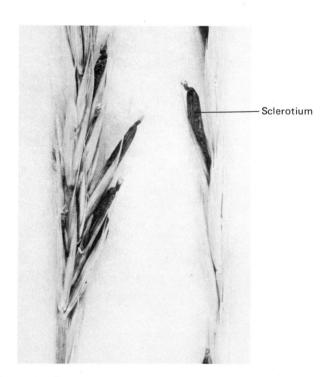

Figure 14-18. This figure shows *Claviceps purpurea* (ergot) on a wild rye plant.

Claviceps purpurea

Claviceps purpurea is one of the most interesting and frightening of all fungi. This ascomycete is parasitic on many grasses (Fig. 14-18), including cultivated rye, and causes the disease known as *ergot of rye*. This disease is responsible for reduced yields in many rye-growing regions of the world. However, the loss of crops is not the primary importance of this fungus. Claviceps infects the flower of the rye plant and replaces the developing seed with a large dark-colored *sclerotium* (Fig. 14-18). This structure is a mass of hardened hyphae that functions in overwintering the fungus. The chemical ergotine is produced by the cells of the sclerotium, and this substance is extremely toxic to

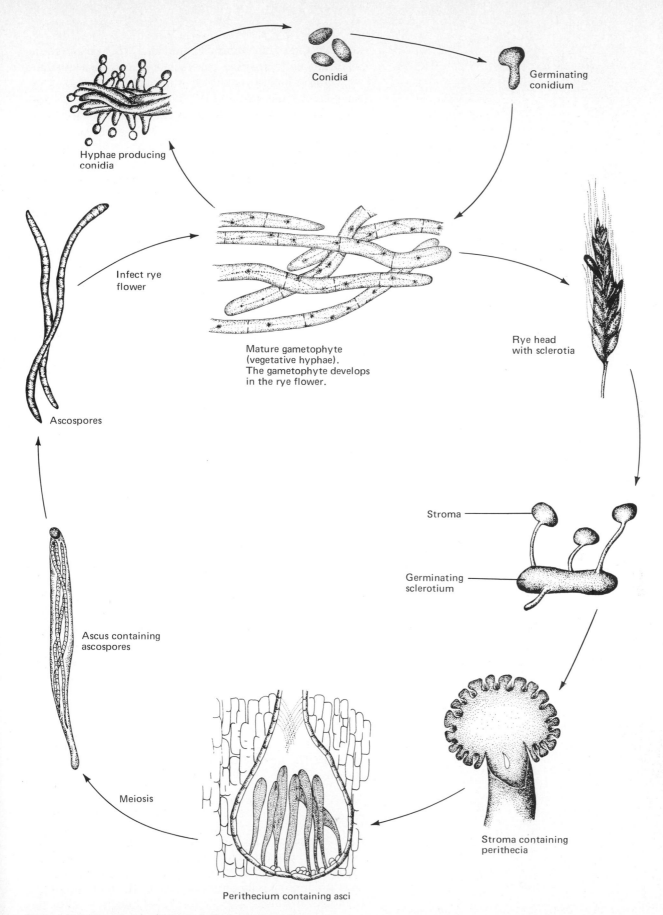

Conidia

Germinating
conidium

Hyphae producing
conidia

Infect rye
flower

Mature gametophyte
(vegetative hyphae).
The gametophyte develops
in the rye flower.

Rye head
with sclerotia

Ascospores

Stroma

Germinating
sclerotium

Ascus containing
ascospores

Meiosis

Stroma containing
perithecia

Perithecium containing asci

Figure 14-19. Disease cycle diagram of *Claviceps purpurea*.

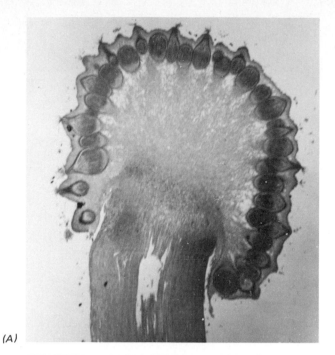

(A)

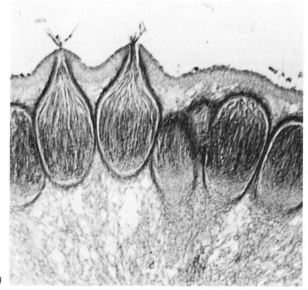

(B)

Figure 14-20. *Claviceps purpurea* overwinters by producing hard-ened masses of hyphae known as sclerotia. The following spring stroma develop from the sclerotia and produce perithecia. These sections through the stroma of *Claviceps purpurea* show an entire stroma containing numerous perithecia (A) and an enlargement of A showing several perithecia (B).

humans and their domestic animals. When ingested in small amounts, ergotine causes blood-vessel constriction, and cattle or humans eating ergot-infected rye often develop gangrene of the extremities. In greater amounts, acute poisoning results and hallucinations and death are often imminent. The hallucinations apparently always deal with fire and burning. Ergot poisoning in humans is also termed *Saint Anthony's Fire* since superstition has it that only prayers to St. Anthony can relieve its pain and suffering. Incidentally, ergotine is chemically similar to the halluci-nogen, LSD, which is also extremely toxic.

Since ergotine causes blood-vessel constriction, this substance has been used to restrict bleeding at childbirth and in hemorrhage patients. Ergotine is also a strong abor-tive agent. Because of the present and potential medicinal uses of this substance, several pharmaceutical companies are currently trying to develop methods to induce high levels of *Claviceps* infection and sclerotium formation under controlled situations.

Ascospores of *Claviceps* are produced in the spring at the same time rye and other grass flowers develop. These spores blow about on wind currents and some of them con-tact rye flowers where they germinate to cause infection. A mycelium develops slowly in the ovary of the flower until near the time when the grains mature. Conidia are pro-duced throughout the period that may infect other rye flowers (Fig. 14-19). In the early fall, the mycelium de-velops rapidly and becomes hardened on its outer surface to form the sclerotium. This structure falls to the ground after a time and the fungus overwinters in this state. Dur-ing the following spring, several *stroma* arise from the sclerotium (Figs. 14-19 and 14-20). These structures are composed entirely of fungal tissue and are stalked with a spherical head. Numerous perithecia are imbedded within this head (Fig. 14-20). The typical ascomycete sexual cycle occurs within them to produce ascospores. These spores are long and needle-shaped and often fragment into several pieces. They are released as the rye flowers are being pollinated and blow about on air currents to infect other rye plants.

A cycle such as that described above is often termed a *disease cycle* rather than a life cycle since it involves many steps in the development of a plant disease.

Claviceps is closely related to several other pathogenic Ascomycetes, some of which are responsible for large losses of crops.

Powdery mildews

Powdery mildews include several genera of Ascomycetes that are serious plant pathogens of a wide variety of hosts, including many cereal grains and other food crops. These fungi produce mycelia on the surface of the host plant that often appear white. Specialized hyphal branches penetrate

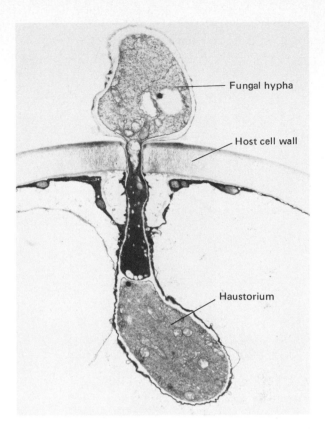

Fungal hypha

Host cell wall

Haustorium

(A)

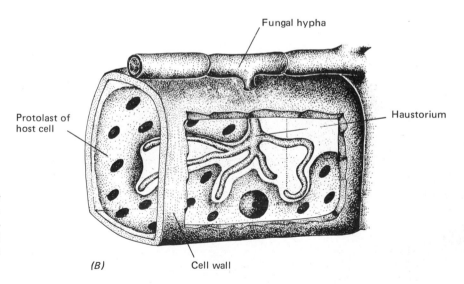

Fungal hypha

Protolast of
host cell

Haustorium

Cell wall

(B)

Figure 14-21. Many parasitic fungi penetrate host cells and produce structures known as *haustoria*, which are organs of absorption. *A* shows an electron micrograph of a cross section through a fungal hypha and haustorium and *B* is a diagram of a haustorium. (*A* is courtesy of Charles Bracker, Purdue University.)

the epidermis of the host and *haustoria* (Fig. 14-21) develop in many epidermal cells. Haustoria are specialized structures that absorb water and nutrients from host cells, often without killing penetrated cells for long periods of time.

Conidia develop along hyphae (Fig. 14-22) and are released on air currents to infect other host plants. These are produced in such large numbers that they often make an infected plant appear as though it had been dusted with

(A)

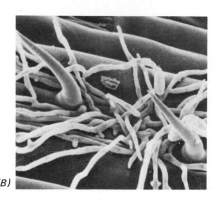

(B)

(C)

Figure 14-22. These electron micrographs show stages in infection of the powdery mildew *Erysiphe graminis* on wheat. *A* shows a germinating conidium. A mycelium develops from the conidium and spreads over the surface of the host *(B)*. As the mycelium matures, it produces more conidia in the characteristic chains *(C)*, which make the host appear powdered and gives the fungus its common name. (Courtesy James V. Allen, Brigham Young University.)

talcum powder, giving these fungi their common name. When a conidium contacts a suitable host cell, it germinates to produce a short hypha. This soon penetrates the epidermis in order to obtain nutrients and then continues growth. Conidia will be produced by the new mycelium as it approaches maturity.

SUMMARY

The class Ascomycetes contains some of the most important plants known to humans. These plants are both beneficial and harmful in their effects. Some are responsible for the degradation of large amounts of organic matter and thus are important recycling organisms. Others are serious plant pathogens, and still others are responsible for certain diseases of animals and humans. Antibiotics are derived from several Ascomycetes, and some are useful both directly and indirectly as food.

Ascomycetes are similar in many respects to some Rhodophyta. No motile cells of any kind are produced, and the female gametangium often exhibits a trichogyne. Ascomycetes differ from red algae by having a high proportion of chitin in their cell walls and by lacking chlorophyll of any kind.

Ascomycetes may be parasitic or saprophytic. Parasitic species are chiefly plant or insect parasites, but may infest a wide range of hosts. Saprophytic forms occur on a wide variety of substrates including animal dung and humus.

FURTHER READING

DENNIS, R. W., *British Cup Fungi and Their Allies,* The Ray Society, London, 1960.

GRAY, W. D., *The Relation of Fungi to Human Affairs,* Henry Holt and Co., New York, 1959.

KAVALER, LUCY, "Mold-Made Flavors," *Natural History,* 75(4):62–66, 1966.

ROSE, A. H., "Yeasts," *Scientific American,* 202(2):136–146, 1960.

YARWOOD, C. E., "Powdery Mildews," *Botanical Review,* 23:235–300, 1957.

Also see the list of general works on fungi at the end of Chapter 12.

15 Division Mycophyta, Class Basidiomycetes

Amanita muscaria, a hallucinogenic mushroom. (Courtesy of Walter Hodge.)

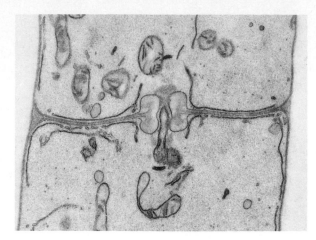

Figure 15-1. The filaments of basidiomycete mycelium are septate, and the septae have elaborate pores known as dolipores, one of which is illustrated in this electron micrograph of *Rhizoctonia solani*. (Courtesy of Charles Bracker, Purdue University.)

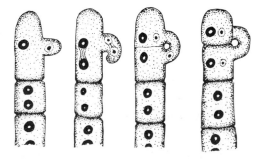

Figure 15-2. The mycelium of some Basidiomycetes is extensive and can be seen easily. The mycelium in this photograph appears white at the center of this decaying log.

Figure 15-3. Clamp connection division is characteristic of many Basidiomycetes. In this type of division one nucleus of the dividing dikaryotic cell is carried backward into the second daughter cell in a cytoplasmic protrusion known as a clamp.

Basidiomycetes are perhaps the fungi best known to most people, since this class contains the common conspicuous fungi. Mushrooms, "toadstools," polypores, earth stars, puff balls, shelf fungi and others are all Basidiomycetes. In addition, this class contains several very serious plant pathogens, particularly the rusts and smuts. About 14,000 species of Basidiomycetes are presently known.

GENERAL CHARACTERISTICS

Basidiomycetes have a well-developed filamentous thallus that penetrates the substrate and absorbs nutrients. Hyphae are septate, and the septum has a peculiar type of pore known as a *dolipore* which is unique to Basidiomycetes (Fig. 15-1). It has a thickened rim and is closely associated with a double membrane in the cytoplasm on each side of the septum. Individual hyphae are invisible without a microscope, although mycelia are often readily apparent. Thus, rotting logs may often be split open to expose a white mat of fungi tissue which is usually a basidiomycete mycelium (Fig. 15-2). Hyphae of many Basidiomycetes pass through three stages. When a haploid spore germinates, it produces *primary hyphae*. Cells of these hyphae are haploid and generally uninucleate. After a brief period of time, two primary hyphae of different mating strains come into contact. A pore is dissolved between them, and nuclei are exchanged to form dikaryotic cells. These cells then divide rapidly to form *secondary hyphae*.

Most of the life cycle of a typical basidiomycete occurs in the secondary hyphae stage. When proper ecological conditions develop, secondary hyphae become organized to form a fruiting body or *basidiocarp*. Hyphae comprising the fruiting body are *tertiary hyphae*.

Cell division in most Basidiomycetes is unique throughout the secondary and tertiary hyphae stages. A small protrusion develops on the side of the cell, and one nucleus moves into it. Each nucleus then divides mitotically, and a new septum forms to divide the mother cell into two daughter cells. However, this septum separates three nuclei in the terminal daughter cell and only one nucleus in the second daughter cell. The "extra" nucleus remains in the protrusion, which is known as a *clamp*. The clamp grows "backwards" to contact the uninucleate cell and deposit its nucleus there (Fig. 15-3). Thus the dikaryotic nature of the cells is maintained. This type of division is known as *clamp connection division*. It is similar in many respects to the development of the ascus-forming hook in Ascomycetes. Several mycologists see this as good evidence that Basidiomycetes arose from Ascomycetes.

Asexual reproduction is well developed among many Basidiomycetes. It occurs by fragmentation of the mycelium or by the production of conidia or other asexual spores. Asexual reproduction of some species may occur

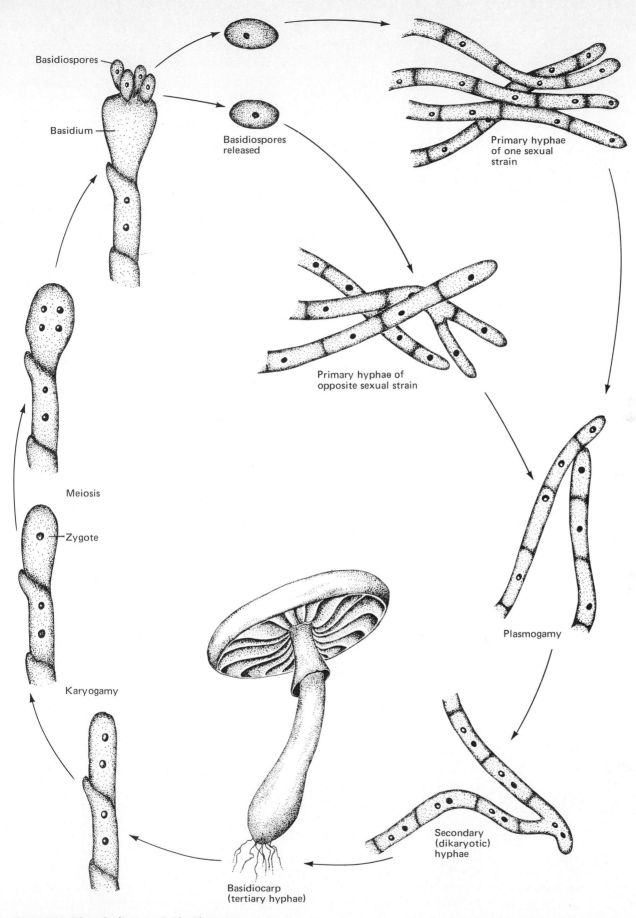

Basidiospores

Basidium

Basidiospores
released

Primary hyphae
of one sexual
strain

Primary hyphae of
opposite sexual strain

Meiosis

Zygote

Karyogamy

Plasmogamy

Secondary
(dikaryotic)
hyphae

Basidiocarp
(tertiary hyphae)

Figure 15-4. Life cycle diagram of a basidiomycete.

(A) Spring *Amanita*

(B) Common *Psathyrella*

(C) Horse mushroom
(*Agaricus arvensis*)

(D) Clitocybe sp.

(E) Inky cap
(*Coprinus* sp.)

(F) Clean *Mycena*

(G) Leopard *Tricholoma*

(H) Morgan's *Lepiota*

(I) Fly agaric
(*Amanita muscaria*)

Figure 15-5. The classification of Basidiomycetes is largely dependent on shape, color, and other characteristics of the basidiocarp. Several basidiocarps are illustrated in this figure.

rapidly, although many Basidiomycetes apparently rely more on sexual rather than asexual reproduction to propagate the species.

Similar to Ascomycetes and red algae, Basidiomycetes do not produce motile cells of any kind. From this and other evidence, many botanists propose that the series Cyanophyta—Rhodophyta—Ascomycetes—Basidiomycetes represents a direct evolutionary line.

Sexual reproduction occurs in almost all Basidiomycetes (Fig. 15-4), although a few species reproduce chiefly asexually. Sexuality has never been observed in a few probable Basidiomycetes which are therefore placed at least temporarily in the form class Fungi Imperfecti.

Many Basidiomycetes produce a basidiocarp at the initiation of the sexual process. The shape of this structure varies greatly among different species (Fig. 15-5) and is perhaps the single most important feature used in the classification of these fungi. Some Basidiomycetes never produce a basidiocarp in sexual reproduction.

A hymenial layer develops on gills, pores, or other parts of the basidiocarp. This hymenium is similar to the same layer produced on or within an ascocarp since it contains meiospore bearing cells and paraphyses. However, the spore-bearing cells in Basidiomycetes are known as *basidia* (Figs. 15-4 and 15-6) rather than asci. A basidium is a club-shaped structure which develops from the terminal cell of a tertiary hypha and bears external meiospores known as *basidiospores*. These spores are generally produced in fours. They develop on short *sterigmata* that occur on the surface of the basidium (Figs. 15-4 and 15-6). The two nuclei in this dikaryotic cell fuse to form a 2N zygote (Fig. 15-4). This zygote divides immediately by meiosis to produce four haploid basidiospores that then migrate through the sterigmata to their external position.

Basidiospores are either passively released or forcibly ejected from basidia. These spores then blow about until

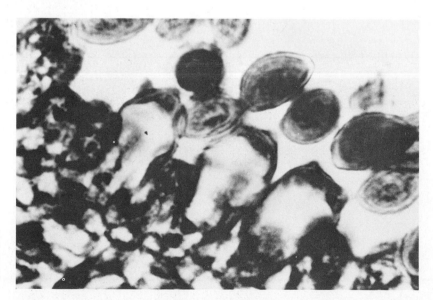

Figure 15-6. Basidiospores are produced externally on a swollen club-shaped basidium. Three basidia are illustrated at the center of this light micrograph. Note the basidiospores attached and released from the basidia.

Figure 15-7. Some fungi are luminescent. For instance, the mushrooms in this figure were photographed in their own light. (Courtesy of Milton Cormier, University of Georgia.)

they settle out of the air. Some of them land in a suitable environment and germinate to produce primary hyphae (Fig. 15-4).

It is significant that gametangia are never produced by Basidiomycetes, representing another distinction between these fungi and Ascomycetes. Other distinctions include the external meiospores, the dolipore septum, clamp connection division and the longevity of the dikaryotic (secondary hyphae) stage in Basidiomycetes, all of which are absent from Ascomycetes.

One interesting characteristic of some Basidiomycetes is their bioluminescence (Fig. 15-7). This has led to much superstition and misunderstanding of these organisms, particularly since bioluminescent forms often occur in swamps or graveyards.

OCCURRENCE AND ECOLOGY

Both parasitic and saprophytic Basidiomycetes are common. Parasitic species mostly attack vascular plants, although saprophytes occur on a wide variety of substrates. Many Basidiomycetes are mycorrhizal, and most will only become associated with a specific kind of tree. Thus, mushroom hunters frequently search for one species in one type of forest and for another species in another type of forest.

Saprophytic Basidiomycetes may be collected from many habitats where adequate moisture and organic matter exists. The forest floor is a common basidiomycete habitat from which many different species can be obtained in a day of collecting, particularly following a rain storm. Some species are common in lawns or even in the cracks of sidewalks (Fig. 15-8).

Figure 15-8. A mushroom in an unusual habitat is shown here pushing up through the asphalt of a parking lot. (Courtesy of A. Krikorian, State University of New York.)

(A)

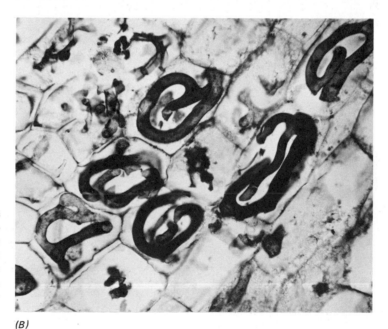

Figure 15-9. Fungi are often associated with the roots of higher plants to form mycorrhizae. A is an enlargement of a pine root which is completely covered with a dense growth of fungal hyphae. B is a cross section through the root of yellow popular showing hyphae within the cells. (Courtesy of Edward Hacskaylo, U. S. Dept. of Agriculture. Used by permission of the American Institute of Biological Science.)

(B)

IMPORTANCE

Basidiomycetes decompose many organic products, particularly wood. They are essential in this role in nature, although they may cause large economic losses when they attack commercially valuable species.

Mycorrhizal Basidiomycetes (Fig. 15-9) are beneficial in their association with some higher plants. It has been shown that many vascular plants, including several economically important trees, will not grow well unless associated with a mycorrhizal fungus. The reasons for this are

Figure 15-10. Some fungi are commercially grown and marketed. This figure shows a bed of *Agaricus campestris*, the common edible mushroom of commerce. (Courtesy of Campbell Soup Company.)

not well understood, although the fungus probably aids the vascular plant in the uptake of nutrients.

Several Basidiomycetes are plant pathogens, causing some of the most serious of all plant diseases. These pathogens are especially significant to humans since many attack the critically important cereal grains. For instance, black stem wheat rust and several common smut diseases are a constant menace to farmers throughout the world. The annual worldwide budget expended on controlling these and related diseases is immense but completely necessary to protect our present rate of food production.

Some Basidiomycetes, especially mushrooms, are used as food. A large industry exists to produce the common pizza or mushroom soup mushroom, *Agaricus campestris* (Fig. 15-10). This mushroom is widely eaten and is a surprisingly good source of several vitamins and minerals. Many mushrooms are prized for their flavor, although they are gathered from natural habitats rather than cultivated. Caution must be used when eating these mushrooms since some species are violently poisonous.

Some poisonous mushrooms cause hallucinations or drunkenness when ingested in small amounts. These mushrooms have been used in the religious rites of primitive peoples. Several of these hallucinogenic species have probably been a part of black magic and witchcraft for hundreds of years.

Some species are currently being investigated for medicinal uses including the treatment of tumors. Others are under investigation because of the allergenic reactions their spores cause in some people.

FOSSIL BASIDIOMYCETES

As with all fungi, Basidiomycetes do not fossilize well and hence are poorly represented in the fossil record. However, a few fossil fungi have been found. These are generally preserved as mycelium in petrified wood or compressed leaves. They are definitely Basidiomycetes since clamp connections on the hyphal strands have been observed.

Several examples of mushrooms are also known from archaeological records. For instance, many mushrooms carved in stone have been collected from Central and South America, indicating that these people were familiar with and probably used Basidiomycetes. Several of these mushroom statues are carved with a tripod base and may have been used as signs or markers of some sort.

CLASSIFICATION

The class Basidiomycetes is divided into two subclasses based on the type of basidium. Each subclass contains several orders based on similarities and differences in basidiocarps and life histories. Representatives of four Basidiomycete orders will be discussed.

REPRESENTATIVE BASIDIOMYCETES
ORDER POLYPORALES

Fungi of the order Polyporales include the coral fungi, shelf and bracket fungi, hounds tooth fungi, and others. Polyporales are characterized by the hymenial layer lining the outside of protrusions or the inside of pores. Basidiospores are exposed throughout their maturation period and are forcibly discharged when they mature.

Polyporus

Polyporous is one of the shelf fungi commonly found growing on living or dead trees. This fungus is usually flat and shelf-like, and its undersurface contains thousands of very small pores. The hymenial layer lines the inside of these pores with the basidia protruding into the pore so that spores may be readily discharged as they mature.

Polyporous and related genera including *Fomes* (Fig. 15-11*A*) are common in most forests of the world. Some cause damage to lumber crops. Some of these fungi produce tremendous numbers of spores. For instance, a large basidiocarp may produce over five trillion spores during a period of a few months. These spores are continuously produced and liberated throughout the growing season so that at least some of them will fall into favorable habitats and germinate to produce a new *Polyporous* plant.

ORDER AGARICALES

The order Agaricales contains the mushrooms and "toadstools." There is actually no difference between these two except that some people have come to call poisonous mushrooms "toadstools." Agaricales are among the most common of all Basidiomycetes and may be collected almost anywhere in the world, except in extremely cold environments. The color, shape, and form of these fungi place them among the most beautiful of all the plants on earth. All who enjoy the beauty of nature should look for mushrooms whenever they are walking in the forest, along a stream, or simply out of doors. In fact, I have photographed some of my favorite mushrooms in deserted cow barns or pastures, and even in the middle of city lawns.

Agaricus

The fungus *Agaricus campestris* is the common edible mushroom so widely used in cooking. It is commercially grown in many regions of the world. Mushroom cultivation on a commercial basis is somewhat difficult, and great care is exerted to insure that no poisonous fungi grow with the edible *Agaricus*. Many people in recent years have begun to grow their own *Agaricus*. Several stores and mail-order

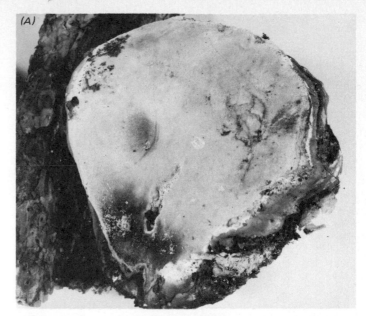

Figure 15-11. Several basidiomycete fruiting bodies reach rather large sizes such as this 6-in. "conk" of the genus *Fomes* (A), and this giant puffball (*Lycoperdon*) which is over 1 ft. in diameter (B). C shows a smaller puffball growing on a decaying log. The largest basidiocarp in C is about 1-in. in diameter.

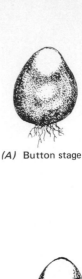

(A) Button stage

(B) Stipe elongates universal veil ruptured

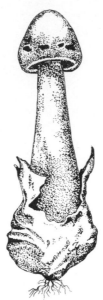

(C) Stipe fully elongated, pileus not expanded

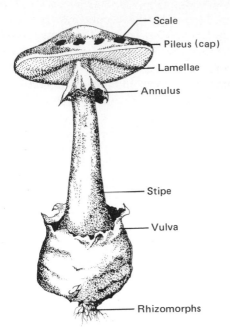

Scale

Pileus (cap)

Lamellae

Annulus

Stipe

Vulva

Rhizomorphs

(D) Mature basidiocarp

(E) Button stage

(F) Stipe elongates, universal veil absent

(G) Stipe fully elongated, pileus not expanded

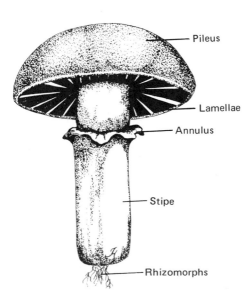

Pileus

Lamellae

Annulus

Stipe

Rhizomorphs

(H) Mature basidiocarp

Figure 15-12. The young basidiocarp of *Amanita* is enclosed by a universal veil *(A, B)* while that of *Agaricus (E, F)* is not. The universal veil forms the scales and vulva *(D)* characteristic of *Amanita*.

houses now sell preprepared flats containing *Agaricus* mycelia which may be tended to produce a nice group of mushrooms.

Many people are also finding pleasure and great eating in collecting wild *Agaricus* and related mushrooms. These people have developed the skill necessary to recognize edible fungi and follow proper rules of collection and identification. Several good books are presently available for the amateur mushroom hunter. A good rule of thumb is to only eat a very small portion of any new mushroom, even if it has been properly identified, to determine how it affects you. Some mushrooms that are perfectly edible and choice to one person may make another person ill or intoxicated.

Agaricus is similar in form to most other mushrooms. It has a cap or *pileus* with *gills* or *lamellae* on its undersurface (Fig. 15-12*H*). The hymenial layer covers these gills. When mature, spores are violently discharged from the basidia. The pileus is supported by a stalk or stipe. Many mushrooms, including *Agaricus,* have an *annulus* which is a ring of tissue surrounding the stipe below the pileus (Fig. 15-12*H*). The annulus results from the pileus enlarging and tearing away from the stipe leaving a ring of tissue where it was once attached. Several root-like structures attach the basidiocarp of some species to the mycelium. These structures are known as *rhizomorphs* or *mycelial connections.*

A good deal of variation in form exists among the basidiocarps of different mushrooms (Fig. 15-6). Some species lack an annulus while others produce one that is wide and conspicuous. Some have widely separated lamellae while others have closely packed lamellae. Some produce scales or scurf on the pileus while others are smooth or even sticky. These and other features are used in the specific classification of Agaricales.

Amanita

Amanita is one of the most infamous of all fungi. This genus is similar in most respects to *Agaricus,* and so is also placed in the order Agaricales. When *Amanita* is young, the entire basidiocarp is enclosed within a sheath of tissue known as the *universal veil* (Fig. 15-12*A–D*). This structure ruptures as the basidiocarp enlarges, leaving more or less prominent scales on top of the pileus and a definite cup or *vulva* at the base of the stipe (Fig. 15-12*D*).

This genus contains the most poisonous of all fungi. *Amanita verna* is commonly called the "destroying angel" and is so deadly that one basidiocarp contains sufficient toxin to kill several people. Horrible accounts of death have come from descriptions of *Amanita* poisoning. Vomiting, diarrhea, muscle cramps, seizures, horrible pain, and finally death occur over a 3–4-day period. Again, do not eat any "wild" mushroom unless you are competent to know it is safe.

Figure 15-13. White pine blister rust does millions of dollars worth of damage yearly. This photograph shows a pine tree infected with this rust *(Cronartium ribocola).* (Courtesy of the U. S. Department of Agriculture.)

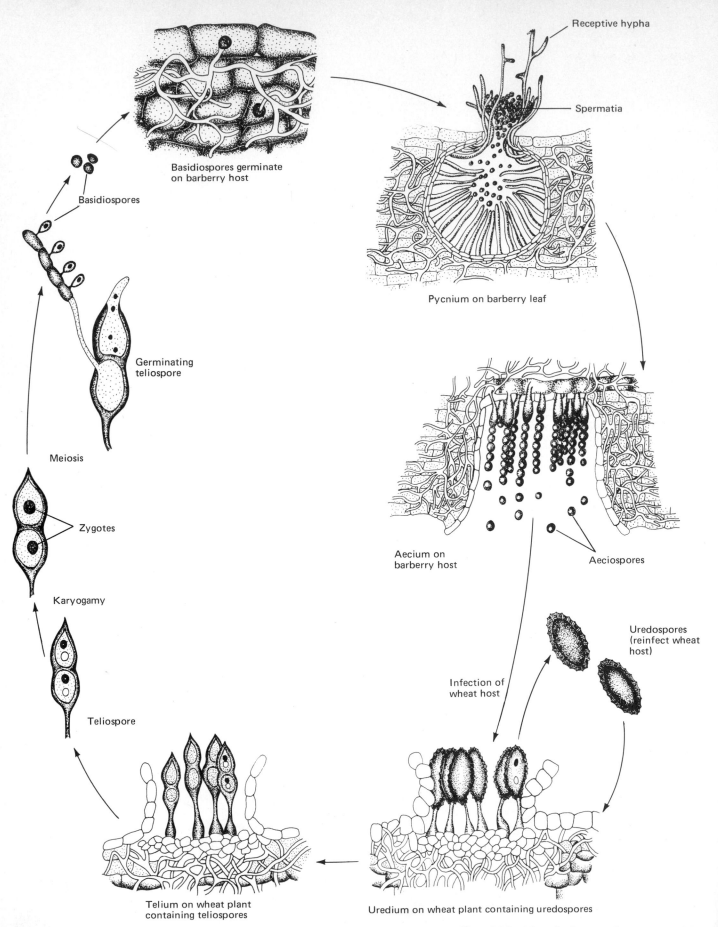

Receptive hypha

Spermatia

Pycnium on barberry leaf

Basidiospores germinate
on barberry host

Basidiospores

Germinating
teliospore

Meiosis

Zygotes

Karyogamy

Teliospore

Aecium on
barberry host

Aeciospores

Uredospores
(reinfect wheat
host)

Infection of
wheat host

Telium on wheat plant
containing teliospores

Uredium on wheat plant containing uredospores

Figure 15-14. Life cycle diagram of *Puccinia graminis*.

Rusts are among the most destructive of all fungi, causing great economic losses each year to such diverse plants and crops as pine (Fig. 15-13), coffee, cultivated ornamentals, and cereal grains. Perhaps the most serious aspect of these losses is that to individual farmers who may have their entire crops destroyed. Some rusts are such serious pathogens that they may significantly lower worldwide production of certain crops.

Puccinia graminis

Black stem wheat rust is caused by *Puccinia graminis* variety *tritici*. This disease has been known since wheat was first cultivated. Early Romans were plagued by wheat rust and held ceremonies to insure the favor of the rust gods Robigus and Robigo who were thought to be responsible for its control.

Puccinia graminis differs from other fungi we have studied since it completes its life cycle on two separate hosts (Fig. 15-14). That is, a portion of its life cycle occurs on a common barberry plant, and another part occurs on a wheat plant. Rusts which complete their life cycles on two different host species are *heteroecious* rusts. If they complete their life cycle on a single host, they are *autoecious* rusts. Both types are common.

Five different types of spores are produced by *Puccinia graminis*. In the spring, a basidiospore lands on the upper surface of a barberry plant and germinates to cause infection. Soon after infection, a small pustule or *pycnium* (Figs. 15-14 and 15-15B) develops on the upper surface of the leaf. This pycnium contains special haploid hyphae known as *receptive hyphae* as well as numerous small haploid *pycnospores* or spermatia. These pycnospores are imbedded in a droplet of sticky liquid which is extruded from an ostiole in the pycnium. This liquid attracts flies or other insects which carry thousands of pycnospores from pycnium to pycnium. Some pycnospores thus are carried to receptive hyphae of a different sexual strain. The nucleus of the pycnospore then passes into a cell of the receptive hyphae creating a dikaryotic cell. This cell divides rapidly producing dikaryotic hyphae which grow toward the lower surface of the barberry leaf where a second, larger, open pustule develops (Fig. 15-15B). This structure is the *aecium* which produces dikaryotic *aeciospores*. These spores are produced in rows within the aecium, and are released to be distributed on air currents.

When an aeciospore lands on the leaf or stem of a wheat plant, it germinates to penetrate the epidermis. The resultant hyphae are dikaryotic and obtain nourishment directly from cells of the wheat host. Soon after initial infection, some hyphae become organized beneath the epidermis to form a *uredium* (Fig. 15-14, 15-15C, and 15-16). This structure produces numerous dikaryotic,

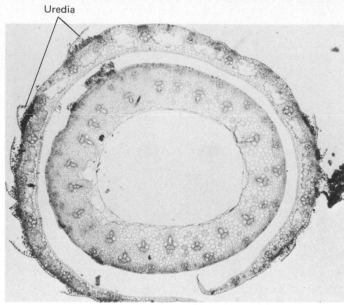

Uredia

(C)

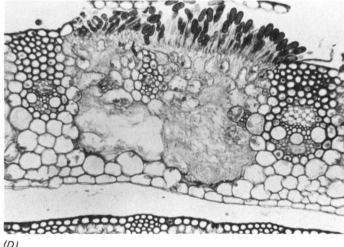

(D)

Pychium

(A)

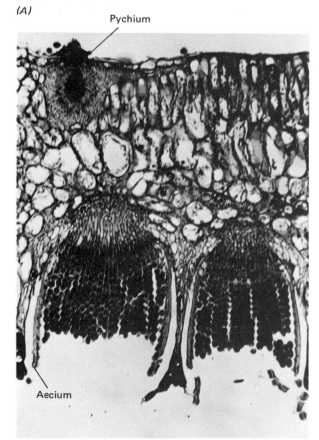

Aecium

(B)

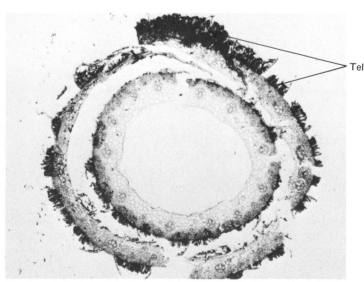

Telia

(E)

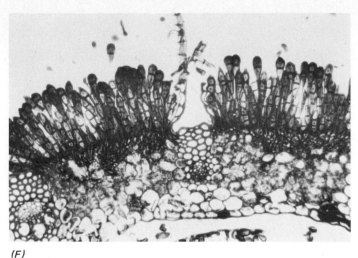

(F)

Figure 15-15. Black stem wheat rust *(Puccinia graminis)* is a very important plant pathogen. *A* shows several clusters of aecia on the lower surface of barberry leaves. *B* is a section through a barberry leaf showing a pycnium at the top of the leaf and two aecia projecting from the bottom. *C* shows a wheat stem producing uredospores (reinfection spores), and *D* is a closeup view of a uredium. *E* is a wheat stem with telia and *F* is a closeup of two telia.

asexual *uredospores* that are released and may be blown to reinfect the same or a new wheat plant. The uredium-uredospore stage of the fungus is usually responsible for the rapid spread of the disease since uredospores are viable for several days and may travel several hundred miles on air currents to infect new wheat plants. This stage lasts throughout the active growing season of the wheat plants.

As the wheat matures, the uredium ceases to produce uredospores. Instead, thick-walled, dark-colored *teliospores* are produced in the old uredium which is now known as a *telium* (Fig. 15-14 and 15-15E). Teliospores are two-celled dikaryotic spores. When they are produced in large numbers on the wheat plant, they cause dark colored streaks or patches and are thus responsible for the common name black stem wheat rust.

Teliospores are resistant to low temperatures and function in overwintering the rust. Either in the fall or the following spring karyogamy occurs in both cells of the teliospore to produce two zygote nuclei (Fig. 15-14). During the spring a peg-like projection develops from each cell and the zygote nuclei soon divide meiotically, each producing four haploid nuclei. These nuclei migrate into the projections on the teliospore and eventually become basidiospores, which are produced externally on sterigmata.

Basidiospores are released early in the spring and are carried on air currents to barberry leaves. These germinate to begin the life cycle of the rust again.

This life cycle is complex, and was extremely difficult for plant pathologists to unravel. In fact, even though *Puccinia* was investigated extensively in the late 1800's, it was not definitely shown to be heteroecious until the late

(A)

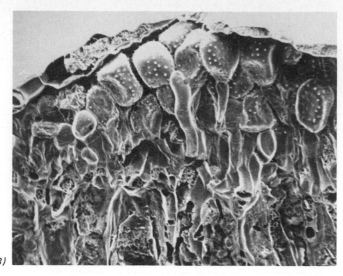
(B)

Figure 15-16. Uredospores of most rusts are produced beneath the epidermis of their host in tight groups. As they mature, they cause the host epidermis to rupture. Uredia of two rust fungi are shown in this figure. A is *Puccinia graminis* (wheat rust) and B is *Uromyces* (bean rust). (Scanning electron micrographs courtesy of Elizabeth Parsons, University of Bristol. Used by permission of the *Journal of Microscopy*.)

(A)

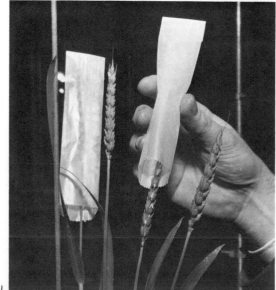
(B)

1920's. At that time a large movement was begun in America by the United States Department of Agriculture and later by the Civilian Conservation Corps to eradicate common barberry plants from wheat-growing regions where it was commonly planted as a hedge plant and for landscaping. Such measures helped in controlling the disease, although they did not eliminate it as was predicted since the uredospore stage of the fungus is so effective in infection. Thus, winter wheat grown in the southern United States serves as a source of uredospores that may infect young spring wheat planted hundreds of miles north in the wheat belt.

The most effective means of combating this and many other serious plant diseases is by developing resistant strains of host plants. That is, some strains of wheat are more resistant to black stem wheat rust than are others. Plant breeders constantly select and cross these plants with others in an effort to produce new more-resistant strains (Fig. 15-17). Such programs are currently underway in all agricultural countries of the world and are constantly being improved and expanded. It has been estimated by some plant pathologists that if resistant crops were not produced continually, as much as 25–50 percent of world food crops could be lost in one year. The results of such a loss would certainly be catastrophic.

Figure 15-17. The researcher at A is removing male parts of the wheat flower so that it cannot pollinate itself. He will then pollinate the flower with pollen from another desired plant and cover the flowers (B) so that no further pollination can occur. Seeds will subsequently be gathered and grown in experimental gardens to determine their degree of disease resistance. (Courtesy of U. S. Department of Agriculture.)

(A)   (B)

Figure 15-18. Smut diseases often cause very obvious symptoms. A shows a corn plant infected with the smut *Ustilago maydis*. The structure at the right is a corn ear that has been completely destroyed by the fungus. B is a smut-infected brome grass.

ORDER USTILAGINALES—THE SMUTS

Fungi of the order Ustilaginales are commonly known as smuts. These fungi are similar to rusts in their significance since they cause widespread damage to many crops including cereal grains. Smuts are responsible for great economic losses each year. Similar to rusts, smuts are obligate parasites of vascular plants, particularly grasses. However, many species occasionally overwinter saprophytically in plant parts discarded at harvest time and left in the field.

Ustilago maydis

Several smuts cause great damage to their hosts. For instance, *Ustilago maydis,* which causes corn smut (Fig. 15-18*A*), often causes stunting of infected corn plants and may destroy entirely the fruit of the plant. Almost anyone who has lived in corn-growing regions has seen smut-infected corn plants. Generally the disease causes large, grey-colored galls on stems or ears, and often these galls completely replace the ears. Although this disease is probably not as common as it once was, several recent outbreaks have caused serious damage to local corn crops throughout the world. The most effective control of corn

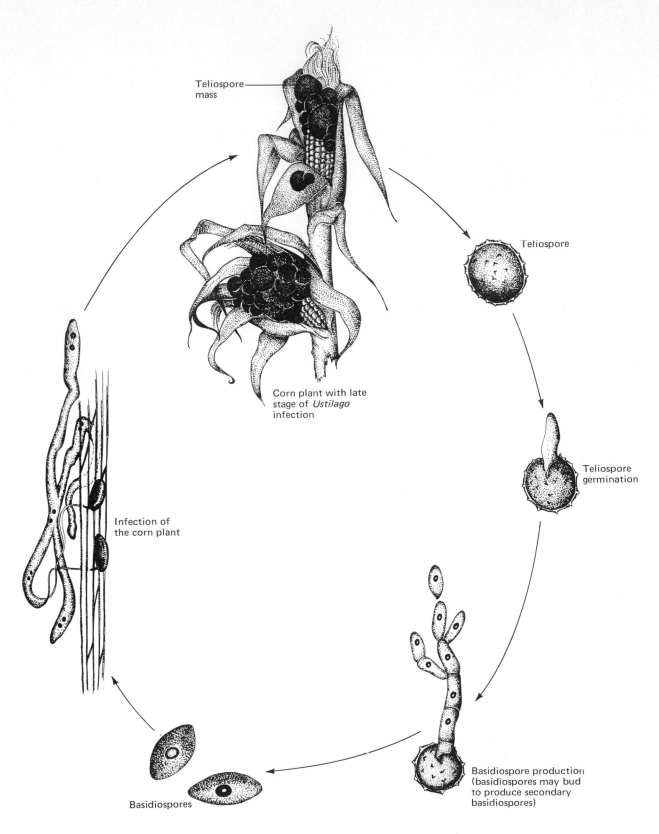

Teliospore mass

Teliospore

Corn plant with late stage of *Ustilago* infection

Teliospore germination

Infection of the corn plant

Basidiospore production (basidiospores may bud to produce secondary basidiospores)

Basidiospores

Figure 15-19. Disease cycle of *Ustilago* (corn smut).

smut is the development of resistant strains through plant-breeding programs.

The life cycle of *Ustilago maydis* and other smuts is simpler than that of most rusts. All smuts complete their life cycles on a single host; and pycnospores, receptive hyphae, uredospores, and aeciospores are never produced.

In the life history of *Ustilago maydis* (Fig. 15-19), spring winds carry a basidiospore to a corn flower, leaf, or stem. This spore germinates to produce haploid hyphae which infect the plant. Two cells of opposite mating strains soon come into contact and fuse to form a dikaryotic hypha within the host. Either or both the haploid or dikaryotic hyphae may produce conidia which act as reinfection spores.

The fungus lives throughout the summer in this hyphal state within the host. At the onset of cool weather, the dikaryotic hyphal cells encyst to form thousands of spherical, one-celled teliospores (Fig. 15-20). These are generally ash grey to black and are produced in tremendous numbers, often replacing each kernel of corn in an ear to form a large smutty or sooty gall (Figs. 15-18 and 15-19). Teliospores are the chief means of overwintering the fungus. Karyogamy occurs within them the following spring, and the resultant zygote nucleus divides meiotically to produce four basidiospores. These are released to infect new corn plants and continue the life cycle of the disease.

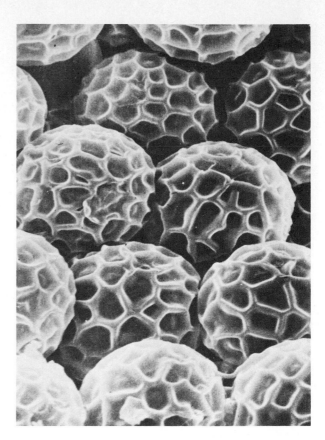

Figure 15-20. Scanning electron micrograph of the teliospores of a common wheat smut fungus. (Courtesy of James V. Allen, Brigham Young University.)

SUMMARY

Basidiomycetes are very important plants. They are responsible for degrading organic matter, particularly wood, and some are among the most serious plant pathogens known to humans. Many species are sought as food, and *Agaricus* is the common mushroom of commerce.

Many botanists feel that Basidiomycetes arose from Ascomycetes, and the two classes are alike in many features. However, they differ since Basidiomycetes produce external meiospores, a dolipore septum, and a relatively longer dikaryotic stage. In addition, Basidiomycetes lack well-defined gametangia.

FURTHER READING

CHRISTENSEN, C. M., *Common Fleshy Fungi,* Burgess Publishing Co., Minneapolis, 1965.

GRAHAM, VERNE, *Mushrooms of the Great Lakes Region,* Chicago Academy of Natural Sciences, Reprint 1970, Dover Pub., New York, 1944.

MILLER, ORSON, *Mushrooms of North America,* E. P. Dutton & Co., New York, 1972.

SMITH, A. H., *The Mushroom Hunters Field Guide,* 2nd ed., The University of Michigan Press, Ann Arbor, 1963.

SMITH, A. H., *Mushrooms in Their Natural Habitats,* Sawyers Inc., Portland, 1949.

YOUNG, PEGGY, "Mushrooms," *Natural History,* 79(6):66–71, 1970.

16 Lichens

Spanish moss (a foliose lichen) growing on bald cyprus trees in Louisiana. (Courtesy of Walter Hodge.)

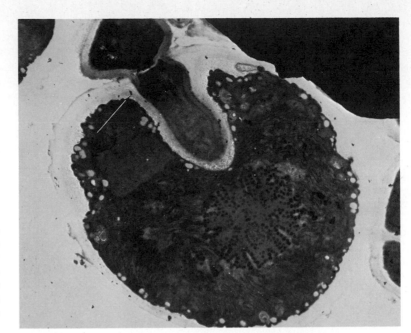

Figure 16-1. It was long thought that all lichens were mutualistic associations where both the alga and the fungus benefitted. This has shown to be incorrect for some lichens which seem to be more like a fungus with an internal algal garden. This electron micrograph shows an algal cell penetrated by a fungal cell which absorbs nutrients. (Courtesy of Vernon Ahmadjian, Clark University.)

Lichens are fascinating plants that grow in nearly any conceivable environment. These plants are actually an association of two plants, an alga and a fungus. However, this association is intimate so that the lichen appears to be a single plant. Furthermore, individual lichen thalli are distinctive and characteristic, and about 15,000 different species have been named. Not all botanists agree that the lichen should be given a separate name and argue that the fungus and alga should be named independently. In fact, some botanists have stated that the fungus is the most important member of the lichen and prefer to place lichens in the division Mycophyta.

GENERAL CHARACTERISTICS

Generally the fungus of a lichen is an ascomycete, although in four genera it is a basidiomycete. Lichens often develop fruiting bodies that are produced directly by the fungal component. For instance, it is common to collect lichens with well-developed, highly-colored apothecia.

The algal component of the lichen is a species of either Cyanophyta or Chlorophyta. Lichen algae are of several different genera, including both filamentous and non-filamentous species. Several common lichen algae include *Nostoc, Scytonema, Gloeocapsa, Palmella,* and *Trebouxia.* Both the alga and fungus of a particular lichen species are apparently specific to that lichen.

Several theories concerning the association between the fungus and alga in lichens have been proposed. Classically, this association has been described as a mutualism where both members are benefited. According to this explanation, the alga provides the fungus with simple foods from photosynthesis, and the fungus in turn provides the alga with water and mineral nutrients. This type of association

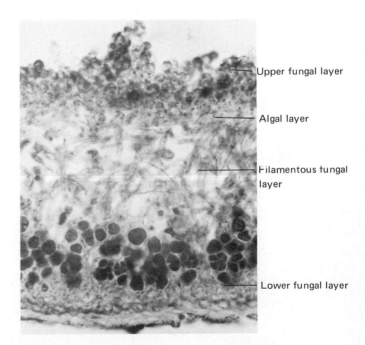

Upper fungal layer

Algal layer

Filamentous fungal layer

Lower fungal layer

Figure 16-2. The lichen thallus is often constructed of distinct algal and fungal layers. This cross section of a lichen thallus clearly shows these layers.

(A)

(B)

(C)

Figure 16-3. The diversity in form and color among lichens is great. Even so, three broad categories of lichen types are recognized. *A* is a crustose lichen. Note the trumpet-shaped apothecia arising from the thallus. *B* is a foliose lichen, and *C* is a fruticose lichen.

probably does occur in some lichen species. Another theory is that the fungus is parasitic on the alga, and thus literally grows an "algal garden" for its own benefit. Recent studies of some lichen species with the electron microscope tend to support this theory since the fungus has been observed penetrating algal cells in much the same manner as other definite plant parasites (Fig. 16-1). Probably several types of association between the alga and fungus occur among different lichen species.

The lichen thallus is often constructed so that a tightly woven layer of fungal hyphae comprises the upper surface (Fig. 16-2). A layer of algal cells occurs immediately beneath this surface. These algae are often near the surface and are usually intermingled with a loosely woven fungal layer. Tightly woven hyphae constitute the lowermost layer of the lichen. Some variation in this pattern exists among various lichens, particularly concerning the position of the algae. For instance, in some lichens algae are more or less scattered throughout the fungal hyphae and do not form a localized layer.

Three general types of lichen thalli are common, although they intergrade to some extent. Many lichens appear to be little more than colored crusts or spots on rocks or trees. These are termed *crustose* lichen (Fig. 16-3). They are very common and may be collected in a wide variety of habitats. The second type, or *foliose* lichen, is leaf-like and is attached to the substrate only by rhizoids (Fig. 16-3). *Fruticose* lichens are deeply divided or lacy in appearance (Fig. 16-3). They may be rigid or soft, and are common epiphytes on trees and shrubs in areas of high moisture such as tropical or temperate rain forests.

Asexual reproduction is common among lichens, often occurring by fragmentation. In addition, specialized structures for asexual reproduction are produced by many species. These are *soredia* (Fig. 16-4) that are small marginal outgrowths of the thallus containing both algal and fungal cells. These become separated from the thallus and are scattered to become new lichen plants.

Other species of lichens reproduce asexually by *death from behind*. This process involves the progressive branching growth of one portion of the thallus accompanied by the progressive death of older portions of the thallus. In other words, as the lichen grows forward, it dies from behind, separating branches of the lichen which become individual plants.

Lichens as individual plants do not reproduce sexually. However, the lichen fungus individually often produces ascospores by the typical ascomycete sexual cycle. These spores are released and carried about by the wind. Many of them apparently germinate and grow for a short period of time. If the resultant hypha contacts algal cells of the proper species, it may continue to grow to produce a new lichen. However, if the hypha does not contact the proper algal cells it usually dies.

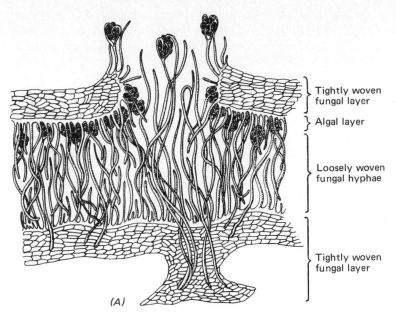

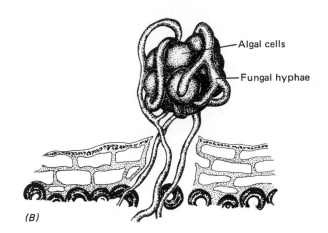

Figure 16-4. Many lichens reproduce by producing soredia, which are small bodies containing both algal and fungal cells. These drawings of a lichen thallus show the separate layers (A) and a close view of a soredium (B).

OCCURRENCE AND ECOLOGY

Lichens occur in a tremendous number of widely diverse habitats from the tropics to the Artic. They are perhaps the most widely distributed of all plants with a few possible exceptions. In tropical and temperate regions with high rainfall, lichens occur in large numbers as epiphytes. For instance, the "Spanish moss" characteristic of oak trees of the American West Coast and Gulf Coast (see beginning of chapter) is actually a fruticose lichen.

Lichens are also plentiful in cool and cold climates. All three types of lichens are present in these regions, although crustose lichens are probably most common. These lichens grow on almost any surface often under the most rigorous circumstances. In cold, northern arctic regions with nine months of winter, lichens are among the most common plants, often forming a spongy layer several inches thick on the ground. Lichens are also frequent in alpine habitats, often growing on bare stone surfaces where nothing else could possibly grow. In harsh environments lichens are very slow growing, are easily destroyed, and often take a very long time to become reestablished. Since these plants represent such an important part of alpine and Arctic ecosystems, caution must be used in developing natural resources in such regions to avoid significantly damaging those fragile environments.

Lichens also inhabit harsh dry habitats. Thus, they are common desert dwellers and often represent the dominant vegetation in certain local desert ecosystems. Desert lichens are mostly crustose species that are often brilliantly colored, and are particularly beautiful in the spring and just after a rainfall.

Figure 16-5. Lichens are often important in the process of primary succession since they may act as pioneer plants and aid in the formation of soils. This photograph was taken on the central Utah desert and shows a cover of lichens (arrow) on a part of a stone face. Mosses also inhabit this face as seen in the middle and right of the photograph.

IMPORTANCE

Some lichens, particularly those of high northern latitudes, are important in the food chains of higher animals, especially when other foods are scarce. Some studies indicate that certain lichens, such as "reindeer moss," are used extensively for food by Arctic caribou. However, the nutritional value of most of these lichens is apparently quite low.

A few lichen species are important as a source of litmus, which is a pH indicator. Litmus is red in acid solutions and blue in basic solutions, as any beginning chemistry student is well aware. Other lichens are used in manufacturing perfumes, dyes, cosmetics, and even certain antibiotics.

Perhaps the primary role of lichens is in *plant succession*. This process occurs when one species of plant occupies a given habitat and prepares it through certain changes to be occupied by a different species. Thus, lichens often grow on bare rocks where no other form of life could long exist (Fig. 16-5). As these hardy plants live there for many years, they eventually erode the rock surface and collect dust. In addition, the lichens contribute a certain amount of organic material and a very primitive soil may begin to develop. This process continues until enough soil accumulates for moss or other plants to inhabit the spot where previously only lichens could grow (Fig. 16-5). The moss may then grow there for some time and contribute to building a better soil. Succession may continue until eventually several plants may inhabit the area simultaneously. Such an association of plants in a particular habitat is known as a *plant community*. The process of successive habitation in previously uninhabited areas is *primary plant succession*. *Secondary* plant succession is theoretically a similar process, although colonization is actually a recolonization process since the area colonized was formerly inhabited by a plant community which was disturbed in some manner such as by burning or erosion. The study of plant communities and their development is part of the science of plant ecology.

SUMMARY

Lichens are symbiotic plants that are an association of a fungus and an alga. This association may be a mutualism or a parasitism of the alga by the fungus.

Lichens reproduce asexually by fragmentation, death from behind, and soredia. Sexual reproduction of the lichen as an entire plant does not occur although the fungal component may reproduce sexually by itself. Sexual spores of the fungus will not develop into a new lichen plant unless the germinating hypha contacts and entraps an algal cell.

Crustose, foliose, and fruticose lichens are so named because of outward characteristics on their thalli. In all three types, the fungus generally forms a tightly woven upper and lower layer, and the alga occurs between the two.

FURTHER READING

AHMADJIAN, VERNON, "The Fungi of Lichens," *Scientific American,* 208:122–131, 1963.

AHMADJIAN, VERNON, "Lichens," *Annual Review of Microbiology,* 19:1–20, 1963.

AHMADJIAN, VERNON, *The Lichen Symbiosis,* Blaisdell Publishing Co., Waltham, Mass., 1967.

HALE, MASON, *The Biology of Lichens,* American Elsevier Publishing Co., New York, 1970.

HALE, M. E., *Lichen Handbook,* Smithsonian Institute Press, Washington, D.C., 1968.

HALE, M. E., *How to Know the Lichens,* Wm. C. Brown Co., Dubuque, Iowa, 1969.

LAMB, I. M., "Lichens," *Scientific American,* 201:144–156, 1959.

SMITH, DAVID, "The Biology of Lichen Thalli," *Biological Reviews,* 37:537–570, 1962.

17 Life on the Land

Cacti such as these have developed means of survival in very arid conditions. (Courtesy of Walter Hodge.)

EVOLUTIONARY TRENDS IN THE PLANT KINGDOM

Several evolutionary trends that serve as unifying principles among the separate divisions are evident in the plant kingdom. These trends have been mentioned occasionally in this text but will be discussed here in greater detail. Generally speaking, algae and fungi comprise the "lower plants," while mosses, ferns, conifers, flowering plants, and others make up the "higher plants."

The organization of the thallus of lower plants is always rather simple since it usually lacks tissues and always lacks organs such as roots, stems or leaves. On the other hand, higher plants are more complex, containing specialized food and water conductive tissues and definite organs (Figure 17-1). The great complexity of many higher plants is often amazing, since several parts of these plants may be modified to perform specialized functions. Likewise, many are highly modified for living in specialized habitats. As a whole, higher plants demonstrate much more diversity in size, shape, and form of the plant body than lower plants.

With some noteworthy exceptions, lower plants have dominant gametophyte and reduced sporophyte generations. In fact, many lower plants produce a single-celled sporophyte that is represented only by the zygote. Conversely, the sporophyte generation of higher plants predominates and most often the gametophyte plant is highly reduced.

All lower plants exhibit unicellular gametangia, whereas higher plants produce gametangia with an outer, sterile protective layer (Fig. 17-2). This is a significant advancement since multicellular gametangia effectively protect the developing gametes, and subsequently the developing embryo.

A much greater proportion of the entire plant is used in reproduction of lower plants than of higher plants. For instance, it is common for every cell of an algal filament to enter into reproduction. Conversely, a very small portion of the total tissue of most higher plants enters the reproductive process. Thus, a mature oak or maple tree is often very large, and only a small amount of tissue located in the flowers functions in reproduction. This is the evolutionary result of the greater efficiency in reproduction among higher plants and the greater likelihood for offspring survival.

The zygote of a lower plant germinates directly either by meiosis to produce meiospores, or by mitosis to produce a sporophyte plant. The sporophyte plant rarely is associated with the gametophyte during this process. Conversely, the zygote of a higher plant germinates to produce a juvenile sporophyte plant or *embryo* which is always closely associated with and often dependent on the gametophyte plant. This is a significant evolutionary step which has been partly responsible for the great success of the higher plants. Some botanists propose that this is perhaps the most important difference between lower and

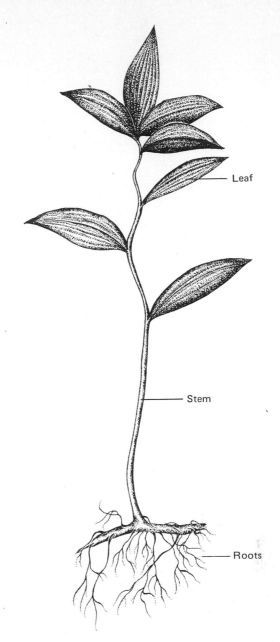

Figure 17-1. This figure shows an example of a "higher plant" (skunk cabbage) that illustrates a good deal of complexity including organs such as leaves and stems which contain food and water conductive systems.

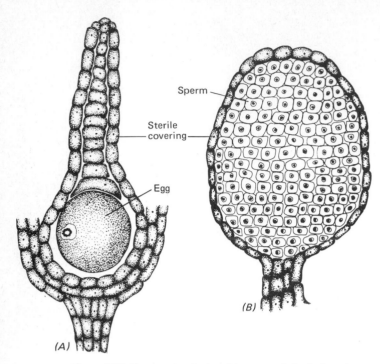

Figure 17-2. The gametangia found in many of the higher plants are larger and generally more complex than those in the lower plants. Both the archegonium (A) and antheridium (B) are composed of an outer layer of sterile tissue that surrounds the gamete producing tissue.

higher plants. It is why plant scientists of the late 1800's and early 1900's recognized only two divisions of plants, the *Thallophyta* and *Embryophyta*. Some botanists still use this classification system, although most now use these two words descriptively rather than as terms of classification.

Green algae (Chlorophyta) have exactly the same pigmentation as all of the higher plants. Starch is stored by green algae as well as all higher plants. This and other evidence suggests that chlorophyta perhaps were the ancestors of the higher plants. Even though the fossil record of algae is poor, evidence does indicate that algae were among the first living plants on earth. It is certainly possible that green algae gave rise to primitive land plants that eventually gave rise to our modern higher plants.

LIFE ON THE LAND

Since environmental conditions on land are generally more extreme than those in water, several basic modifications of plants were necessary as they began to inhabit the land. This is especially true for plants which were to inhabit extreme habitats such as the desert or tundra.

One of the most critical modifications was the development of a system to protect against water loss. This was accomplished by the formation of an *epidermis,* which is a specialized layer of cells on the outer surface of the plant forming a more or less watertight seal. (Fig. 17-3). Thus, a minimum of water is lost from the cells of the interior of the plant. In most land plants another protective layer, the *cuticle,* has also developed. It is a layer of waxy material secreted externally by the cells of the epidermis to further aid in sealing the plant against water loss (Fig. 17-4).

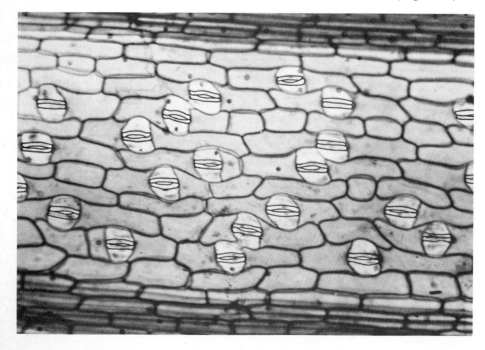

Figure 17-3. A face view of the epidermis of a leaf of a vascular plant (*Tradescantia*) is shown here. Note the tight arrangement of the cells and the stomates. Each stomate is outlined by two guard cells.

Since a plant living on the land is sealed from the atmosphere by the epidermis and cuticle, a method of exchanging gases with the atmosphere is necessary. That is, carbon dioxide must be obtained by the plant for photosynthesis and oxygen must be released. *Stomata* (stomates) have evolved for this purpose. These are "holes" in the epidermis bordered by specialized *guard cells* (Figs. 17-3 and 17-5). These cells are able to swell or contract under different environmental stimuli, and therefore may open or close the stomate. Generally stomata are open during the day when the active photosynthesis occurs and closed during the night.

Cells of plants restricted to aquatic habitats usually absorb water directly from their environment. Land plants lack this ability, since parts of the plant often are far removed from a source of water. For example, even though root cells of an oak tree grow in soil that may have a high water content, cells of the leaves of the same plant are far distant from that water. Thus, a mechanism for transporting water to isolated parts of the plant has developed. Likewise, some parts of many plants, such as roots, are nonphotosynthetic, and the cells of these tissues must obtain food from other parts of the plant. A system for transporting foods to such parts of the plant has also evolved. Any tissue functioning in transporting water or food materials is referred to as a *vascular tissue*. All the higher plants except Bryophyta have well-developed, highly-specialized vascular tissues (Fig. 17-6).

Land plants are usually subject to more physical or structural stresses than aquatic plants. Thus, mechanisms for adding strength and support have developed in these plants that allow them to grow upright, often to great heights. Vascular tissue aids in lending strength since cells of the water conducting system are generally thick walled and rigid. In addition other kinds of cells have evolved in many terrestrial plants which add strength and support to the plant.

Since some parts of higher plants such as roots are non-photosynthetic, the photosynthetic parts must produce enough sugars for the entire plant. In response to this need for efficiency, expanded, flattened leaves have developed (Fig. 17-7). These are very efficient in photosynthesis and are connected to the remainder of the plant by vascular tissue that transports the photosynthetic products.

Land plants also must have an efficient mechanism for attaching the plant and absorbing water and mineral nutrients from the soil. This need has been filled by the development of roots that are efficient in both roles (Fig. 17-1).

All of the evolutionary developments common in higher plants are not necessarily incorporated into each species. For instance, the division Bryophyta contains the mosses and liverworts that are somewhat transitional between

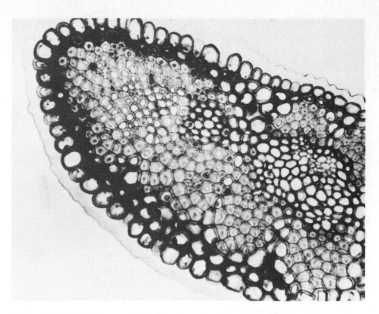

Figure 17-4. This photograph of the edge of a *Yucca* leaf illustrates a very thick cuticle on the outside. This cuticle is effective in protecting against excessive water loss.

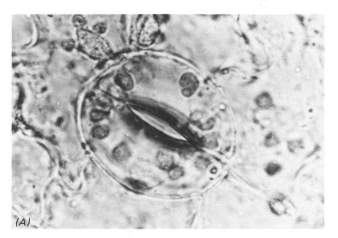

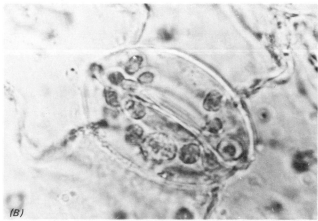

Figure 17-5. Stomata are effective in the exchange of gasses between the plant and the atmosphere. The stomate of geranium shown here is open at A and closed at B.

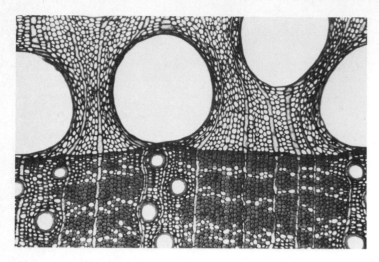

Figure 17-6. Many woody plants have very extensively developed vascular tissue. The cross section through the vascular tissue of an oak stem shown here demonstrates large round "holes" which are vessel elements and are efficient in water transport.

terrestrial and aquatic plants. Bryophytes are generally found in damp or wet habitats and have not developed well-defined vascular tissue. Therefore, these plants lack true roots, stems, or leaves, but demonstrate many other characteristics of higher land plants. Also, some higher plants have become readapted for living in aquatic environments. Such plants often have reduced root systems and vascular tissues. The study of the adaptations of plants to their environment is exciting and often furnishes valuable clues concerning the function of plant structures.

FURTHER READING

ANDREWS, H. N., "Evolutionary Trends in Early Vascular Plants," *Cold Springs Harpor Symposia Quantitative Biology,* 24:217–234, 1959.

BANKS, H. P., "Major Evolutionary Events and the Geological Record of Plants," *Biological Review,* 45:451–454, 1970.

BANKS, HARLAN, *Evolution and Plants of the Past,* Wadsworth Publishing Co., Belmont, Calif., 1970.

BOWER, F. O., *The Origin of a Land Flora,* Macmillan Co., London, 1908.

BOWER, F. O., *Primitive Land Plants,* Macmillan Co., London, 1935.

CHALONER, W. G., "The Rise of the First Land Plants," *Biological Review,* 45:353–377, 1970.

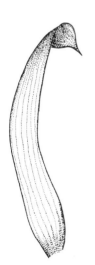

Figure 17-7. Leaves are very efficient at photosynthesis, producing enough sugars for the entire plant, including even the cells which are nongreen. Note the different pattern of leaf veins (vascular tissue) in each of the three leaves illustrated here.

18 Division Bryophyta

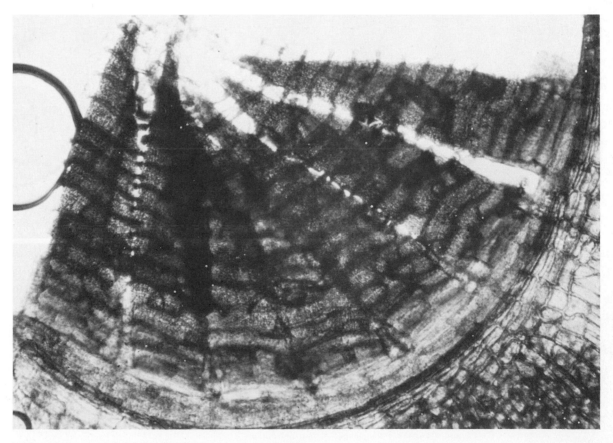

Portion of the spore dispersal apparatus (peristome) of the moss *Funaria.*

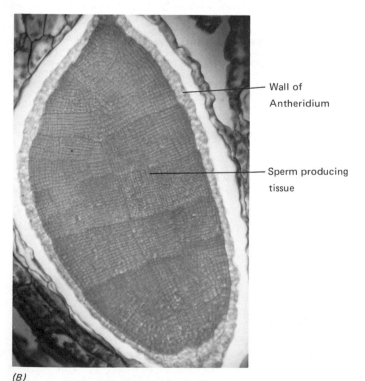

Neck

Venter

Egg

(A)

Wall of
Antheridium

Sperm producing
tissue

(B)

Figure 18-1. The gametangia of Bryophyta are large, multicellular structures. This figure shows the archegonium (A) and antheridium (B) of the common bryophyte *Marchantia*.

The Bryophyta is a rather large division containing about 20,000 species of plants which are placed into three distinct classes. The class Hepaticae contains the "liverworts," the class Musci contains the mosses, and the class Anthocerotae contains the "hornworts."

Generally speaking, bryophytes exhibit characteristics intermediate between lower and higher plants. They are primitive since the gametophyte generation is prominent as among lower plants, and probably the majority are more or less restricted to moist habitats throughout their life cycle. However, bryophytes are advanced since they produce embryos and the sporophyte generation is more complex than that of most lower plants.

Bryophyte gametangia are multicellular. The female gametangium, or *archegonium,* is divided into two regions. The *venter* is the swollen basal portion containing the egg. An elongate *neck* extends upward from the venter (Fig. 18-1). Sperm enter the neck when the egg matures and swim through a canal to the venter where fertilization occurs.

The male gametangium is the antheridium (Fig. 18-1). The bryophyte antheridium is more complex than that of lower plants, since a layer of sterile, protective cells comprises the outer wall. Motile sperm are produced by bryophyte antheridia.

CLASS HEPATICAE

GENERAL CHARACTERISTICS

The gametophyte plants of Hepaticae, or liverworts, are prostrate and have a definite upper and lower surface (Figs. 18-2 and 18-3). Such plants are *dorsiventral.* The upper surface is generally a deep green color since the cells of this region contain a high concentration of chloroplasts. The lower surface is often yellowish or brownish green since the cells of this region contain relatively few chloroplasts and function largely in storage.

Rhizoids are produced from the lower surface of the thallus (Figs. 18-3 and 18-4). These attach the plant to the soil and also absorb water and mineral nutrients to some extent. More than one kind of rhizoid may be found on some liverworts. (Fig. 18-4), and these are often used in species classification.

Two basic types of liverworts are recognized. The first resembles a foliose lichen and is known as a *thalloid liverwort.* These liverworts are unique in form and never bear structures resembling leaves or stems. They were originally named liverworts by medieval herbalists using the *doctrine of signatures.* This doctrine refers to the common Middle Age medical practice of using plants resembling body parts for treating the ailments of those same parts. Thus, since some thalloid Hepaticae resemble the liver, these plants were commonly used for the treatment of liver ailments and were named liverworts.

Some Hepaticae produce leaf and stem-like structures but are similar to thalloid liverworts in reproduction. These plants are commonly known as leafy liverworts. Even though they appear to have stem and leaves, they lack vascular tissues and are prostrate and dorsiventral with the lower surface producing numerous rhizoids.

The sporophyte plant of all Hepaticae is attached to the gametophyte plant, depending on it for water and nourishment throughout its life. Many botanists consider the sporophyte plant to be parasitic on the gametophyte. The sporophyte is nonphotosynthetic at maturity and is always smaller and usually less conspicuous than the gametophyte. It is composed of an absorbing organ or *foot* that is imbedded within gametophyte tissue, a *seta* or stalk that arises from the foot, and a meiospore-producing *capsule* (Fig. 18-5). The capsule occurs at the apex of the seta and is generally swollen and spherical. *Elaters* often are formed within the capsule along with meiospores. These structures have differentially-thickened cell walls that absorb atmospheric water at different rates. This causes the elaters to bend rapidly in response to changes in humidity, thus separating the mass of spores in the capsule. The wall of the capsule ruptures at maturity, and the spores are shed with the aid of the elaters.

OCCURRENCE AND ECOLOGY

Most liverworts grow in very moist, shady habitats, such as the floor of the forest, the edge of streams or the spray zone of waterfalls. Some species are epiphytic and many occur on fallen logs or other plant detritus. A few species survive under relatively dry conditions although this is rare, since these organisms do not have particularly efficient methods for absorbing and transporting water from the soil.

IMPORTANCE

Hepaticae have little direct importance to humans, although a few are found in the food chains of animals. Some species are used for botanical research, and a few are used as decorative plants. The class Hepaticae is one of the largest plant groups that is relatively economically unimportant to humans.

REPRESENTATIVE GENERA
THALLOID LIVERWORTS

Marchantia

Marchantia is widely distributed throughout North America and Europe. It is most likely to be found in moist habitats on silty or clay soils. Streamsides are very good places to collect this common liverwort.

Figure 18-2. Liverworts occupy several different habitats, including the aquatic. This figure shows a surface view of several plants of the liverwort *Ricciocarpus* floating on water. (Courtesy of William C. Steere, New York Botanical Garden.)

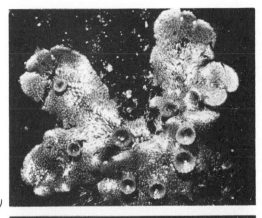

(A)

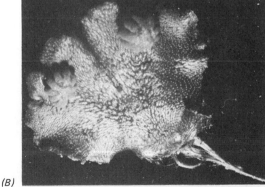

(B)

(C)

Figure 18-3. One of the most common liverworts is *Marchantia*. A is a dorsal view of a plant with several gemmae cupules. B is a dorsal view of a female plant with two developing receptacles. C is a ventral view illustrating the numerous rhizoids.

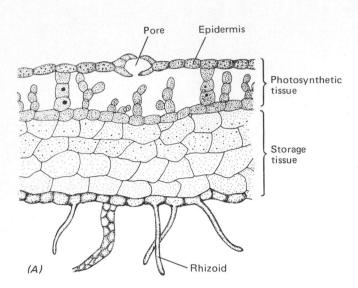

(A)

Pore — Epidermis

Photosynthetic tissue

Storage tissue

Rhizoid

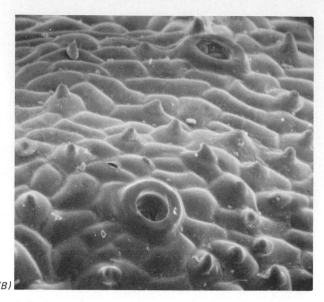

(B)

Figure 18-4. The thallus of some liverworts is rather complex. A is a longitudinal section through *Marchantia* and B is a scanning electron micrograph of the surface of *Marchantia* showing an air pore. (B is courtesy of James V. Allen, Brigham Young University.)

Foot

Calyptra

Seta

Elater

Capsule

Meiospores

(A)

(B)

Figure 18-5. The sporophyte of *Marchantia* is illustrated in this longitudinal section (A) and face view (B). Note the spores and elaters in A and the shedding spores in B.

The gametophyte plant of *Marchantia* is a branched, rather ribbonshaped, prostrate thallus (Fig. 18-3). The branches range up to about 1 inch wide and are deep green in color on their upper surface. On close examination, the upper surface of a branch often appears to be composed of many small polygonal units. An air pore occurs in the center of each of these units (Fig. 18-4). The pores open internally into a chamber bounded on all sides by small cells containing numerous chloroplasts (Fig. 18-4). These form a loosely packed layer one to several cells thick, which is responsible for most of the photosynthesis of the plant. The pores allow the exchange of gases necessary for photosynthesis, and may close if the plant is subjected to water deficiency.

The gametophyte plant is attached to the ground by conspicuous, whitish rhizoids that generally occur in two rows on the lower surface of each branch. These rhizoids are of two types (Fig. 18-4). One functions mostly in attachment, and the second in absorbing and transferring water.

Asexual reproduction occurs in *Marchantia* by three methods. Occasionally the thallus is fragmented. This may happen, for instance, when an animal eats a part of the thallus or when a flood breaks up a plant at the edge of a stream. Death from behind also occurs. The third asexual method is more specialized. Several small cupules often develop on the upper surface of a thallus. These are *gemmae cupules* that produce *gemmae* (Figure 18-3). Gemmae are multicellular, disk-shaped structures (Fig. 18-6) that are released from the cupule and grow to become a new thallus. During a rainstorm, raindrops splash into the cupule displacing the gemmae, often splashing them several inches away from the old thallus.

Marchantia is self-sterile in sexual reproduction since the gametophyte plants are unisexual. At the initiation of the sexual process, small upright branches develop at the tips of some prostrate branches. These specialized upright branches are termed *gametophores* since they bear gametangia in receptacles at their apices. The young receptacle of a female plant is disk-shaped and produces several archegonia on its upper surface. However, this receptacle becomes inverted as it matures, so that the archegonia hang from what is now its lower surface (Fig. 18-7). As the receptacle inverts, several *rays* develop on its surface so that the entire mature structure somewhat resembles a miniature palm tree (Figs. 18-7 and 18-8).

The male receptacle is also disk-shaped. Unlike the female receptacle, it does not become inverted or produce rays. Antheridia are imbedded within the receptacle near its upper surface (Fig. 18-7). Sperm are liberated from the antheridia during periods of very high humidity or rain. A sperm then must travel to a female plant to fertilize an egg in the venter of archegonium. This often occurs during a rain storm when sperm may be splashed by raindrops onto the female receptacle. A sperm then enters the neck of the archegonium and swims through the neck canal to the egg where fertilization occurs.

The resultant zygote divides several times to form an embryo which develops in place within the venter of the archegonium (Fig. 18-7). During the growth of the embryo, the venter enlarges by cell division to form a protective covering or *calyptra* surrounding the young sporophyte plant (Fig. 18-9). The sporophyte plant completes its development from the undersurface of a female receptacle. This plant is generally whitish or yellow in color and is nonphotosynthetic. Meiospores intermingled with elongate elaters are produced within the capsule of the sporophyte (Fig. 18-10).

Meiospores are released from the capsule with the aid of elaters and are dispersed by the wind. Spores landing in a favorable environment germinate to produce new gametophyte plants. These mature with time to produce gametangia, and the life cycle of the species is repeated.

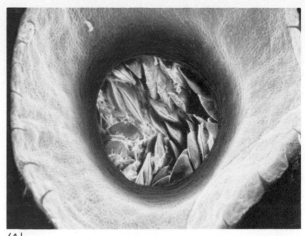

(A)

(B)

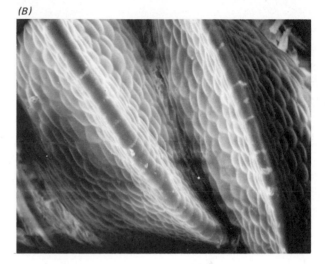

Figure 18-6. One of the common methods of asexual reproduction among many liverworts is the production of gemmae. *A* shows a gemmae cupule of *Marchantia* with several gemmae within. *B* is an enlargement of two gemmae. (Scanning electron micrographs courtesy of James V. Allen, Brigham Young University.)

"LEAFY" LIVERWORTS

Porella

Porella is a common genus containing "leafy" liverworts. These occur throughout the United States. Most species grow in moist habitats, although some are found in quite dry or even arid environments.

The gametophyte plant of *Porella* and other "leafy" liverworts is branched or unbranched, and it has con-

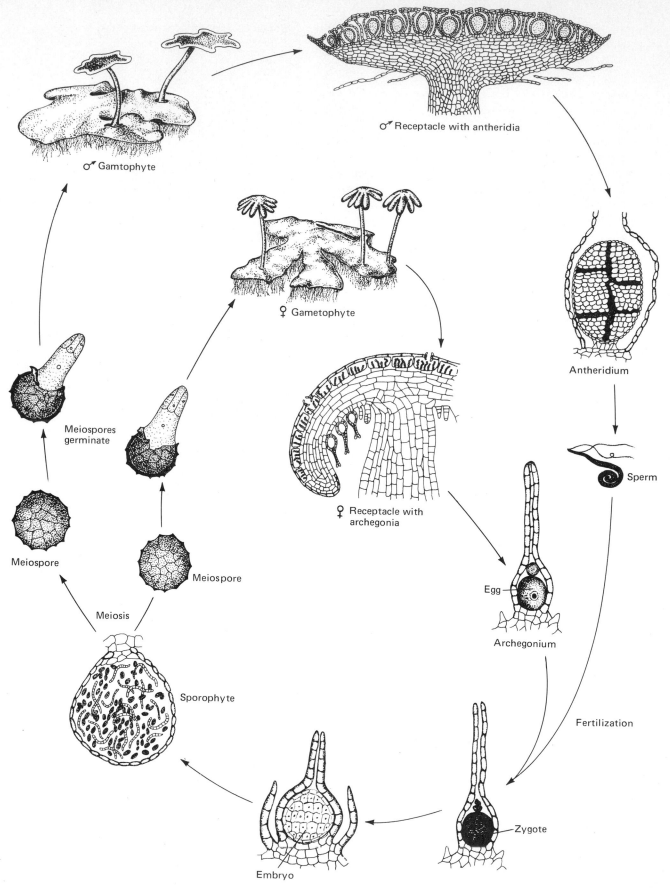

♂ Receptacle with antheridia

♂ Gamtophyte

Antheridium

♀ Gametophyte

Sperm

Meiospores
germinate

Meiospore

♀ Receptacle with
archegonia

Egg

Meiospore

Archegonium

Meiosis

Fertilization

Sporophyte

Embryo

Zygote

Figure 18-7. Life cycle diagram of *Marchantia*.

(A)

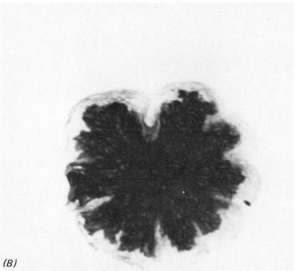

(B)

Figure 18-8. Face views of the female (A) and male (B) receptacles of *Marchantia*. Note the prominent rays on the female receptacles giving it a characteristic "palm tree" configuration. (Courtesy of W. M. Hess, Brigham Young University.)

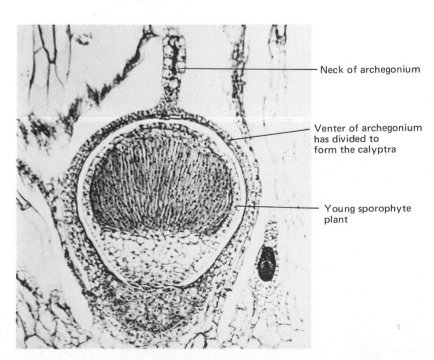

— Neck of archegonium

— Venter of archegonium has divided to form the calyptra

— Young sporophyte plant

Figure 18-9. The young sporophyte plant of *Marchantia* shown here is protected by a calyptra.

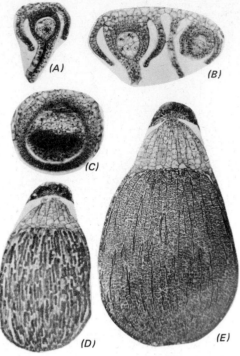

Figure 18-10. Several stages in the development of the sporophyte of *Marchantia* are illustrated in this figure. *A* shows an archegonium containing a zygote. The zygote divides several times to produce a small embryo (*B*). The embryo continues development to become an immature sporophyte shown in two stages at *C* and *D*. *D* contains meiospore mother cells. A fully mature sporophyte is shown at *E*. This mature plant contains meiospores and elaters.

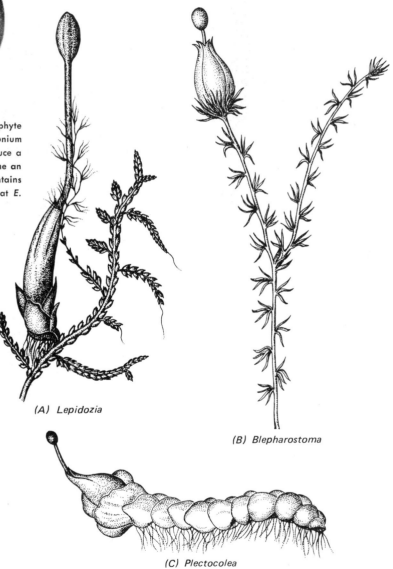

(A) *Lepidozia*

(B) *Blepharostoma*

(C) *Plectocolea*

Figure 18-11. Many liverworts are closely related to *Marchantia*, even though they appear leafy. These three "leafy" liverworts show both the gametophyte and attached sporophyte generations.

spicuous false leaves or *phyllidia* in two rows along a false stem or *caulidium* (Fig. 18-11). A third row of small phyllidia is often evident on the ventral side of the plants. Small, smooth-walled rhizoids also occur on the ventral surface. They function mostly in attaching the thallus since the ventral phyllidia are thought to absorb water. The dorsal phyllidia are arranged so that they channel water to the ventral surface where it may be retained often for long periods of time for use by the plant under dry conditions.

Antheridia and archegonia are produced at the tips of separate short branches that develop in the axils of some phyllidia. Two or three phyllidia often fuse around an archegonium to form a protective covering. Sperm are released from the antheridia when mature and make their way to the archegonium where fertilization occurs.

The zygote germinates to produce an embryo in place within the venter which expands to form a protective calyptra. The sporophyte plant is composed of a foot, seta, and capsule. The seta elongates rapidly as the spores begin to mature to push the capsule through the calyptra (Fig. 18-12). Mature spores are released from the capsule with the aid of elaters (Fig. 18-13). These spores are disseminated by wind and germinate under proper conditions to become new gametophyte plants.

Other "leafy" liverworts

About 4,000 different species of "leafy" liverworts are known and can be found almost anywhere in the world. They are among the most interesting and beautiful of all plants as anyone with a hand lens and a few minutes to spare in the forest or at the edge of a lake knows. Conard's book on *How to Know the Mosses and Liverworts* (see Further Reading section) identifies many of these plants and discusses where they grow.

Leafy liverworts prevalent in the United States include *Calypogeia, Frullania, Pellia* and *Lepidozia*.

CLASS ANTHOCEROTAE

Species of Anthocerotae are commonly known as *hornworts* or *horned liverworts*. This is a small class containing about 500 species.

GENERAL CHARACTERISTICS

Characteristics of Anthocerotae are largely based on *Anthoceros,* which is the most common hornwort genus. The following discussion concerns *Anthoceros* in particular, as well as the class as a whole.

The gametophyte plant of *Anthoceros* is simpler in structure than that of many liverworts. It is a flattened, branched, often disk-shaped thallus occasionally reaching

Figure 18-12. The sporophyte of the "leafy" liverwort *Porella* is shown here. Note the gametophyte at the base of the sporophyte.

(A)

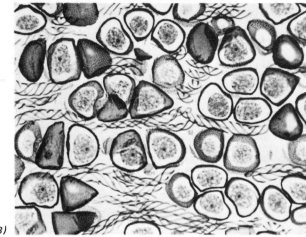

(B)

Figure 18-13. These sections through the capsules of two "leafy" liverworts show the characteristic spores and elaters in both. A is *Pelia* and B is *Porella*.

2 or 3 inches in diameter. Unlike *Marchantia, Anthoceros* lacks a specialized arrangement of tissues into a photosynthetic region. A storage region and air pores also are absent. All cells of the *Anthoceros* gametophyte are essentially alike, containing one chloroplast with a single large pyrenoid. This pyrenoid is unique among embryophyta and functions in producing starch just as in lower plants. A few simple pores or slits are common on the ventral surface of the thallus. Often these become filled with mucilage and are commonly inhabited by colonies of the blue-green alga *Nostoc. Nostoc* colonies apparently fix nitrogen, which then becomes available for use by the *Anthoceros* gametophyte.

Whitish rhizoids occur along the ventral surface of the gametophyte. They are unicellular and function in attachment and to some extent in absorption.

Asexual reproduction is by fragmentation, which often occurs when animals eat part of the thallus. In addition, the thallus dies from behind as it grows, thus separating branches which develop into separate plants. Anthocerotae do not produce gemmae. However, thickened regions on the margin of the gametophyte plant of several species are resistant to adverse conditions, and may grow into a new thallus. This is especially common in the spring since the remainder of the gametophyte plant is often killed by cold temperatures in winter.

Sexual reproduction occurs in all species (Fig. 18-14). The gametangia of Anthocerotae are imbedded within the gametophyte thallus. Antheridia develop individually or in small groups within specialized antheridial chambers. They produce large numbers of sperm which are released at maturity during conditions of high moisture.

Archegonia of *Anthoceros* are not well delineated from the gametophyte thallus as in other bryophytes because the venter is poorly developed (Fig. 18-14). At maturity the neck of the archegonium is open at the surface of the thallus.

Most species of Anthocerotae are self-fertile. A sperm swims to the neck of the archegonium and migrates down the neck canal to fuse with the egg. The zygote resulting from this fusion divides to become the embryo which is retained within the gametophyte thallus (Fig. 18-14). The embryo is completely dependent on the gametophyte plant for nourishment as it develops into a mature sporophyte plant.

The mature sporophyte is easily distinguished from that of other Bryophyta. The foot is imbedded within the gametophyte plant and occasionally develops rhizoids that may penetrate the gametophyte. A seta is lacking from the *Anthoceros* sporophyte. The capsule is elongate and horn-shaped, and is continually produced throughout the growth season by a *meristematic tissue* at its base. Because of the active division of the cells of this meristem, the sporophyte plant of *Anthoceros* is long lived and continues to produce

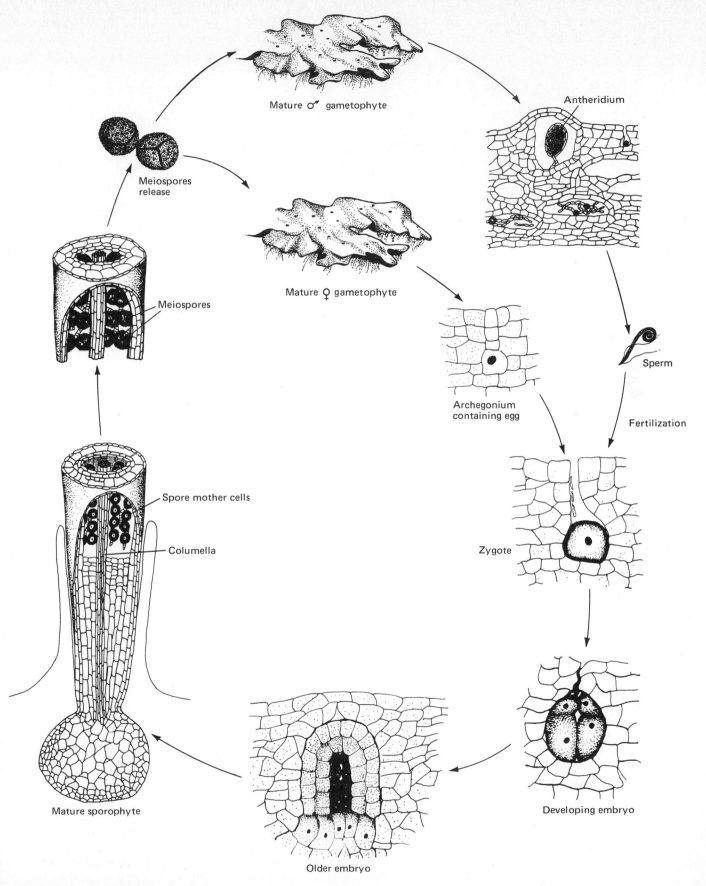

Mature ♂ gametophyte

Antheridium

Meiospores release

Meiospores

Mature ♀ gametophyte

Sperm

Archegonium containing egg

Fertilization

Spore mother cells

Columella

Zygote

Mature sporophyte

Older embryo

Developing embryo

Figure 18-14. Life cycle diagram of *Anthoceros.*

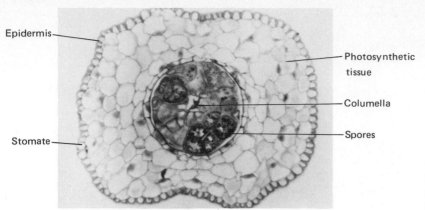

Epidermis

Photosynthetic tissue

Columella

Spores

Stomate

Figure 18-15. The sporophyte of *Anthoceros* contains a considerable amount of sterile tissue, as illustrated in this cross section through the capsule.

spores for several months throughout the spring and summer. This prolonged maintenance of the sporophyte generation is a significant advancement over lower plants.

The *Anthoceros* sporophyte plant is photosynthetic and thus contributes to its own nourishment. It has been shown by experimental evidence that a sporophyte removed from a gametophyte will continue to grow for an extended period though it is less vigorous than when attached to the gametophyte thallus.

A well-defined epidermis which is often coated externally by a cuticle is present on the *Anthoceros* sporophyte (Fig. 18-15). Numerous stomata are scattered throughout this epidermis. The capsule has a central sterile columella throughout its entire length. Spores and elaters are produced in the region surrounding this columella. A zone of photosynthetic cells occurs external to the sporogenous tissue (Fig. 18-15). The top of the capsule splits at maturity, and spores are released throughout the growing season as they mature. Each spore has the capacity to germinate and produce a new gametophyte plant.

OCCURRENCE AND ECOLOGY

Anthocerotae are worldwide in occurrence. However, most species are tropical, and hornworts are often conspicuously absent from very cold habitats, preferring moist shady environments such as those inhabited by Hepaticae.

IMPORTANCE

Anthocerotae are of very little direct economic importance. However, some botanists feel that this class is important in understanding possible evolutionary relationships between the lower plants and the higher plant divisions.

POSSIBLE EVOLUTIONARY DEVELOPMENT OF HIGHER PLANTS

The gametophyte generation of most lower plants (thallophyta) is predominant, and the sporophyte plant is often

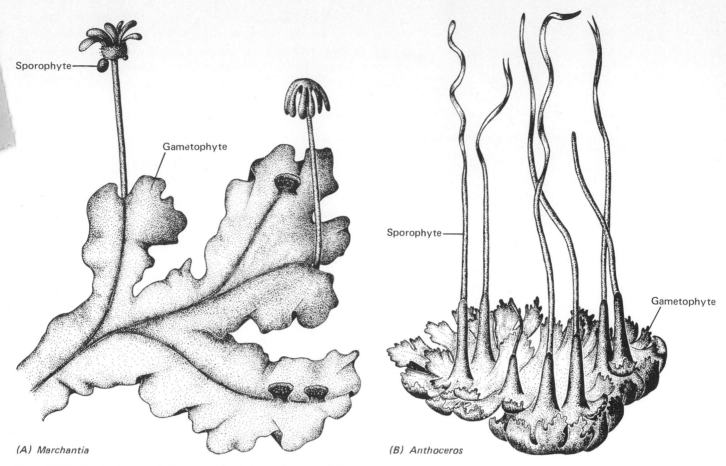

Sporophyte

Gametophyte

Sporophyte

Gametophyte

(A) Marchantia

(B) Anthoceros

Figure 18-16. *Marchantia* and *Anthoceros* sporophytes and gametophytes are compared in this figure. The sporophyte is larger and more complex in *Anthoceros*.

reduced. Conversely, higher plants (embryophyta) are normally opposite in this tendency. Bryophytes are intermediate between higher and lower plants in this regard, since both generations are similar in complexity and duration in the life cycle. *Anthoceros,* in particular, produces a gametophyte and sporophyte generation that are equally long lived in the life cycle. In addition, the *Anthoceros* sporophyte plant is quite complex   (Fig. 18-16) with the epidermis producing a cuticle and stomata, and with a meristematic region at the base of the capsule. Because of these significant advancements, some botanists feel that an *Anthoceros*-like plant may have been ancestral to the higher plants.

Further evidence for this hypothesis is seen in the fact that a comparatively large portion of the *Anthoceros* sporophyte is sterile and never enters into reproduction. Furthermore, this sporophyte is photosynthetic and often produces rhizoids that may penetrate the gametophyte to absorb nutrients directly from the substrate. Thus, the *Anthoceros* sporophyte demonstrates a good deal of independence.

This evidence does not indicate that *Anthoceros* itself was the ancestor of higher plants, but rather that a plant

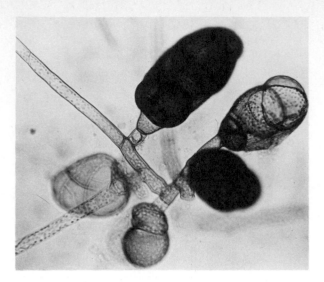

Figure 18-17. The upright gametophyte of a moss develops from buds along the prostrate gametophyte or protonema. Several buds attached to a protonema are illustrated here.

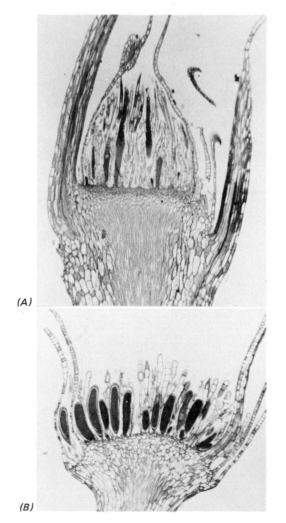

(A)

(B)

Figure 18-18. Most mosses, such as *Mnium* illustrated here, produce unisexual gametophytes. *A* shows a female plant with several long-stalked archegonia mixed with paraphyses. *B* shows a male plant with antheridia and paraphyses. The gametangia of both sexes are surrounded by a tight whorl of false leaves for protection.

with similar characteristics perhaps filled this role. Whether or not this theory is correct, it is important to remember that *Anthoceros* is intermediate in many respects between lower plants and higher plants. Some botanists feel that bryophytes as a group are similar in theoretical position in the plant kingdom to the amphibians of the animal kingdom.

CLASS MUSCI

Plants of the class Musci are the mosses. They are very common in the earth's flora, and one can rarely walk through a moist shady area without finding several different species. Mosses are especially beautiful, although they are small and so often overlooked by the nonobservant. Over 14,000 species of mosses have been identified.

GENERAL CHARACTERISTICS

The mature gametophyte plant of a moss is usually upright and symmetrical. Most species are not dorsiventral, but rather produce a central caulidium bearing spirally arranged phyllidia. The gametophyte is attached to the ground by large multicellular rhizoids that absorb water and nutrients as well as attach the plant to the soil.

A moss spore germinates to produce a filamentous immature gametophyte known as a *protonema* (Fig. 18-17).

This structure produces rhizoids and is green and self-supporting. After a period of time, the protonema develops buds that grow to become the upright gametophyte plants (Fig. 18-17). As the upright gametophyte develops, the protonema generally dies and decomposes. A few species of green algae may be easily mistaken for moss protonemata except that cells of a protonema have oblique rather than straight end walls and the algae never produce buds.

Most mosses do not have well-defined methods for asexual reproduction. In most species this only occurs by fragmentation of the protonema. A few mosses produce gemmae although usually not in gemmae cupules.

Sexual reproduction is very successful among mosses and represents the only prominent method of reproduction for most species. Nearly all mosses are self-sterile and most are unisexual, with one plant bearing antheridia and a second bearing archegonia (Figs. 18-18 and 18-19). Sperm are released from the antheridia at maturity often during a rainstorm, and swim or are splashed to the archegonia. One sperm swims down the neck canal of the archegonium to the venter where fertilization occurs.

The zygote develops in place within the venter to form an embryo (Fig. 18-19). This embryo continues development in place, and a sporophyte plant develops upward from the tip of the female gametophyte (Fig. 18-20). The

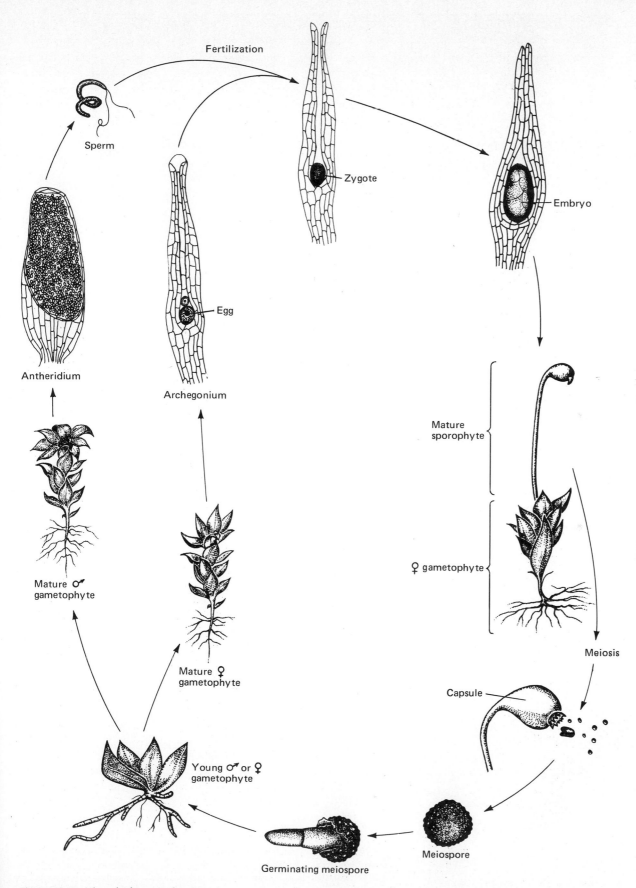

Fertilization

Sperm

Zygote

Embryo

Antheridium

Archegonium

Egg

Mature sporophyte

♀ gametophyte

Mature ♂ gametophyte

Mature ♀ gametophyte

Meiosis

Capsule

Young ♂ or ♀ gametophyte

Germinating meiospore

Meiospore

Figure 18-19. Life cycle diagram of a moss.

(A)

(B)

(C)

Figure 18-20. The gametophytes and sporophytes of several mosses are demonstrated in these photographs. A shows *Tortula ruralis,* B shows *Polytrichum,* and C shows *Funaria.* Notice that in all cases, the sporophyte arises from the gametophyte. (*A* is courtesy of Macmillan Science Company—Turtox/Cambosco.)

young sporophyte plant is protected by a calyptra which develops from the archegonial venter. The calyptra may remain attached to the gametophyte plant as the sporophyte matures or it may tear free and remain as a protective covering surrounding the sporophyte capsule.

The moss sporophyte is somewhat similar to that of a liverwort since it is composed of a foot, seta, and capsule. However, the moss capsule is more complex and highly specialized. It often contains more sterile tissue than that of other bryophytes because the central columella and photosynthetic regions are large (Fig. 18-21). The sporogenous tissue surrounds the columella and produces meiospores but no elaters. An epidermis with stomata comprises the outermost layer of the capsule. The moss sporophyte is dependent on the gametophyte plant during its development but is partly self-supporting as it approaches maturity.

The sporophyte capsule is very efficient in spore dissemination. Most species have an *operculum* or lid at the capsule apex which is shed at maturity (Figs. 18-22 and 18-23). Several tooth-like structures arranged in a ring and known as *peristome teeth* project into the capsule below the operculum (Fig. 18-23). The walls of peristome cells are unequally thickened and thus absorb atmospheric moisture differentially. This causes the peristome teeth to move, and, when moving, to open or close the apex of the capsule. Spores are shed when the peristome apparatus is open and are disseminated by air currents. Some spores fall into favorable environments and germinate to produce new protonemata. The characteristics of the three classes of Bryophyta are compared in Table 18-1.

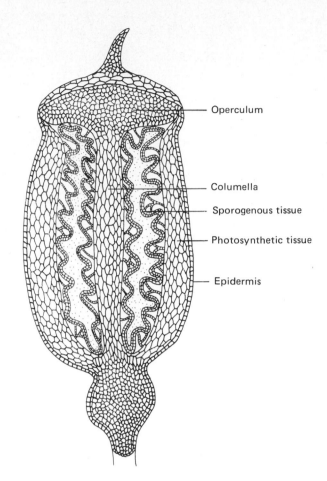

Figure 18-21. This longitudinal section through the capsule of a moss shows the large amount of sterile tissue and the complexity of the sporophyte which is characteristic of the mosses.

Labels: Operculum · Columella · Sporogenous tissue · Photosynthetic tissue · Epidermis

TABLE 18-1. A Comparison of the Classes of Bryophyta.

	Anthocerotae	*Hepaticae*	*Musci*
Gametophyte plant	Thallose; always dorsiventral.	Thallose or "leafy"; always dorsiventral.	"Leafy"; often not dorsiventral.
Sporophyte plant	Complex; meristematic region present; elaters present; columella present.	Complex; meristematic region absent; elaters present; columella absent.	Complex; meristematic region absent; elaters absent; columella present; peristome present.
Method of asexual reproduction	Thickened margins on thallus; gametophyte fragmentation; death from behind.	Gemmae in cupules; gametophyte fragmentation; death from behind.	Generally no specialized method; occasionally gemmae but not in cupules; protonema fragmentation.

OCCURRENCE AND ECOLOGY

Mosses occur in a wide variety of habitats ranging from flowing stream bottoms to desert soils. Many species are

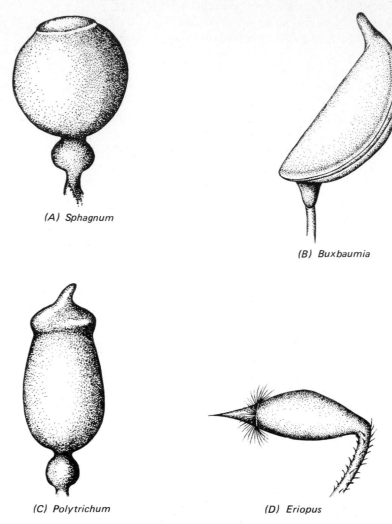

(A) Sphagnum

(B) Buxbaumia

(C) Polytrichum

(D) Eriopus

Figure 18-22. These drawings illustrate four different moss capsules. The capsule differs widely among the different mosses and is used in their identification. Most moss capsules are equipped with an operculum at their apex which is evident here in A, C, and D.

epiphytes and some exist on bare rocks or sand. Mosses are worldwide in distribution and are common from the tropics to the high latitude tundras. In certain habitats they are especially prevalent, often forming pure stands. In other habitats they occur with a wide variety of other plants, often in the understory under low light conditions.

IMPORTANCE

Musci are sometimes very important pioneers in plant succession. They are often the first plants to grow on the surface of rocks, or on freshly exposed soil. These plants then change the environment to some extent to allow other plants to inhabit these regions. After extended periods of time, they may cause a good soil to be formed and the nature of the flora to be changed.

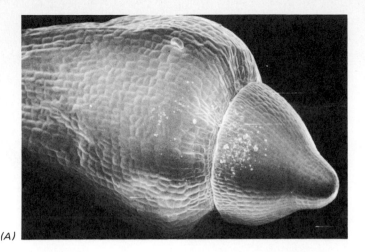
(A)

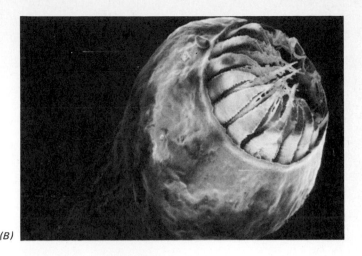
(B)

(C)

Figure 18-23. These scanning electron micrographs show well the sporophyte capsules of the moss *Mnium*. The operculum is still in place in *A* and has been lost at *B,* exposing the peristome apparatus. C is a closeup of the peristome. (Courtesy of James V. Allen, Brigham Young University.)

Peat moss is widely used in gardening and landscaping and in packing delicate objects for shipping. *Sphagnum* (Fig. 18-24), the most important peat moss, often accumulates in large amounts along the shores of acidic lakes where it is harvested and dried for use. If these peat deposits remain in place for many years, they often form a low-grade coal often known as *bog* or *peat coal.* This material is burned as a fuel in some parts of the world, and in some countries it is considered a very useful natural resource.

SUMMARY

The division Bryophyta contains plants with heteromorphic alternation of generations in which the sporophyte and gametophyte generation are of nearly equal importance and duration. However, in most cases the sporophyte

Figure 18-24. *Sphagnum* is one moss with commercial importance since it is harvested and sold as peat moss. *A* shows three *Sphagnum* gametophytes with several attached sporophytes. *B* shows an enlargement of a single gametophyte. *C* and *D* are closeup views of a single "leaf" showing dense (appearing black) photosynthetic cells and large dead cells. These large cells have unequal wall thickenings that appear as bands across the cells and are efficient for water storage.

plant is dependent on the gametophyte throughout all or much of its life.

Bryophyte gametangia are multicellular and may be produced on specialized gametophores or directly on or within the gametophyte thallus. Water is necessary for the fertilization of all species. Embryos develop in place on the female gametophyte plants which nourish the developing sporophyte plants.

Except for peat mosses, bryophytes are of little direct economic importance. Anthocerotae are of theoretical significance since some botanists postulate that plants similar to these were ancestral to the higher plants.

FURTHER READING

CONARD, H. S., *How to Know the Mosses and Liverworts,* Wm. C. Brown Co., Dubuque, Iowa, 1956.

DOYLE, W. T., *The Biology of Higher Cryptograms,* The Macmillan Co., New York, 1970.

GROUT, A. J., *Mosses With a Hand-lens,* Published by the author, Newfane, Vermont. Sold by the Chicago Natural History Museum.

LACEY, WILLIAM, "Fossil Bryophytes," *Biological Review,* 44:189–205, 1969.

WATSON, E. V., *The Structure and Life of Bryophytes,* Hutchinson and Co., London, 1971.

19

Introduction to the Vascular Plants: Division Psilophyta

The underside of a leaf of the Victoria water lily. The reticulations are veins of vascular tissue that are used to transport water and dissolved substances. (Courtesy of Walter Hodge.)

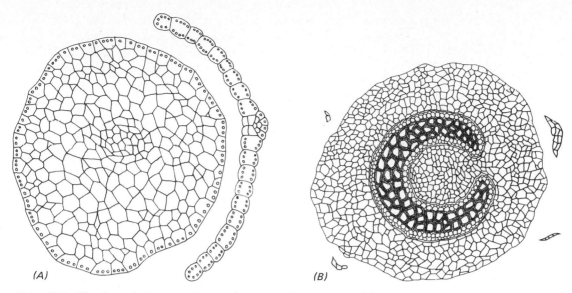

Figure 19-1. All advanced plants contain vascular tissue. This comparison between the caulidium of a moss (A) and the stem of a fern (B) shows that the fern stem is more complex, largely because of the ring of vascular tissue near the center of the stem. This vascular tissue is efficient in transporting water and other substances.

The unifying characteristics of all vascular plants is their possession of vascular tissue (Fig. 19-1). This tissue is used to transport water and mineral nutrients as well as photosynthetic food products in the plant (Fig. 19-2).

Specifically, two types of vascular tissue, each with characteristic kinds of cells, are present in vascular plants. *Xylem* (Fig. 19-2) is the tissue chiefly responsible for transporting water and dissolved mineral nutrients from the soil through roots to all parts of the plants. Most of

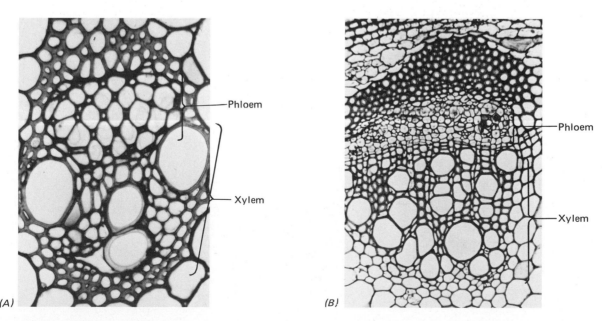

Figure 19-2. Most of the common plants we see around us, such as corn (A) and sunflower (B) contain copius vascular tissue in characteristic patterns.

the water is transported to the leaves where it is used in photosynthesis.

Xylem tissue is composed of several different types of cells. Two main types are responsible for conduction, *tracheids* and *vessel elements*. Tracheids are elongate cells with hollow interiors, pointed ends, and conspicuous pits or thin spots in their side walls (Fig. 19-3). These cells overlap and the pits of opposing tracheids are aligned so that water and minerals may pass through them. Vessel elements are barrel-shaped with a single large or several smaller ports in their end walls (Fig. 19-3). They are aligned end to end to form a continuous *vessel*. The flow of water and nutrients through a vessel is rapid and free, which is an advantage to the plant. Primitive vascular plants usually contain only tracheids in their xylem, while more advanced vascular plants generally contain vessel elements.

Both types of xylem cells are thick-walled, relatively rigid, and dead when they are mature and functioning. They add strength and rigidity to the plant as well as conduct water. These cells are very well adapted to perform both functions, and some vascular plants grow more than 200-ft high.

Phloem is the tissue responsible for transporting the photosynthetic products of the plant. This tissue normally transports simple sugars away from the leaves, into the stem, roots, and other nonphotosynthetic tissues. Two types of cells are responsible for most of this transport in the phloem. The *sieve cell* is the more primitive of these two and is somewhat analogous to a tracheid (Fig. 19-4). They are elongate and tend to be narrowed at both ends.

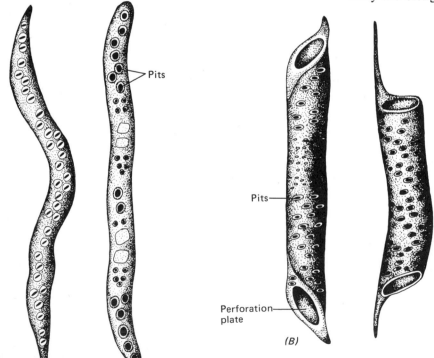

Pits

Pits

Perforation plate

(A)

(B)

Figure 19-3. Conductive cells of the xylem are tracheids (A) and vessel elements (B). The vessel element is the more advanced of the two and is often found aligned end to end in advanced plants to form vessels.

Groups of small pits or *sieve areas* occur on the side walls. Food materials being transported pass from cell to cell through these areas.

The second conductive cell, or *sieve tube element* (Fig. 19-4), is somewhat analogous to the vessel element. These cells are shorter and wider than sieve cells and are aligned end to end. The ends of a sieve tube element are perforated by large pores forming a *sieve plate*. Generally, sieve cells are found in primitive vascular plants, while sieve tube elements are characteristic of more advanced species.

Both types of phloem cells are living when mature, although the sieve tube element usually lacks a nucleus. It lies close to another, smaller phloem cell known as a *companion cell* (Fig. 19-4), which has a nucleus and is thought to control the sieve tube element.

In addition to conductive cells, xylem and phloem also contain other cells with different functions. *Parenchyma* cells are rather large and isodiametric and often function in storage. Parenchyma occurs in nearly all other tissues within a plant and is the most prevalent cell type in many plants. Elongate, thick-walled *fibers* are also common in vascular tissues, functioning to support and protect the plant. Various other cells play a minor role in the vascular tissues of some species.

In some primitive vascular plants, the xylem forms a central strand in the stem and is surrounded by phloem. In other species, xylem and phloem may be closely associated to form individual strands or *vascular bundles*. Often these bundles are characteristically arranged within a plant, and this arrangement is one basis for classification. The arrangement of vascular tissues among different vascular plants will be further discussed in later chapters.

Since vascular plants contain vascular tissue, they have true roots, stems, and leaves. These organs are fundamentally similar throughout the plant kingdom but have reached a great level of specific diversity among different plants.

The gametophyte generation of all vascular plants is much reduced and does not contain vascular tissue except in a few rare cases. This gametophyte plant is usually inconspicuous and is never as complex as the sporophyte. In advanced vascular plants the gametophyte is completely dependent on the sporophyte, reversing the trend of many of the lower plants and bryophytes. The female gametophyte plant of a Pinyon pine, for instance, is a small, spherical, white structure about ¼ in. in diameter. It develops entirely within the seeds produced by the sporophyte plant. Likewise, the gametophyte generation of a flowering plant is extremely reduced, comprised of merely a few cells produced within the flower.

DIVISION PSILOPHYTA

The division Psilophyta is one of the smallest of all living plant divisions. Less than ten species in the two

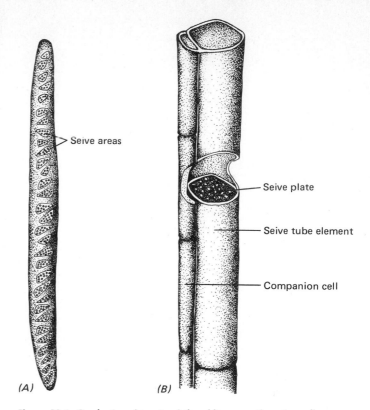

Figure 19-4. Conductive elements of the phloem are the seive cell (A) and seive tube element (B) shown in these drawings. The seive tube element is the more advanced of the two and is closely associated with a companion cell.

genera *Psilotum* and *Tmesipteris* are currently placed in this division. However, many more species of Psilophyta occurred during the geologic past. It should be mentioned that some botanists no longer recognize this division, prefering to place *Psilotum* and *Tmesipteris* with ferns and the fossil plants in several other divisions.

For our purposes, the division Psilophyta can conveniently be divided into two orders. The Psilotales contain all living representatives, and the Psilophytales (Rhyniales) contain all extinct species. This latter order is considered by some botanists to be distinct enough from the Psilotales to be a separate division, the Rhyniophyta. However, these two plant groups will be considered here as orders of the division Psilophyta.

ORDER PSILOTALES
GENERAL CHARACTERISTICS

The sporophyte plant of *Psilotum* and *Tmesipteris* is conspicuous, although small (Figs. 3-3 and 19-5). It is composed of a branched or vine-like stem with equal and opposite branches (*dichotomous branching*). This upright stem arises from an underground stem or *rhizome* that grows parallel to the ground a few inches below the surface. The rhizome produces many rhizoids that absorb water and mineral nutrients from the soil for transport through vascular tissue to the upright portions of the plant. True roots are absent among Psilophyta.

The psilophyte stem is simple in construction. Both the aerial stem and the rhizome exhibit an *apical cell* at the terminus. This cell functions as a *meristem* since it continually divides to produce new cells and is thus responsible for stem elongation.

A region of thin-walled parenchyma cells occurs just inside the epidermis of the stem (Figs. 19-6 and 19-7). These cells function in photosynthesis in the upright stem and storage in the rhizome. This layer of parenchyma tissue is the *cortex*. The innermost layer of the cortex is composed of a ring of closely appressed cells lacking intercellular spaces. These cells have thickened regions along their walls which form a seal that does not allow water or other substances to pass except directly through a cell by osmosis or active uptake. This ring of cells is the *endodermis* (Fig. 19-7) which regulates the uptake of water and mineral nutrients, particularly in roots and rhizomes.

The vascular tissue of the stem occupies the central position. It is arranged with the xylem at the center generally forming an X or irregular star shape. Phloem surrounds the xylem. The total vascular tissue of a stem or root comprises the *stele* (Figs. 19-6 and 19-7). When the stele is a solid core of vascular tissue as in Psilophyta, it is said to be a *protostele* which is the simplest and most primitive of all stelar types. The protostele is found in the stems of the most primitive vascular plants and the roots of nearly all vascular plants.

(A)

(B)

Figure 19-5. *Psilotum nudum* is a primitive vascular plant thought by some to be related to certain ancient forms. *A* is a mature plant showing the branching habit. *B* is a small group of branches illustrating the prominent synangia.

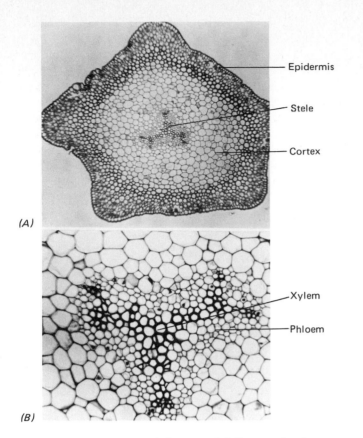

(A)

(B)

Figure 19-6. A cross section of the stem of *Psilotum nudum* is shown here. *A* shows the entire stem and *B* is an enlargement of the protostele showing the central tracheids.

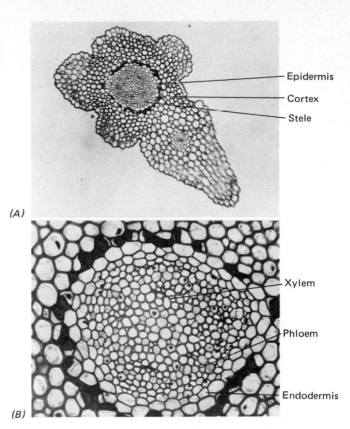

(A)

(B)

Figure 19-7. One other living genus of Psilophyta is *Tmesipteris,* which is illustrated in this figure. *A* shows an entire stem at the point of branch formation. *B* is an enlargement of the protostele.

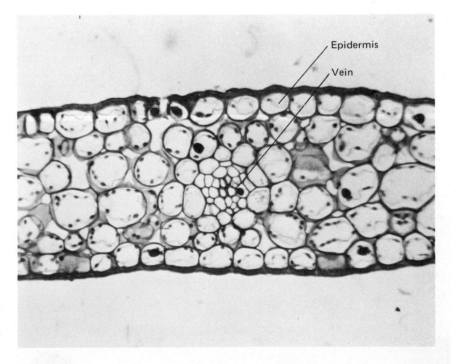

Figure 19-8. A cross section of the leaf of *Tmesipteris* illustrating the central vascular bundle (vein) is shown here. The vein is composed of a few central tracheids surrounded by a region of phloem.

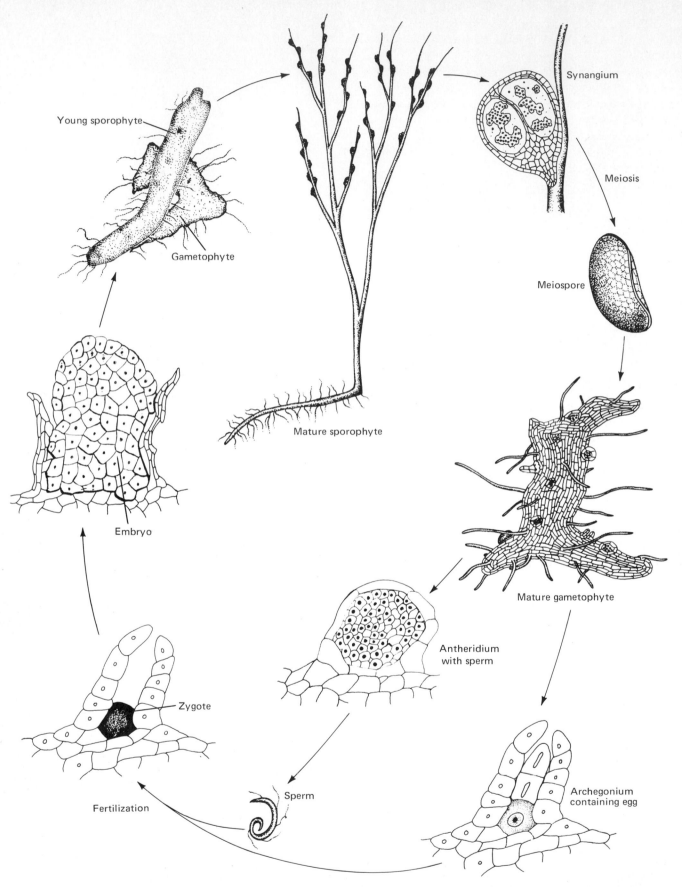

Young sporophyte

Gametophyte

Synangium

Meiosis

Meiospore

Mature sporophyte

Embryo

Mature gametophyte

Antheridium
with sperm

Zygote

Sperm

Fertilization

Archegonium
containing egg

Figure 19-9. Life cycle diagram of *Psilotum*.

The xylem of Psilophyta is composed only of tracheids and parenchyma. The phloem contains sieve cells rather than sieve tube elements.

Most characteristics of Psilophyta such as those discussed above are primitive. It is the opinion of most botanists that these are the most primitive of all vascular plants. However, a few botanists postulate that these plants are not as primitive as they seem, and that their lack of complexity is the result of an evolutionary simplification. Whichever theory proves to be accurate, there is no question that Psilophyta are the least complex of all vascular plants.

Psilotum and *Tmesipteris* produce leaves along the upright stem. They are usually only about 1 mm long in *Psilotum* and contain vascular tissue only in their bases. Leaves are considerably larger in *Tmesipteris* and contain vascular tissue throughout their length (Fig. 19-8). Some botanists prefer to call the small leaves of *Psilotum* scales since they contain such a limited vascular system.

Sporangia occur along the stems in the axils of some leaves (Figs. 19-5 and 19-9). They are formed in groups of three in *Psilotum* and of two in *Tmesipteris* and fuse as they mature. Such a group of fused sporangia is termed a *synangium* (Figs. 19-9 and 19-10). Synangia are easily distinguished along the stem, since they are large and often bright yellow at maturity. Spore mother cells are produced within the synangia. These cells are $2N$ in chromosome number and each divides to produce four meiospores which are released at maturity and are blown about by the wind. These spores germinate under favorable conditions to produce the gametophyte plant.

The gametophyte plant of *Psilophyta* is colorless or slightly yellow and is generally subterranean. It is usually less than 5-mm long and is cylindrical with swollen or branched ends. Many rhizoids are produced by this plant. Since the gametophyte plant is nonphotosynthetic, it derives its nourishment from a mycorrhizal association. One interesting feature of the gametophyte of *Psilotum* is that it may contain a few central tracheids surrounded by a small amount of phloem.

Both antheridia and archegonia are produced on the same gametophyte plant (Fig. 19-9). These gametangia are multicellular, but are less complex than those of Bryophyta. Sperm are released from the antheridium at maturity and swim to the archegonium. The archegonium usually loses its neck prior to the release of the sperm, allowing easy access to the egg in the venter.

The zygote develops in place on the gametophyte plant. During the initial period of growth, the young sporophyte plant is dependent on the gametophyte for nutrients which are absorbed through a foot. The sporophyte plant soon develops a mycorrhizal rhizome and soon afterward an upright stem. It becomes independent from the gametophyte on the production of the rhizome. The gametophyte

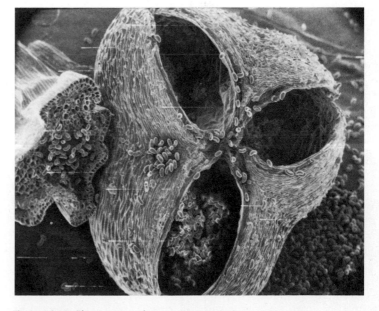

Figure 19-10. This scanning electron micrograph shows a close view of a synangium of *Psilotum* that has ruptured and is spilling spores. Note the stem at the left. (Courtesy of James V. Allen, Brigham Young University.)

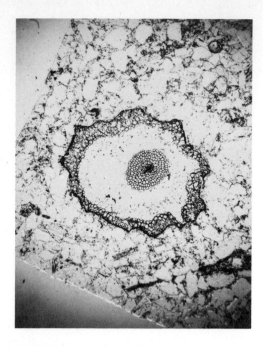

Figure 19-11. Many of the earliest vascular plants such as *Psilophyton,* illustrated here, were very simple plants rather similar to the modern *Psilotum* in stem structure. (Courtesy of Francis Hueber, Smithsonian Institution.)

plant generally dies and disintegrates during maturation of the sporophyte.

OCCURRENCE AND ECOLOGY

Tmesipteris is presently restricted in occurrence to Australia and a few Pacific Islands and is often epiphytic on trunks of tree ferns and palms. On the other hand, *Psilotum* is widely distributed in tropical and subtropical regions of the world. It occurs commonly in the southeastern United States as far north as South Carolina, and is especially common in parts of Florida. Psilophyta prefer moist habitats in sand or light soil.

IMPORTANCE

Psilophyta are of very little direct economic importance. However, they may be important in determining relationships and evolutionary progressions in the plant kingdom.

ORDER PSILOPHYTALES (RHYNIALES)

The Psilophytales are an extinct order of plants similar in many respects to modern Psilotales. This order was named for the fossil *Psilophyton* (Fig. 19-11) which was collected in the middle 1800's and was thought to be very similar to

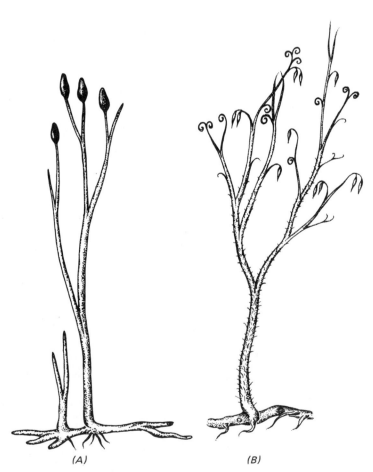

(A)   (B)

Figure 19-12. In the past, several genera of fossil plants were placed in the division Psilophyta. However, the present trend of many paleobotanists is to place these plants into several other divisions. The two plants illustrated here, *Rhynia major* (A) and *Psilophyton princeps* (B), are two fossil plants often placed in the Psilophyta.

Psilotum. This similarity indeed is evident, although some botanists recently have proposed that these two orders are not as similar as once thought. Since *Psilophyton* was discovered, several other closely related genera have been discovered and named (Fig. 19-12). These fossil plants made up the largest portion of the earth's terrestrial flora during the Devonian period. Most botanists consider them to be very primitive plants and some have suggested they evolved from bryophyte-like plants. It has been proposed that Psilophytales (Rhyniales) are ancestral to the higher vascular plants, a matter that is currently open to debate.

FURTHER READING

GENERAL REFERENCES ON
VASCULAR PLANTS

BIERHORST, D. W., *Morphology of Vascular Plants,* The Macmillan Company, New York, 1971.

BOLD, HAROLD, *Morphology of Plants,* Third edition, Harper & Row, New York, 1973.

EAMES, ARTHUR, *Morphology of Vascular Plants: Lower Plants,* McGraw-Hill Book Co., New York, 1936.

FOSTER, A. S. and E. GIFFORD, *Comparitive Morphology of Vascular Plants,* Second edition. W. H. Freeman and Co., San Francisco, 1974.

MacROBBIE, ENID, "Phloem Translocation," *Biological Reviews,* 46:421–481, 1971.

SPORNE, K. R., *The Morphology of Pteridophytes,* Hutchinson University Library, London, 1970.

STEWART, W. N., "More About the Origin of Vascular Plants," *Plant Science Bulletin,* 6(5):1–5, 1960.

ZIMMERMAN, MARTIN, "How Sap Moves in Trees," *Scientific American,* 208(3):132–142, 1963.

20 Division Lycophyta

Stumps of *Lepidodendron* in growth position. These stumps are located in Glasgow and are preserved in a public park. (Courtesy of S. A. J. Oldham, Director of Parks, Glasgow.)

Lycophyta are commonly known as club mosses, quill worts, or lycopods. These plants may be collected in many areas throughout the world. Over 1000 species of lycopods presently live on the earth, and many extinct fossil species are known as well.

GENERAL CHARACTERISTICS

The lycopod sporophyte plant is green and conspicuous and seldom over 18-in. tall. A few species are considerably larger and resemble small shrubs. In some areas certain lycopods are termed ground pine since they bear numerous sharply pointed leaves that somewhat resemble the leaves or needles of a pine.

The lycopod stem is terminated by a region of several meristematic cells that divide to produce new tissues and cause stem elongation. This group of cells is the *apical meristem*. Nearly all higher vascular plants exhibit a multi-cellular apical meristem, which is an advancement over a single-celled apical meristem.

The apical meristem divides to produce cells on the outer surface of the stem. These become the epidermis (Fig. 20-1). The epidermis is well formed among lycopods and contains numerous stomata. A cortex occurs immediately inside the epidermis. This tissue is composed largely of parenchyma cells with numerous chloroplasts. The innermost layer of the cortex is the endodermis (Fig. 20-1). The central portion of the stem is occupied by a protostele. This stele is slightly more complex than that of *Psilotum* since the xylem often forms several deeply divided lobes with bundles of phloem occurring between them. Most lycopods produce only tracheids in the xylem, although some species of the common genus *Selaginella* have vessel elements in their xylem (Fig. 20-2). A region of thin-walled parenchyma cells generally occurs between the vascular tissue and the endodermis. These cells comprise the *pericycle* which may function as a meristematic tissue. Generally, when a pericycle is present in a stem, it has little function. However, when it occurs in roots, this tissue is almost always responsible for the formation of branches.

Leaves develop from small protrusions formed at the apical meristem. These structures are relatively small and contain one vascular bundle or leaf *vein*. The vascular tissue of this vein is connected directly to the vascular tissue of the stem so that a continuous system for translocation throughout the plant is established. No break in vascular tissue, or *leaf gap*, is produced above the point of attachment of the leaf vein to the stele of the stem, although this occurs in most higher vascular plants. A leaf of this type is small, contains a single vein, produces no leaf gap, and is known as a *microphyll* (Fig. 20-1). The position on a stem where a leaf is produced is known as a *node*. That portion of the stem between two nodes is an *internode*.

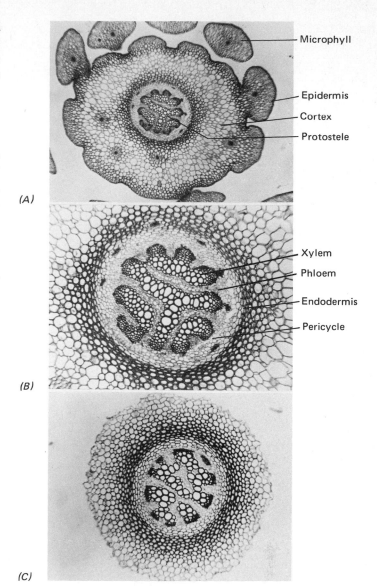

(A)

(B)

(C)

Figure 20-1. The vascular tissue in *Lycopodium* stems is arranged in a characteristic pattern as illustrated here. A is an aerial stem with evident microphylls surrounding the stem. B is a closeup of A. C is a rhizome and lacks the microphylls.

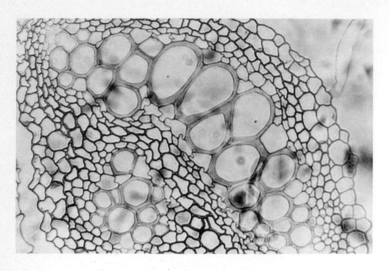

Figure 20-2. This closeup of a stele of *Selaginella* shows the vascular tissue. The large thick-walled cells at the center of the photograph are vessel elements.

Roots are produced at the base of the stem. These structures lack chlorophyll, as do all roots, and may range from unbranched to highly branched. They terminate in an apical meristem covered by a cap of cells known as the *root cap*. Root caps are produced by most roots and function in protecting the apical meristem of the root. Cells of the root cap are slimy and are continually sloughed off as the root grows through the soil. As this occurs, new cells are produced by the root meristem to replace them. The vascular tissue of a lycopod root is very similar to that of the stem. Together with the pericycle, it forms a protostele that is surrounded by an endodermis.

Sporangia occur in the axils of certain leaves along the stem. Leaves which subtend sporangia are known as *sporophylls,* which literally means "sporangium bearing leaf." These sporophylls may be scattered along the stem but more commonly they are aggregated at the tip of a stem or branch to form a *strobilus* or cone.

Sporangia of some lycopods produce meiospores that are all alike in shape and size (Fig. 20-3). Such species

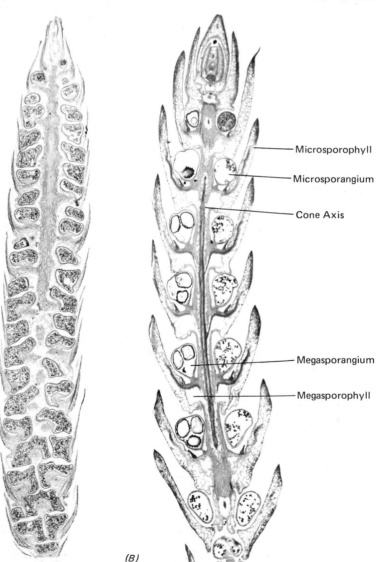

(A)   (B)

Microsporophyll

Microsporangium

Cone Axis

Megasporangium

Megasporophyll

Figure 20-3. Some lycopods are homosporous and others are heterosporous. *A* is the homosporus *Lycopodium* and *B* is the heterosporous *Selaginella.*

are said to be *homosporous*. Other species produce spores of two distinct sizes (Fig. 20-3). The larger spores are termed *megaspores* and are produced in *megasporangia*. *Microspores* are much smaller and are produced in *microsporangia*. A megaspore germinates to produce a female gametophyte plant, while a microspore germinates to produce a male gametophyte plant. Species which produce these two types of spores are *heterosporous*.

The gametophyte plant of all lycopods is small and inconspicuous in comparison to the sporophyte plant. It is entirely free living and is never dependent on the sporophyte for nutrition. Some species produce gametophytes bearing both antheridia and archegonia while others produce unisexual gametophytes.

Two basic types of gametophyte plant development occur among plants. The first type, *exosporic* development, is characterized by the germination of a spore to produce a gametophyte plant completely outside the confines of the spore wall. All gametophyte plants previously studied in this book are of the exosporic type. Heterosporous lycopods and most higher plants exhibit an *endosporic* gametophyte development (Fig. 20-11). This occurs when a spore germinates to produce a gametophyte plant entirely within the confines of that spore wall. The development of the heterosporous habit and endosporic development was essential for the evolution of higher plants and will be discussed again in conjunction with seed development.

CLASSIFICATION

Three orders of lycopods are represented in the present day flora of the earth. All homosporous lycopods are placed in the order Lycopodiales, while the orders Selaginellales and Isoetales contain heterosporous species. The order Selaginellales contains species that are terrestrial in habit and similar in appearance to Lycopodiales. The Isoetales is an aquatic order containing lycopods with very short, swollen stems and long, prominent leaves.

In addition to these modern lycopods, at least three extinct orders are recognized. In the geologic past, these orders made up a very large, prominent portion of the earth's flora.

OCCURRENCE AND ECOLOGY

Lycopods are found in all areas of the earth from the alpine and Arctic tundras to tropical jungles. Perhaps most are tropical in distribution, although many are common in temperate regions. Most species are terrestrial, some are aquatic, and many are epiphytic. Some species are well adapted to desert environments, and in certain of these harsh habitats, they may be among the only surviving plants.

Figure 20-4. Many fossil plants have contributed to the formation of coals. This strip mining operation furnishes coal for a power generating station. (Courtesy of Stan Welsh, Brigham Young University.)

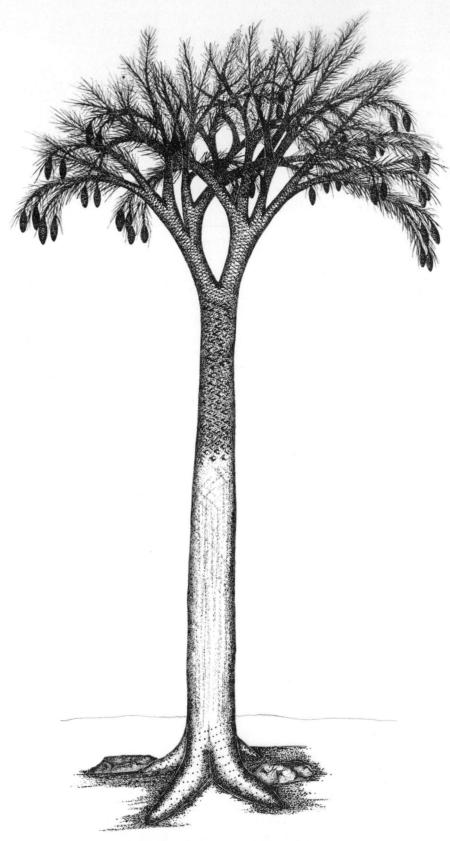

Figure 20-5. Certain ancient lycopods such as *Lepidodendron*, shown here, were large, tree-like plants prominent during the Carboniferous Period. They were important in coal formation.

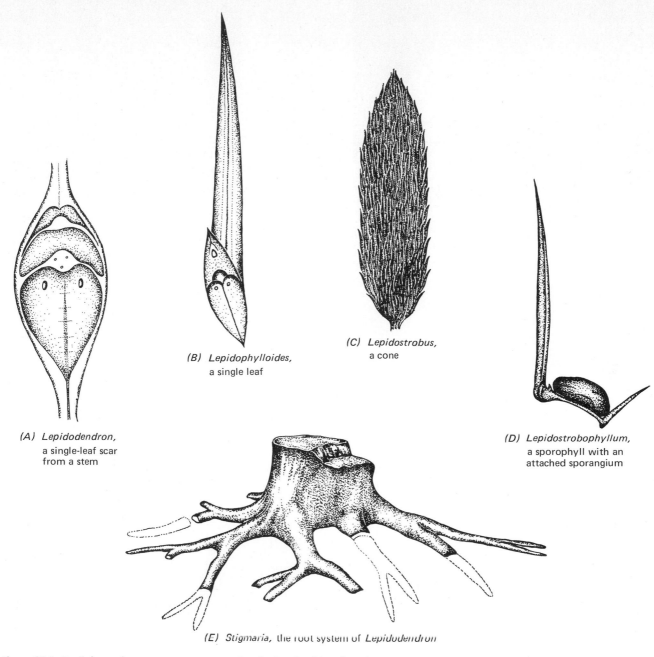

(A) *Lepidodendron,*
a single-leaf scar
from a stem

(B) *Lepidophylloides,*
a single leaf

(C) *Lepidostrobus,*
a cone

(D) *Lepidostrobophyllum,*
a sporophyll with an
attached sporangium

(E) *Stigmaria,* the root system of *Lepidodendron*

Figure 20-6. Fossil lycopods are common at several collecting localities throughout the world. Some of the most common of these fossils are illustrated here.

IMPORTANCE

Modern lycopods are of relatively little direct economic importance. Some are grown in nurseries and sold as ornamental house plants, while others are used as Christmas decorations. Many of the small wreathes common at Christmas time are made of *Lycopodium* or *Selaginella* rather than of a conifer.

Spores of some lycopods were used in the past as flash powder in photography and for lubrication. The electronic

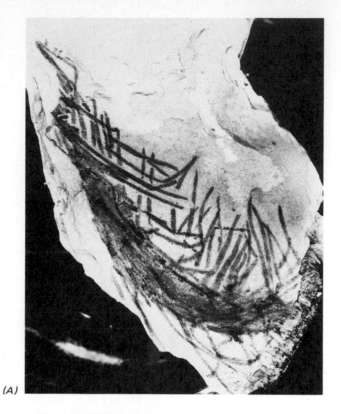

(A)

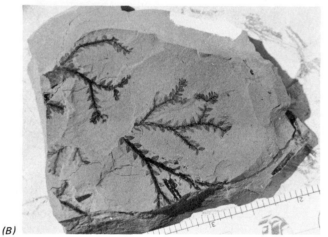

(B)

(C)

Figure 20-7. Three fossil lycopod stems, *Baragwanathia longifolia* (A), *Selaginella anasazia* (B) and *Lepidodendron* (C) are illustrated in these photographs. (The photograph of *Baragwanathia,* one of the oldest vascular plants known, is courtesy of Francis Hueber, Smithsonian Institution. B is courtesy of Sidney Ash, Weber State College.)

flash gun has largely eliminated their use in photography, although they are still used in specialized lubrication applications.

Many fossil lycopods were among the earth's most important coal-forming plants. While admittedly their contribution occurred in the past, it should not be overlooked. Coal is one of the most useful natural resources on earth (Fig. 20-4). It is a versatile product that can be used in the synthesis of other organic products or for direct combustion to provide energy. Much of the power generated in the United States and elsewhere for the many years to come will probably derive from coal-fired generating stations similar to the huge plants currently under construction in the southwestern United States. Coal represents radiant energy from the sun which was trapped by photosynthesis millions of years ago and stored as fossil plant remains.

FOSSIL LYCOPODS

Representative Lycophyta have been on the earth since Early Devonian times. During the Carboniferous Period these plants dominated many landscapes. Many of these Carboniferous plants were large and tree-like. For some reason all of these large forms became extinct and only the smaller lycopods have survived.

Lepidodendron is an extinct, tree-like lycopod (Fig. 20-5). It flourished during the Carboniferous Period in the large coal swamps found at that time. Today we usually find only disassociated parts of these once prominent trees. Roots are often collected without evidence of the stem, since the roots usually were fossilized while the stem decomposed or metamorphosed into coal. Conversely, the stem is often collected without evidence of a root system. In addition, leaves and strobili of *Lepidodendron* are often collected disassociated from the rest of the plant. Because of this piecemeal recovery, different names have been given to different parts of the plant, even though they may have come from the same living plant. Thus, the leaves of *Lepidodendron* are known as *Lepidophylloides,* the cones are known as *Lepidostrobus,* the roots are known as *Stigmaria,* and individual sporangia with attached sporophylls are known as *Lepidostrobophyllum* or *Lepidocarpon.* All of these fossils are well known and have been collected from many parts of the world (Fig. 20-6).

The large tree-like lycopods are not the only fossils of this division. For instance, *Lycopodites* is the genus for fossil lycopods resembling the modern *Lycopodium.* Also, *Selaginellites* is the fossil genus for plants resembling living *Selaginella.* Many other fossil lycopods are known as well, from collecting localities throughout the world (Fig. 20-7).

REPRESENTATIVE LYCOPHYTA
ORDER LYCOPODIALES

Lycopodium

Species of *Lycopodium* are the most common of all living lycopods. They are particularly common in damp or humid regions.

The stem of *Lycopodium* is freely branched and covered with numerous microphylls (Fig. 20-8). These stems are usually bright green and may grow up to about a foot in height. Branches of some species are terminated by

(A)

(B)

Figure 20-8. Some species of *Lycopodium* do not produce typical strobili. A shows an entire *Lycopodium* plant except for the root system. Note the branching habit and tightly arranged leaves. B is a closeup of the apex of one of the branches in A. The sporangia and sporophylls are not aggregated to form tight strobili in this species and the sporangia are clearly evident at the base of some sporophylls.

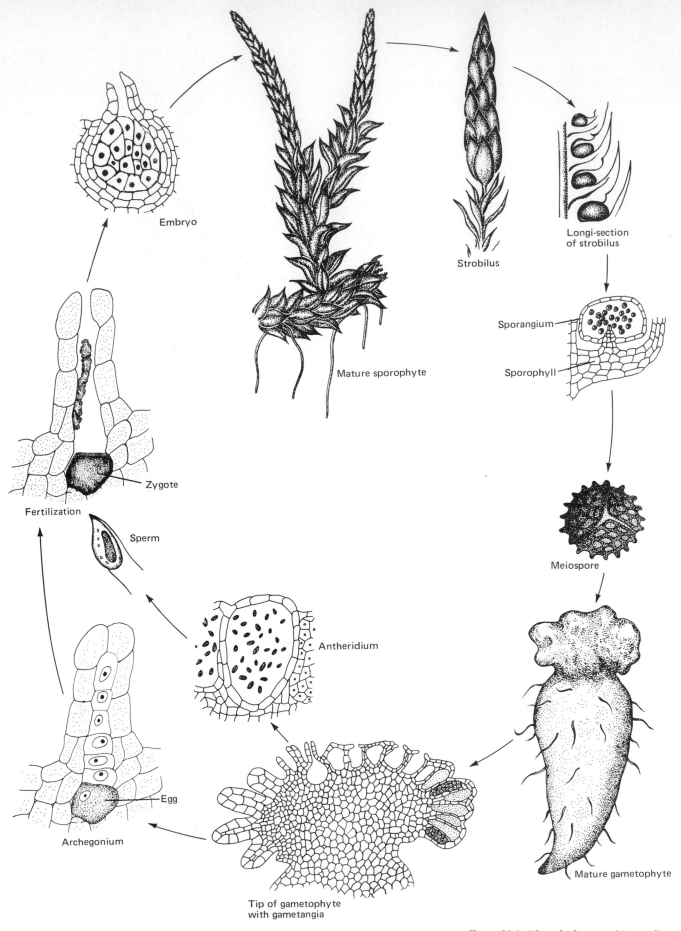

Embryo

Strobilus

Longi-section
of strobilus

Sporangium

Sporophyll

Mature sporophyte

Zygote

Fertilization

Sperm

Meiospore

Antheridium

Egg

Archegonium

Tip of gametophyte
with gametangia

Mature gametophyte

Figure 20-9. Life cycle diagram of *Lycopodium*.

yellowish strobili. Many of these plants are beautiful and are occasionally grown as ornamentals.

Sporangia occur in the axils of sporophylls that are grouped into strobili or scattered along the stem (Fig. 20-8). All species of *Lycopodium* are homosporous. Four meiospores produced by a single mother cell often adhere to each other in tetrads during their maturation. The spores eventually are separated from each other and are released as they mature when the wall of the sporangium ruptures.

A meiospore germinates to produce an exosporic gametophyte. The gametophyte of many species is about 1-in. long and is shaped like a flat-topped carrot or turnip (Fig. 20-9). In other species it is disk-shaped and prostrate. The portion of the gametophyte plant exposed to sunlight often develops chloroplasts, while underground parts of the plant are nongreen and mycorrhizal. Rhizoids occur along the gametophyte, and gametangia develop imbedded in the disk-shaped top.

After fertilization, the embryo develops in place on the gametophyte plant (Fig. 20-9) and depends on it for nourishment through the early stages of sporophyte maturation. Eventually roots develop, a small photosynthetic shoot is pushed above the ground, and the sporophyte becomes free living. The gametophyte plant normally dies and disintegrates after the sporophyte becomes established.

Figure 20-10. A *Selaginella* plant showing leaves and the branching pattern is shown in this figure.

ORDER SELAGINELLALES

Selaginella

Selaginella superficially resembles *Lycopodium* (Fig. 20-10). It differs in habit since the stems of *Lycopodium* are usually upright while those of *Selaginella* are often prostrate. In addition, two sizes of leaves are produced by many species of *Selaginella* with smaller leaves occurring along the upper surface of the prostrate stem and larger leaves along the margins. Both are microphylls.

Sporangia of *Selaginella* are associated with sporophylls which are always grouped into strobili. Microsporangia and megasporangia are generally produced in the same strobilus with the microsporangia above the megasporangia or on different sides of the strobilus (Figs. 20-3 and 20-11). Sporangia are often arranged into four rows within the strobilus, giving it a characteristic square appearance.

A megaspore germinates to produce an endosporic female gametophyte plant (Fig. 20-11). This plant completely fills the cavity of the megaspore, and archegonia are produced at its apex where the megaspore cracks open. Microspores germinate to produce an endosporic male gametophyte plant. This plant produces sperm that swim to the female gametophyte where fertilization occurs. The embryo sporophyte develops in place on the female gametophyte plant (Figs. 20-11 and 20-12) and is dependent

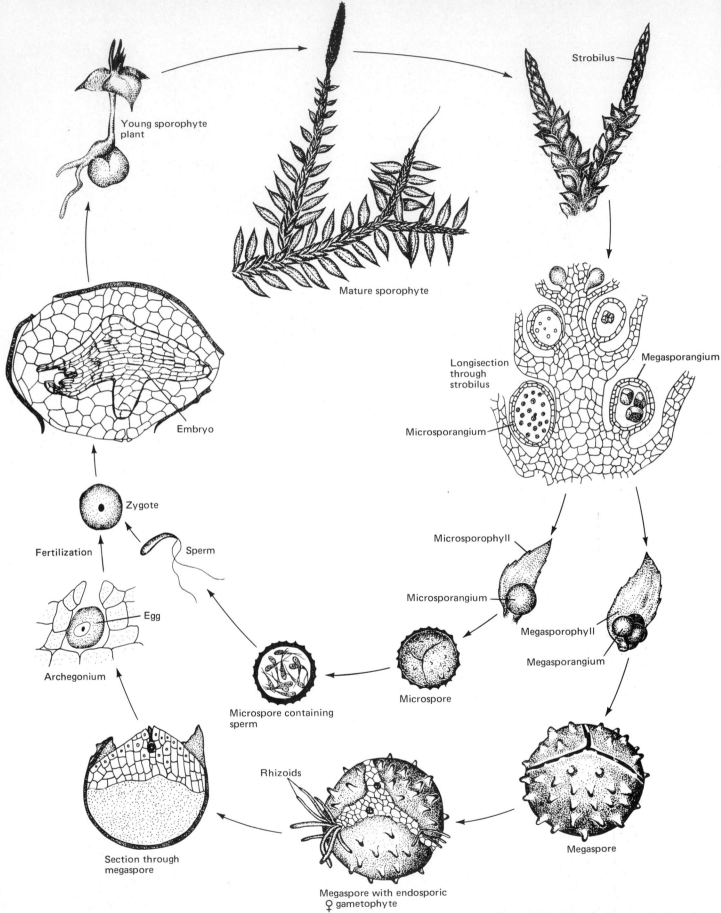

Young sporophyte
plant

Strobilus

Mature sporophyte

Longisection
through
strobilus

Megasporangium

Microsporangium

Embryo

Zygote

Fertilization

Sperm

Microsporophyll

Microsporangium

Megasporophyll

Megasporangium

Egg

Archegonium

Microspore
containing
sperm

Microspore

Rhizoids

Megaspore

Section through
megaspore

Megaspore with endosporic
♀ gametophyte

Figure 20-11. Life cycle diagram of *Selaginella*.

Figure 20-12. The female gametophyte of *Selaginella* is endosporic, and the sporophyte develops attached to the megaspore as illustrated here.

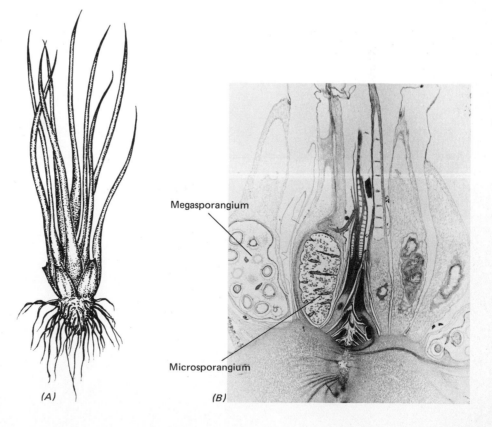

Figure 20-13. *Isoetes* is an unusual grass-like lycopod usually growing as an aquatic plant. *A* shows the entire plant and *B* is a longitudinal section through the stem. *Isoetes* is heterosporous and the two types of sporangia are clearly evident in *B*.

Megasporangium

Microsporangium

(A)   *(B)*

on it for nourishment through its early developmental stages. As the sporophyte plant matures, it becomes self-supporting and the gametophyte plant dies.

SUMMARY

The division Lycophyta has been represented on earth since the early Devonian Period. These plants were among the most prevalent on earth during Carboniferous times. Three orders of lycopods presently occur. These are the *Lycopodiales, Selaginellales* and *Isoetales.*

Lycopods are characterized by lobed protosteles and well-defined roots, stems, and leaves. Sporangia are always associated with sporophylls, which are often aggregated into a strobilus. Some species are homosporous, while others have developed the heterosporous habit. Development of the gametophyte plant may be exosporic or endosporic, but in no case is the gametophyte plant dependent on the sporophyte for nourishment. The sporophyte plant begins development on the female gametophyte plant and is dependent on it during its early growth.

Lycopods are of little direct economic importance although their ancestors contributed greatly to the formation of coal. Some species are presently cultivated for aesthetic reasons.

FURTHER READING

ARNOLD, C. A., *An Introduction to Paleobotany,* McGraw-Hill Book Co., New York, 1947.
BOWER, F. O., *Primitive Land Plants,* Macmillan Co., London, 1935.

Also see the list of general references on vascular plants at the end of Chapter 19.

21 Division Equisetophyta

Two cones of *Equisetum telemia* flanking a short stem section with a whorl of leaves.

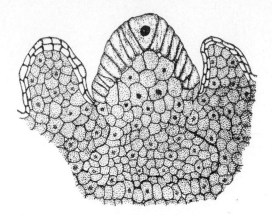

Figure 21-1. Stems of all vascular plants grow longer by the addition of tissue from an apical cell or apical meristem. The stem tip of *Equisetum* illustrated here shows a prominent apical cell.

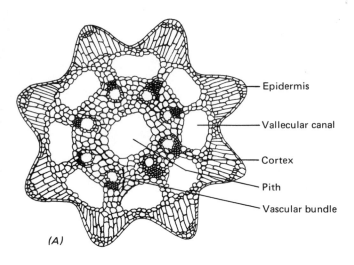

(A)

- Epidermis
- Vallecular canal
- Cortex
- Pith
- Vascular bundle

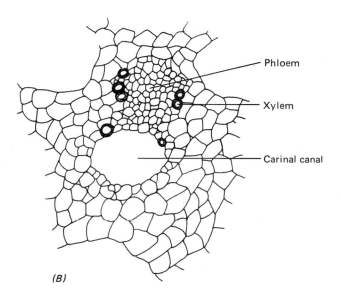

(B)

- Phloem
- Xylem
- Carinal canal

Figure 21-2. The stem of *Equisetum* shows a rather unusual tissue arrangement. A shows an entire stem with a hollow pith and air canals in the xylem and cortex. B is a closeup of a vascular bundle.

Equisetum is the only living representative of the division Equisetophyta. This plant is known by a variety of common names including horsetail, jointgrass, snakegrass, and scouring rush. The division itself has been designated by the names Arthrophyta and Calamitophyta, but since the genus *Equisetum* is the sole modern representative of this division, many authors now prefer to use the divisional name Equisetophyta.

GENERAL CHARACTERISTICS

The following discussion is based primarily on *Equisetum*. As with all vascular plants, the sporophyte plant is large and conspicuous in comparison to the gametophyte.

The stem of *Equisetum* elongates by the activity of an apical cell (Fig. 21-1) and has prominent longitudinal ridges. Conspicuous nodes give the plant a "jointed" appearance. Both aerial stems and rhizomes are similar in construction, although the rhizome often produces *tubers* attached at the nodes. Tubers are modified, swollen stems that function in storing food materials.

Both aerial and underground stems have an epidermis (Fig. 21-2). This tissue is composed of tightly arranged cells and scattered stomata. The epidermal cells of *Equisetum* often become heavily impregnated with silica. Because of this, some of the early American settlers of the West used this plant to clean pots and pans and often called it scouring rush.

The cortex is divided into several regions. Thick-walled cells just inside the epidermis aid in protecting and supporting the plant. A zone of thin-walled, highly photosynthetic cells occurs next. Since the leaves are small these cells are responsible for the bulk of photosynthesis of the plant. A third region of large thin-walled cells occurs inside the photosynthetic tissue. Large air spaces often develop in this tissue. These cavitites are known as *vallecular canals*. They function as air canals to the parts of the plant submerged in water (Fig. 21-2). Similar mechanisms for aeration are common in most plants growing in aquatic or wet habitats. The innermost layer of the cortex is the endodermis which surrounds the stele.

The stele of *Equisetum* is a *siphonostele*. This type of stele is characterized by having a central nonvascular region, or *pith*, composed of parenchyma cells. The siphonostele of *Equisetum* is divided into small, separate vascular bundles that surround the pith. Each vascular bundle contains xylem toward the inside and phloem toward the outside of the stem. The xylem is composed of both tracheids and vessel elements. Some of the earliest formed xylem is stretched and torn as the stem enlarges to form a hollow cavity or canal near the inner edge of each vascular bundle. This cavity, or *carinal canal* (Fig. 21-2), is continuous in each vascular bundle throughout the stem except for interruptions at the nodes. The phloem

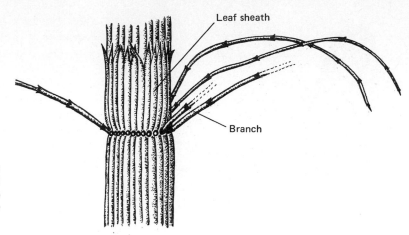

Figure 21-3. This surface view of the stem of *Equisetum* sp. shows whorled leaves and branches. Several branches have been removed in this drawing. Note that the leaves are fused throughout most of their length to form a sheath.

is composed entirely of sieve cells. The siphonostele probably evolved from a protostele when some of the central protostelic tissue became nonvascular.

The pith of *Equisetum* becomes a hollow cavity or *central cavity* as the plant matures (Fig. 21-2). This cavity is not continuous throughout the stem since it is interrupted at each node by a thin partition of tissue. In plants growing in aquatic habitats, the central cavity often becomes filled with water.

Leaves are formed at nodes along the stem. They are small and inconspicuous, and are produced in whorls. The bases of the leaves produced at the same node often fuse together during their development to form a leaf sheath that encircles the stem (Fig. 21-3). Each leaf contains a single vascular bundle or vein connected to the vascular tissue of the stem. No leaf gap is formed and the leaves are thus microphylls.

Branches are formed at the same nodes as leaves. They arise from buds occurring between the leaves and rupture through the leaf sheath as they develop (Fig. 21-4). Branches have essentially the same anatomy as the main stem and often produce leaves in the same manner.

Roots may be simple or branched and occur at nodes on the rhizome. They are protostelic and rarely more

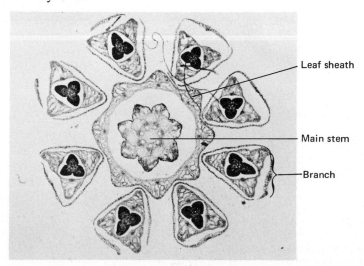

Figure 21-4. This unusual cross section through the stem of *Equisetum* was cut just above a node. The main stem is at the center of the photograph and is surrounded by a leaf sheath. Branches are evident outside of the leaf sheath.

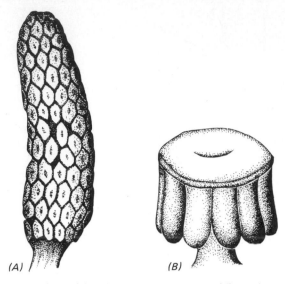

Figure 21-5. The strobilus of *Equisetum* is somewhat different from that of a lycopod. It is composed of tightly compacted, short branch-like structures known as sporangiophores. A shows an entire *Equisetum* cone and B shows a single sporangiophore with attached sporangia.

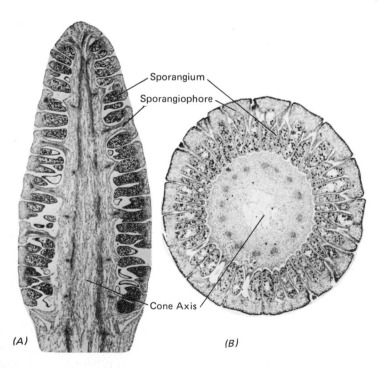

Figure 21-6. The strobilus of *Equisetum* is illustrated in these photographs in longitudinal (A) and cross sections (B).

than 1in long. Each root has a separate apical cell that produces a protostele, cortex, epidermis, and protective root cap.

LIFE HISTORY OF *EQUISETUM*

The life cycles of the modern species of *Equisetum* differ only in small details. The stem of *Equisetum* is terminated by a strobilus just as in many Lycophyta (Fig. 21-5). It is produced at the tip of an ordinary vegetative stem in some species, or at the end of a specialized reproductive stem in others. The strobilus is composed of stalked, sporangium bearing structures known as *sporangiophores* (Fig. 21-6). Sporangiophores are polygonal in face view, and each produces several sporangia attached to the inner surface projecting toward the cone axis.

When immature the strobilus is cone-shaped and tightly arranged. As the sporangia mature, the axis of the strobilus elongates rapidly and the sporangiophores separate (Fig. 21-7). This allows the meiospores to be released when the sporangium splits open.

Meiospores of *Equisetum* are distinctive since each has four elaters attached at one point to its undersurface (Fig. 21-8). These elaters remain coiled around the spore when the humidity is high. When the humidity falls, they uncoil, causing the spore mass to break up and giving the spore more buoyancy in the air for efficient dissemination.

When a spore falls into a favorable environment, it germinates to produce a thalloid gametophyte that may range up to nearly 1 in in diameter. This plant is inconspicuous and irregularly lobed (Fig. 21-9) with many rhizoids on its ventral surface. It is green and completely self-supporting, although often short lived.

Antheridia and archegonia are produced on the upper surface of the gametophyte thallus (Fig. 21-10). Generally both gametangia are produced on the same plant, although unisexual species have been reported. Sperm are large and spirally twisted with numerous whiplash flagella. They are released at maturity as the apex of the antheridium dissolves and are attracted toward the archegonia. One enters the neck and migrates to the venter to fertilize the egg. The zygote develops in place on the gametophyte plant to form an embryo. The embryo is nourished by the gametophyte since it absorbs nutrients through a foot until a functional root and shoot develops and the sporophyte becomes self-supporting. More than one sporophyte may be produced by a single gametophyte plant.

The stem produced from the embryo is small and poorly developed. It soon produces branches that become rhizomes and upright stems. The rhizomes grow parallel to the surface, often only an inch or two beneath the ground. Roots develop from these rhizomes as do more upright stems.

Figure 21-7. These strobili of *Equisetum* are in various stages of maturation. The cone at *A* is mature and is just beginning to elongate to expose its sporangia. The cone at *B* has elongated and spores are being shed, and the cone at *C* has released all of its spores.

(A)   (B)   (C)

OCCURRENCE AND ECOLOGY

Species of *Equisetum* occur in a wide variety of habitats throughout the world, but not in Australia. Most species prefer a moist or even wet environment, although some grow in relatively exposed or dry habitats. Some species may be true aquatics growing emergent from ponds or marshes. Most are small, although *Equisetum giganteum* is a large species with a vine-like stem that may grow to a height of about 25-ft by clinging to other vegetation.

Fossil Equisetophyta apparently shared the preference for a moist or swampy habitat with their living relatives. This is thought to be so because the structure of the stems and roots of these fossils are similar to plants occurring in moist habitats today.

IMPORTANCE

Equisetophyta are of little direct economic importance. However, they were important in the formation of coal in

Figure 21-8. This scanning electron micrograph of the meiospores of *Equisetum* clearly show their characteristic coiled attached elaters. (Courtesy of James V. Allen, Brigham Young University.)

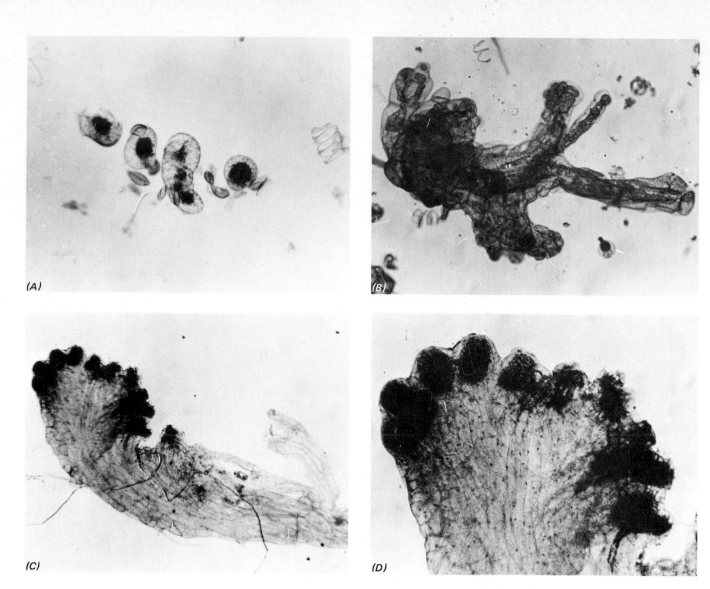

Figure 21-9. Various stages in gametophyte development in *Equisetum* are shown in this figure. A shows germinating spores. B shows a young gametophyte and C is a mature gametophyte bearing antheridia. D is an enlargement of C showing antheridia.

the past. Thus, together with lycopods, Equisetophyta represent the largest group of coal-forming plants of the Carboniferous Period.

Equisetum is occasionally important as a weed in cultivated fields and in some localities it is particularly difficult to control.

CLASSIFICATION

About twenty-five species of *Equisetum* are presently known. These plants are all similar in form and reproduction and represent the only living representatives of this division. All living Equisetophyta are placed in the order Equisetales.

Other orders of Equisetophyta known only from the fossil record include the Hyeniales, Sphenophyllales, and

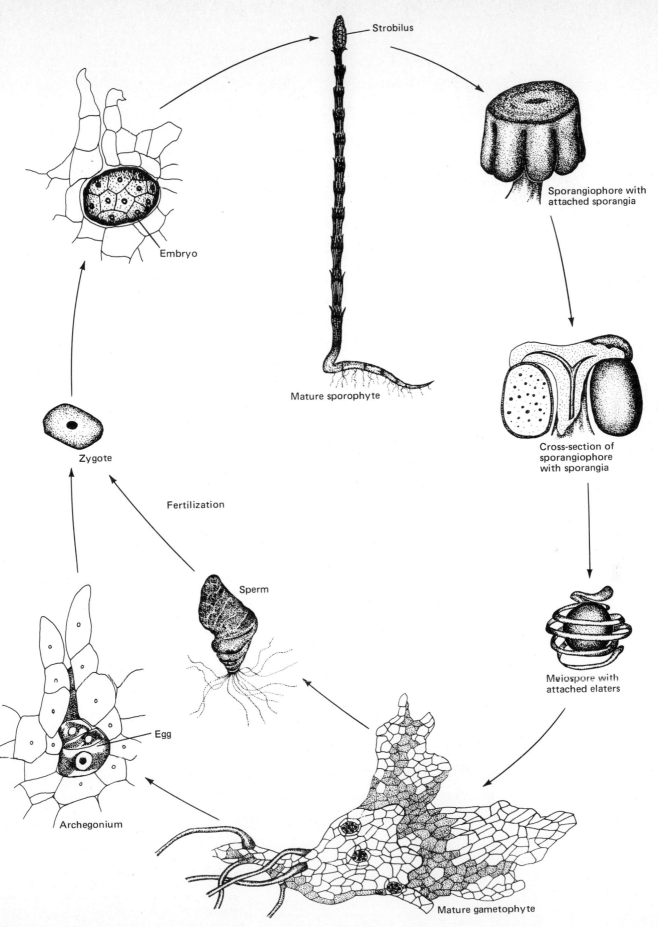

Strobilus

Mature sporophyte

Sporangiophore with
attached sporangia

Cross-section of
sporangiophore
with sporangia

Meiospore with
attached elaters

Embryo

Zygote

Fertilization

Sperm

Egg

Archegonium

Mature gametophyte

Figure 21-10. Life cycle diagram of *Equisetum*.

Figure 21-11. Some fossil Equisetophyta such as *Calamites,* shown in this reconstruction were large, conspicuous plants. All living Equisetophyta are much smaller.

Calamitales. Some species in these orders were very abundant in the geologic past.

FOSSIL EQUISETOPHYTA

Plants of the three extinct orders of Equisetophyta, especially the Calamitales, were prominent members of the earth's flora during the Carboniferous Period. *Calamites* was a large tree-like plant similar to the modern *Equisetum* although much larger than it (Fig. 21-11). *Calamites* had a hollow pith (central canal) that often filled with sediment when the plant fell into a muddy environment. Often this sediment later solidified to form casts of the inside of the stem (Fig. 21-12). *Calamites* also is known from petrified specimens, and its internal anatomy has been well described.

One of the most uniform characteristics of Equisetophyta is the whorled leaves at the nodes. This pattern of leaf production also occurred in fossil members, and detached leaf whorls are commonly collected from some collecting localities. When these leaves are not attached to an identifiable stem, they are placed in the genus *Annularia* (Fig. 21-13) or *Asterophyllites,* depending on their shape and size.

Equisetites is a fossil genus essentially identical to the modern *Equisetum.* Thus, *Equisetum* has apparently been present on the earth for a very long time. It is curious why

Figure 21-12. Pith casts of ancient *Calamites* such as this one are common in many parts of the world.

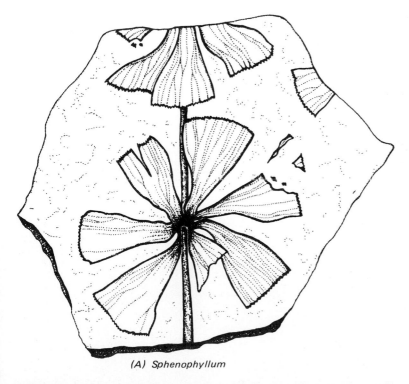

(A) Sphenophyllum

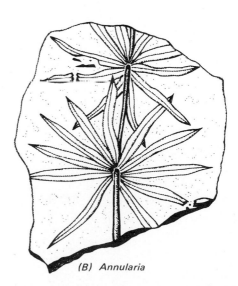

(B) Annularia

Figure 21-13. Leaves of fossil Equisetophyta such as those shown here are characteristically shaped and arranged.

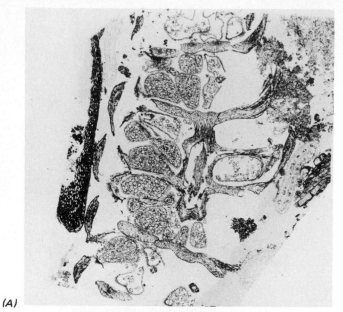

(A)

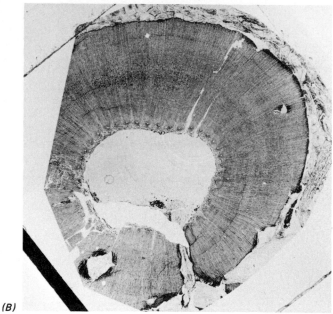

(B)

Figure 21-14. Representitive fossil Equisetophyta are shown in these photos. *A* is a longitudinal section through a strobilus and *B* is a cross section through a stem with copius xylem and a hollow pith.

some fossil plants have become extinct while others have apparently existed almost unchanged throughout long periods of geologic time. The reasons for this are not entirely clear, although it is apparent that when any organism becomes too specialized for its environment, it runs the risk of becoming extinct when that environment changes. Apparently such a fate befell all of the ancient large lycopods and Equisetophyta.

SUMMARY

As it presently occurs on earth, the Equisetophyta is a small division containing only the single genus *Equisetum*. However, this division was much larger in the geological past and constituted one of the great coal forming groups of the Carboniferous Period.

Important characteristics of Equisetophyta include: (1) the leaves are produced in whorls at the nodes; (2) sporangia are produced on stalk-like sporangiophores; (3) spores exhibit attached elaters which aid in dispersal; (4) stems are siphonostelic; and (5) several large air spaces occur in the stems. *Equisetum* usually grows under moist conditions, and is occasionally a serious weed.

FURTHER READING

BARRATT, K., "A Contribution to Our Knowledge of the Vascular System of the Genus *Equisetum*," *Annals of Botany*, 34:201–235, 1920.

BROWNE, I. M., "Anatomy of *Equisetum giganteum*," *Botanical Gazette*, 73:447–468, 1922.

Also see the list of general references on vascular plants at the end of Chapter 19.

22 Division Filicophyta: the Ferns

Large tree ferns (*Dicksonia antarctica*) growing in Australia. (Courtesy of Walter Hodge.)

Figure 22-1. Ferns vary from small, inconspicuous plants to large trees such as those shown here which occur in a temperate rain forest in Tasmania. (Courtesy of Walter Hodge.)

(A)

(B)

Figure 22-2. Some ferns produce upright stems and others produce rhizomes such as those shown here. Leaf scars indicating the points where leaves were attached are clearly evident in B. Note the vascular bundles in these leaf scars.

Almost everyone is familiar with ferns of one kind or another. These plants are common in nearly all parts of the world, and many are cultivated in gardens or grown as house plants for their beautiful foliage. Certainly everyone has seen on a festive occasion a floral display mixed with fern leaves to add color.

The Filicophyta is by far the largest division of vascular plants that does not produce seeds. While the other non-seed vascular plants (Psilophyta, Lycophyta, and Equisetophyta) were much more successful in the geologic past and are now highly restricted on earth, the ferns have not become nearly so restricted. This continuing success is unusual although most pleasant in view of their outstanding beauty.

GENERAL CHARACTERISTICS

The sporophyte plant of a fern is always green and photosynthetic. It is independent of the gametophyte plant at maturity. There is a wide diversity in form of mature fern plants, but all have stems and leaves, and nearly all have roots. Some ferns are large, tree-like plants that grow up to 60-ft tall (Fig. 22-1), while others are tiny, inconspicuous aquatic plants.

The fern stem may be either upright or rhizomatous (Fig. 22-2), and both kinds produce leaves. The anatomy of a rhizome and an aerial stem is similar. It is usually more complex than that of the other lower vascular plants. The anatomy of the stem is variable among different fern species and is often characteristic.

The epidermis of the stem is obscured in some species by a thick mat of epidermal hairs. Other species produce large numbers of *adventitious roots* which may obscure the epidermis.

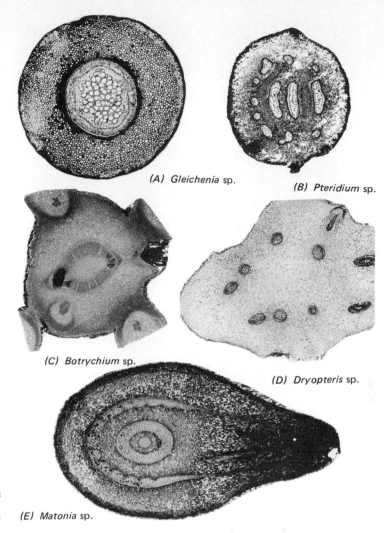

(A) *Gleichenia* sp.

(B) *Pteridium* sp.

(C) *Botrychium* sp.

(D) *Dryopteris* sp.

(E) *Matonia* sp.

Figure 22-3. Fern rhizomes often show characteristic vascular tissue arrangement. Of the rhizomes shown here, A is a protostele and B–E are variations of siphonosteles.

A cortex occurs inside the epidermis. This tissue lacks chlorophyll in a rhizome and often functions in storage or support. The cortex of an aerial stem may be colorless or green and is usually composed of large, isodiametric parenchyma cells.

The stele occupies the center of the stem. Most species produce siphonosteles, although some are protostelic when immature, becoming siphonostelic at maturity. The arrangement of xylem and phloem in the stele is variable among different ferns but is often characteristic for a genus (Fig. 22-3). For instance, in some species a continuous ring of xylem is surrounded by an internal and external ring of phloem. In others a ring of phloem is only external to a ring of xylem. The stele of some species is broken into individual vascular bundles. These bundles may be composed of xylem surrounding phloem, phloem surrounding xylem, or xylem capped by phloem.

The conductive cells of the xylem of most ferns are tracheids, although a few species have vessel elements. Sieve cells are the conductive cells of the phloem. Parenchyma cells often occur in the vascular tissues functioning in storage.

Figure 22-4. This section of a frond of the tree fern *Dicksonia antarctica* shows the arrangement of the pinnae and pinnules. (Courtesy of Walter Hodge.)

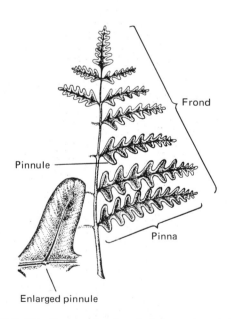

Figure 22-5. This diagram of a fern leaf shows the relationship between pinnules and pinnae on the frond.

The leaves of ferns are generally large and well developed, with a flattened lamina well suited for efficient photosynthesis. A leaf stalk or *petiole* attaches the leaf to the stem. The lamina of many species is divided into small segments or *pinnules*. Several pinnules together on a central stalk or *rachis* comprise a *pinna*. Several pinnae together comprise the entire leaf, which is often referred to as a *frond* (Figs. 22-4 and 22-5).

The petiole contains one or more large vascular bundles or leaf traces that are attached to the vascular tissue of the stem. Unlike other lower vascular plants, a gap in the vascular tissue of the stem is produced above the point of attachment. This *leaf gap* is filled with parenchyma cells that are continuous with those of the pith and cortex. Leaves as described above usually occur quite large and always contain a leaf gap. They are known as *megaphylls*. Such leaves are common to all higher vascular plants.

As it passes into the lamina, the leaf trace divides several times until it becomes dissected into numerous fine bundles or veins. The pattern of these leaf veins is often characteristic and may be open when the veins do not fuse, or reticulate when the veins branch and reunite, to form a network.

Leaves develop from the stem in a rather peculiar manner. Each leaf develops in a "rolled up" position and unfurls as it matures (Fig. 22-6). Such a development is known as *circinate vernation*. The immature leaves in the rolled up position are often known as *fiddleheads*, which may be eaten in some species. This could be hazardous since recent research indicates that consumption may cause serious mutations to unborn offspring in livestock. More research is necessary to determine whether similar effects occur in humans.

Most ferns have a well-developed root system made up of many small fibrous roots. They originate from either the aerial stem, the rhizome, or both and are therefore *adventitious roots*. These roots are similar to those of most other vascular plants since they are formed from an apical meristem, are protected by a root cap, and are protostelic. A few species of floating aquatic ferns lack roots because they are not attached and so absorb nutrients directly through their leaves.

Sporangia are formed on leaves (Fig. 22-7). In many species the sporophylls exactly resemble vegetative leaves, while in others they are modified and scarcely recognizable as leaves. Often certain pinnules of a single pinna are fertile while others are sterile, and the two appear alike. Most species produce sporangia on the lower leaf surface.

Sporangia are relatively large and may be formed singularly or in clusters. When sporangia are clustered they often form a pustule-like structure or *sorus* (Figs. 22-7, 22-8, and 22-9). Sori are often protected by a thin flap of tissue known as an *indusium* (Fig. 22-10), which covers the sporangia. This structure protects the sporangia

(A)

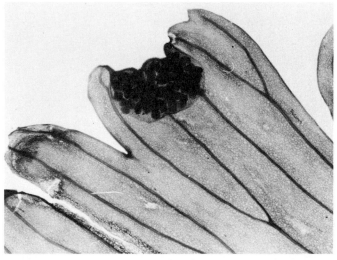

(A)

(B)

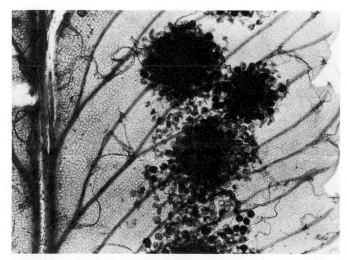

(B)

Figure 22-6. This close view of *Asplenium* illustrates "fiddle heads" near the center of the photograph (A). B is a closeup of the stem of *Dicksonia*. Note the matted epidermal hairs and prominent leaf bases.

Figure 22-7. Enlarged views of leaves of *Adiantum* (A) and *Polystichum* (B) are shown here. Note the prominent veins and the sporangia clustered to form sori. The sorus at A is covered by a false indusium and that at B lacks an indusium.

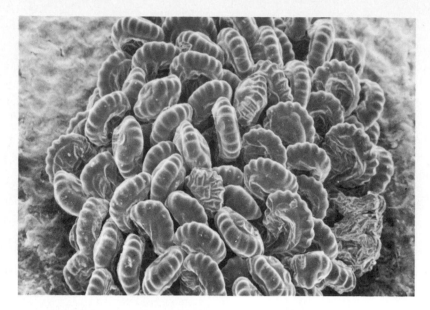

Figure 22-8. This scanning electron micrograph of a sorus of *Polypodium* shows the crowded arrangement of sporangia. Note the vertical annulus on each sporangium. (Courtesy of James V. Allen, Brigham Young University.)

Figure 22-9. Enlarged leaves and attached sori of four ferns are shown here. *A* is *Polypodium*, *B* is *Lygodium*, *C* is *Asplenium*, and *D* is *Adiantum*. Note the false indusia on the *Adiantum* leaf.

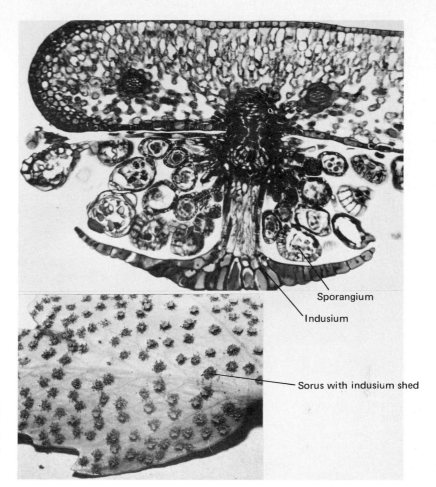

Sporangium

Indusium

Sorus with indusium shed

Figure 22-10. The fern *Cyrtomium* produces sori scattered on the undersurface of sporophylls which resemble vegetative leaves. *A* is a cross section through a *Cyrtomium* leaf, and *B* is a view of the undersurface of a leaf. Some indusia have been shed from the sori in *B*.

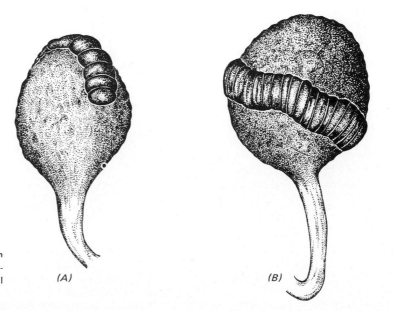

(A)   *(B)*

Figure 22-11. The position of the annulus on a fern sporangium is one feature of importance in classification. The sporangium illustrated at *A* has a vertical annulus while that at *B* has a horizontal annulus.

261

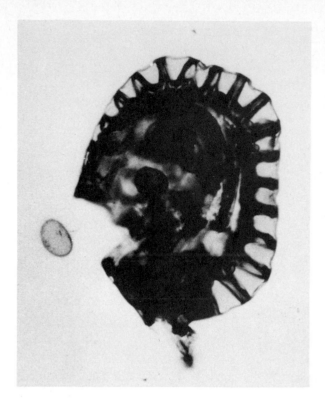

Figure 22-12. This sporangium of the maidenhair fern is just being opened by contraction of the annulus. Note the single spore that has been released.

prior to their maturation but is displaced or shed to allow efficient spore dispersal as the sporangia mature. Some ferns which are nonindusiate produce a false indusium (Figs. 22-7 and 22-9). This occurs when the edge of the pinnule curls downward over the sporangia. Not all ferns produce indusia or false indusia, and their presence or absence is one important characteristic for classification.

The fern sporangium is generally stalked and often has a row of large thick-walled cells on its surface (Figs. 22-11 and 22-12). This row of cells or *annulus* shrinks as the cells lose water and causes the sporangium to rupture at maturity. It is fascinating to watch this process through the microscope. As the annulus dries, it contracts, causing the sporangium to open and the upper half to slowly pull away from the lower (Fig. 22-13). The upper half is eventually pulled back far enough to complete opening the sporangium and expose the spores. The cells of the annulus soon expand again rapidly, causing the top half of the sporangium to launch forward violently catapulting spores into the atmosphere for wind dispersal. This represents one of the exceptionally fine methods of spore dispersal in nature.

Most ferns are homosporous, although a small but significant group of aquatic ferns are heterosporous.

The fern gametophyte plant is small and generally inconspicuous (Fig. 22-14). It is green in most species, although colorless and mycorrhizal in others. The gametophyte plant rarely grows larger than ½ in. in diameter and is often much smaller. Rhizoids are produced on the ventral surface. Antheridia and archegonia are usually produced on the ventral side of the same plant, although often at opposite ends. Even though most ferns are homothallic, often they are functionally heterothallic since antheridia mature prior to most archegonia on the same

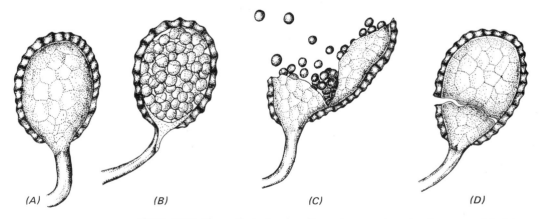

(A)   (B)   (C)   (D)

Figure 22-13. The method of spore dispersal among ferns is often accomplished by the contraction of an annulus. The drawing at A shows a sporangium in side view and B shows the internal spores. The annulus has contracted at C and has ruptured the sporangium. The sporangium then lurches forward violently throwing its spores. D shows the sporangium at completion of the process.

plant. This mechanism fosters outbreeding and thus encourages genetic recombination.

The fertilized egg develops in place within the archegonium and is dependent on the gametophyte plant for its early nourishment (Fig. 22-15). The zygote soon divides to produce an embryo that continues development to become a mature sporophyte plant. Generally the gametophyte plant dies during this maturation process.

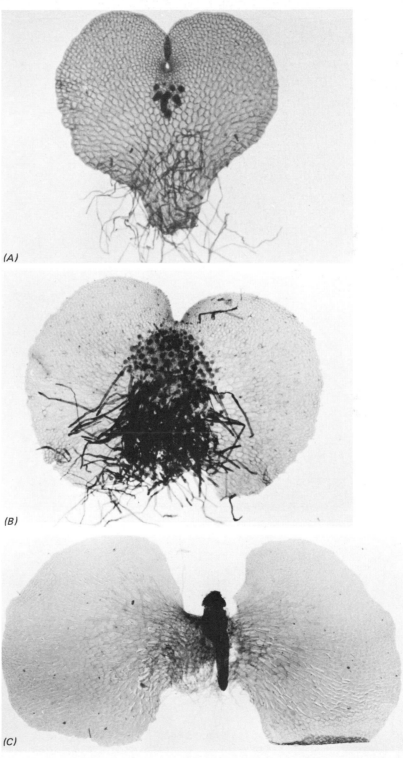

(A)

(B)

(C)

Figure 22-14. The fern gametophytes shown here illustrate gametangia. The top plant (A) demonstrates archegonia and the center plant (B) exhibits several antheridia. The plant at C is a female gametophyte with a young sporophyte attached.

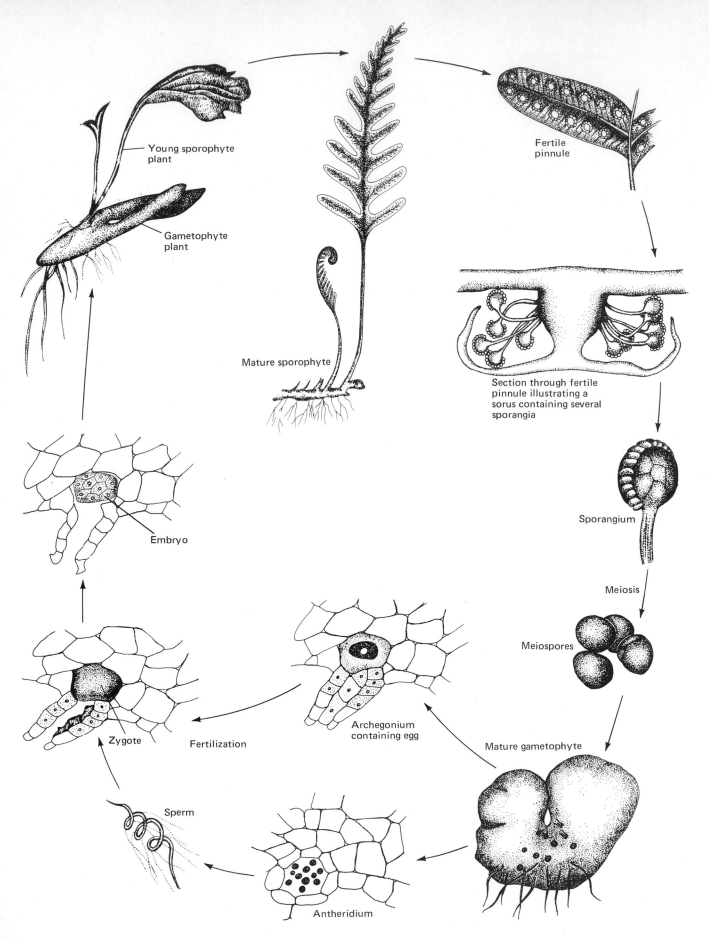

Young sporophyte plant

Gametophyte plant

Fertile pinnule

Mature sporophyte

Section through fertile pinnule illustrating a sorus containing several sporangia

Sporangium

Meiosis

Meiospores

Embryo

Mature gametophyte

Zygote

Fertilization

Archegonium containing egg

Sperm

Antheridium

Figure 22-15. Life cycle diagram of a fern.

OCCURRENCE AND ECOLOGY

Ferns occur in all parts of the world, and are extremely common and important in some habitats. These plants especially prefer moist, shady habitats, although some are adapted for growth on the desert or at high altitudes and latitudes. Probably most ferns presently growing on earth occur in the tropics. Tropical fern forests undoubtedly are among the most beautiful areas of the world. Some ferns in such forests reach heights of 60 ft and the understory contains hundreds of smaller ferns and flowering plants.

A few ferns are true aquatic plants that grow floating in quiet tropical and temperate fresh waters.

IMPORTANCE

Many ferns are used directly by humans. They are grown as house and garden plants and are used in the floral industry. These uses are responsible for a multimillion-dollar business done yearly in the United States alone.

Ferns are also important ecologically. They are primary producers in the food chains of many animals, especially where they comprise a large part of the total vegetation. Ferns are essential in many temperate and tropical watersheds in protecting against erosion and adding large amounts of organic matter to the soil.

CLASSIFICATION

Living and extinct ferns are placed into several orders. Much research is currently being done by paleobotanists to determine if all of these orders are valid. Modern ferns are placed by most botanists into the five orders: Marattiales, Ophioglossales, Filicales, Marsileales, and Salviniales. These orders contain more than 10,000 species, most of which are placed in the Filicales.

REPRESENTATIVE FOSSIL FERNS

Fossil ferns, particularly those from Paleozoic and Mesozoic floras, are among the most prevalent of fossil plant remains. These fossils are placed into several orders based on different morphology. Only a few genera of fossil ferns will be discussed here to illustrate some of the important features of these plants.

Many fern-like plants of the past do not appear to be closely related to the ferns of the present. However, since these fossils exhibit many fern-like characteristics, they are placed in the division Filicophyta. Future work will undoubtedly aid in the accurate classification of such plants.

Zygopteris

Many fossils are only partially complete when they are collected. For instance, the hard parts of a plant, such as

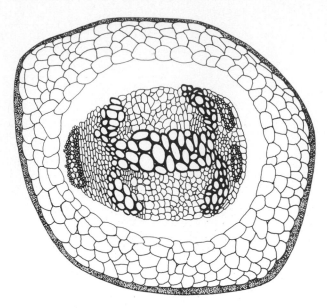

Figure 22-16. This cross section of the petiole of the ancient fern *Etapteris* clearly shows the characteristic *H*-shaped xylem strand in the center of the petiole. *Etapteris* is common from Carboniferous strata.

a stem or petiole, may be well preserved while the readily decomposed laminae are often absent. Such is often the case of *Zygopteris,* which is a fossil fern-like plant stem of the Devonian and Carboniferous Periods. This fossil and several closely related species are identified by their characteristically shaped protosteles. Occasionally *Zygoteris* is collected with petioles still attached. Each petiole contains a single *H*-shaped leaf trace. When these petioles are collected separated from *Zygopteris,* they are known as *Etapteris* (Fig. 22-16).

Cladophlebis

Ferns made up a large portion of the earth's flora throughout the Mesozoic Era. *Cladophlebis* is one of the most common ferns of this era. This genus contains sterile fern leaves resembling certain modern ferns. However, since fern classification largely is based on characteristics of the sporangia and sori, these fern leaves cannot be placed in modern genera. Some specimens of *Cladophlebis* are several inches across and are extremely well preserved so that the pattern of leaf veins and even the actual cells of the epidermis or mesophyll may be studied.

Gleichenites

By Cretaceous times, many of the ferns present on earth were very similar to modern day ferns. Hence, many paleobotanists place such fossils in the living genus or into a fossil genus with the same name but with the suffix *-ites* added to distinguish them from modern ferns of the closely related modern genus.

Gleichenites is one of the most common fossil ferns of the late Jurassic and Early Cretaceous Periods. It grew in profusion in some parts of the world during this time, forming large fern swamps in parts of Europe, Greenland, and North America.

This fern is characterized by large dichotomously branched fronds bearing small rounded or tongue-shaped pinnules (Fig. 22-17).

Matonidum

Matonidum is one of the most beautiful of all fossil ferns. It was widely distributed throughout the world during Jurassic and Early Cretaceous times when it often grew with *Gleichenites*. This fern has a unique frond with several pinnae originating from a common point at the petiole apex and forming a very graceful leaf (Fig. 22-17).

Matonidium is closely related to the modern *Matonia,* which is presently found only on the Malayasian Peninsula. Understanding why a fern with such a wide geographical distribution in the geologic past became so restricted in the present is a fascinating problem. Many such problems

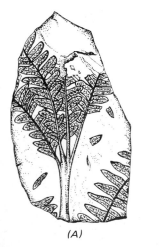 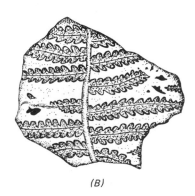

(A)   *(B)*

Figure 22-17. Both *Matonidium* (A) and *Gleichenites* (B) were common Mesozoic ferns and have been collected from many parts of the world.

(A) Sphenopteris

(B) Alethopteris

(C) Mixoneura

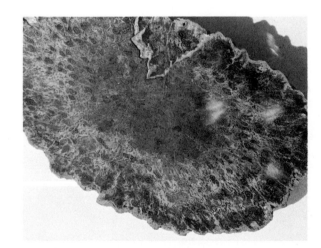

(D) Tempskya

Figure 22-18. During the Carboniferous Period, plants which had fern-like leaves were very prevalent. However, only some of these were true ferns since it has been shown that some producd true seeds, which ferns never do. Thus, when Carboniferous fern-like leaves are collected, they are often called fern-like foliage unless they are collected with reproductive structures attached. This figure shows three examples of fern-like foliage (A–C) as well as the petrified remains of a large true fern, Tempskya (D).

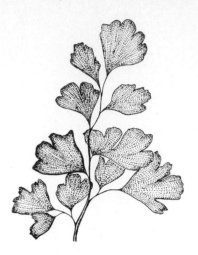

Figure 22-19. Leaves of the maidenhair fern *Adiantum* have many equally branched veins as illustrated in this figure.

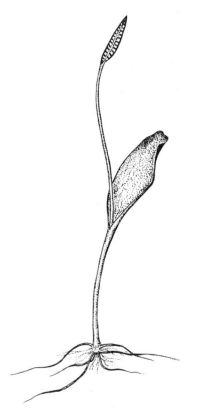

Figure 22-20. Rattlesnake fern (*Ophioglossum*) is a primitive form with the sporangia clustered on a fertile spike and fused to form synangia.

exist in the study of the distribution of fossil and modern plants.

REPRESENTATIVE LIVING FERNS
ORDER FILICALES

The Filicales is by far the largest of all fossil and living fern orders. Nearly all modern plants easily recognized as ferns are placed into this order. Filicales exclude only the aquatic ferns and a small group of primitive ferns. Much of the discussion at the beginning of this chapter was based on *Filicales,* and only a few representative genera are discussed here to illustrate some of the differences between the ferns of this order.

Adiantum

Adiantum is known as the maidenhair fern because of the fine pattern of regularly branched veins in its pinnules (Fig. 22-19). It has large, fan-shaped pinnules attached to a rachis by a short petiole. Pinnules are bright green and the rachis is often shiny black making a very beautiful contrast. The maidenhair fern is a commonly grown house plant because of its striking beauty.

Adiantum prefers rich humic soil in nature and is distributed throughout tropical and temperate regions of the world.

Dryopteris

Dryopteris is one of the largest and most common fern genera of the Northern Hemisphere. This fern grows upright and produces large, shield-shaped, spreading leaves. Because of these characteristics leaves, *Dryoptris* is often called the shield fern. Small, round, indusiate sori are produced on the under surface of *Dryopteris* pinnules along the veins.

Polystichum

The holly ferns are placed in the genus *Polystichum.* These are large, coarse ferns which are mostly north temperate in distribution. *Polystichum* is a large, common genus often collected from moist northern forests.

Polystichum petioles are yellow and the pinnules are yellow-green or bright green. Pinnules are sharply pointed with saw-toothed margins giving these plants the common name holly fern. Sori are large, containing many sporangia and are protected by a prominent, toothed indusium.

Thelypteris

Thelypteris is a large tropical and temperate genus that is widely distributed throughout the world. *Thelypteris* petioles are yellow and the foliage is bright green. Pinnules are tongue-shaped with relatively few simple or sparsely branched veins. Small indusiate sori occur along these veins.

ORDER OPHIOGLOSSALES

The Ophioglossales is a small order containing living ferns with many primitive characteristics. Sporangia particularly are primitive. They differ from those of Filicales in that they are much larger, have thick walls, and produce thousands of spores. In addition, the gametophyte plant differs from that of Filicales since it is colorless, mycorrhizal, and subterranean.

Three genera of ferns are placed in this order. Of these, *Ophioglossum* and *Botrychium* are widely distributed. *Ophioglossum* is commonly known as adder's tongue or rattlesnake fern because of the slight resemblance of the narrow leaves to the tongue of a snake (Fig. 22-20). *Botrychium* is often called the grape fern since the cluster of sporangia on a fertile leaf resembles a miniature cluster of grapes.

Figure 22-21. The aquatic fern *Marsilea* is occasionally mistaken for a four-leaf clover because of its characteristic leaves. This unusual fern is heterosporous.

ORDERS MARSILEALES AND SALVINIALES

Marsileales and Salviniales are aquatic ferns that differ in many respects from other ferns. Both orders contain only heterosporous species.

Marsilea

Marsilea is a common fern of shallow water or marshy area. It produces a rhizome with true roots and upright leaves (Fig. 22-21). These leaves are very characteristic since they resemble long-stalked, four-leaved clovers. These leaves develop in the normal circinate manner common among ferns.

Both microsporangia and megasporangia are produced on the same plant. These sporangia are small and lack annuli. They are produced in sori within a highly modified, hardened, brown sporophyll known as a *sporocarp* (Fig. 22-22). The spores remain viable for long periods of time since sporangia remain inside the sporocarp until it disintegrates, which often takes several years. Eventually, the sporangia shed their spores, and the development of the gametophyte plants is endosporic. Both the male and female gametophyte plant are highly reduced. The maturation of the male gametophyte plant is very rapid (often within one day) and it disintegrates on formation of the sperm. The female gametophyte plant is colorless during its early development, but becomes green and develops rhizoids after fertilization of the egg in order to manufacture food for the developing embryo.

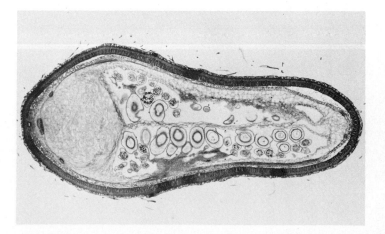

Figure 22-22. *Marsilea* is a heterosporous fern. This section through a fertile leaf shows megasporangia with megaspores and microsporangia with microspores.

Salvinia

Salvinia is a small aquatic fern that lacks roots at maturity. This fern is common in many parts of the world although it is often overlooked because of its small size.

Salvinia is very distinctive in appearance since it bears leaves in groups of three at the nodes. Two of these leaves float on the surface of the water, are broadly oval in shape, and grow up to about 2 cm long. The third leaf grows submerged in the water and is highly dissected and feathery in appearance. This leaf functions to absorb water and nutrients from the water, and hence roots are absent in *Salvinia*. Sporocarps develop on this submerged leaf. They are released from the leaf often on the death of the plant when they sink to the bottom. Microspores and megaspores rise to the surface when they are released, and the life cycle is completed floating upon the surface of the water.

SUMMARY

The ferns are a large and diversified division of nonseed producing vascular plants. They are advanced over the vascular plants previously studied since they produce megaphylls and more complex vascular tissue.

Both homosporous and heterosporous ferns are known, although homosporous species are by far the most common. Spores are produced within sporangia which are often equipped with an annulus to aid in spore dispersal. Sporangia are often clustered into a sorus on the undersurface of a sporophyll. The sorus may be covered by an indusium which functions in protecting the sporangia during their maturation. The indusium is often shed or displaced when the sporangia are mature.

The gametophyte plant of most ferns is small but green and free living. However, the gametophyte of Ophioglossales is nongreen, subterranean, and mycorrhizal.

The embryo is dependent on the gametophyte plant for its early nourishment, but soon it becomes independent. The sporophyte plant has true roots, stems, and leaves, and may be small and succulent or large and tree-like. Ferns are distributed throughout the world, especially in the tropics. They prefer moist shady habitats, although some are adapted to a wide variety of different environments.

FURTHER READING

BOWER, F. O., *The Ferns,* Vols. 1, 2 and 3, Cambridge University Press, London, 1928.

DELEVORYAS, T. (Ed.), "Origin and Evolution of Ferns," *Memoirs of the Torrey Botanical Club,* 21:1–95, 1962.

NAYAR, B. K. and S. KAUR, "The Gametophytes of Homosporous Ferns," *Botanical Review,* 37:295–396, 1971.

WILSON, KENNETH, "Biology of Reproduction in Ferns," *Natural History,* 74(6):52–59, 1965.

Also see the list of general references on vascular plants at the end of Chapter 19.

23

Introduction to the Seed Plants

Closeup of milkweed seeds. (Courtesy of Sheril Burton, Brigham Young University.)

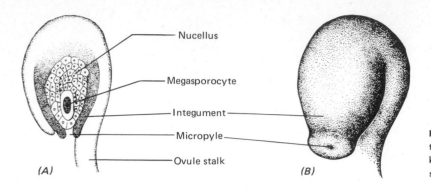

Nucellus

Megasporocyte

Integument

Micropyle

Ovule stalk

(A)   *(B)*

Figure 23-1. The megasporangium of a seed plant together with one or more outer protective layers is known as an *ovule*. This figure shows a longitudinal section (A) and face view (B) of an ovule.

Until now we have discussed only plants that reproduce by spores. In addition, all thalloid plants including algae, fungi, and bryophytes, and most lower vascular plants are homosporous. However, *Selaginella* and some ferns were mentioned as being heterosporous. The development of this heterosporous habit was critical in the evolution of seed-producing plants since all seed plants are heterosporous.

The megasporangium of a seed-producing plant is a highly specialized structure. Together with one or more outer protective layers, this megasporagium is known as an *ovule* (Fig. 23-1). An ovule is usually produced on a megasporophyll, although this structure may be highly modified and scarcely recognizable as a leaf. In conifers, the ovule is produced on a hard and woody cone scale, and in most flowering plants, it is completely enclosed within a protective structure.

The outer layer of an ovule is the *integument* which protects the internal tissues. Immediately inside the integument, a diploid tissue or *nucellus* develops. This tissue is composed of large parenchyma cells. One of the cells of the nucellus enlarges somewhat to become a *megaspore mother cell,* or *megasporocyte* (Fig. 23-2) which divides meiotically to produce four megaspores in a linear tetrad. Generally, three of these megaspores degenerate, leaving one functional megaspore.

This remaining haploid megaspore divides to produce a female gametophyte plant (megagametophyte). Development of this female gametophyte plant is always endosporic within the functional megaspore. The function of any gametophyte plant is to produce gametes, and accordingly, on reaching maturity, the megagametophyte produces one or more eggs. The method of producing the egg varies among the different divisions of seed plants. The megagametophyte of some seed plants produces one or more archegonium which produces the eggs, while the more advanced flowering plants lack archegonia entirely.

The embryo develops within the ovule following fertilization of the egg. As it matures, the other tissues of the ovule mature as well to produce a fully matured ovule or *seed* (Fig. 23-3). The mature seed always contains an embryo. In addition, it always has an external *seed coat,*

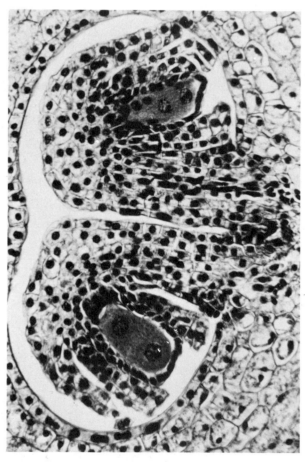

Figure 23-2. These two lily ovules contain a large central megasporocyte which is in the process of meiotic division to produce a tetrad.

which is a protective layer developed from the integument. Most or all of the nucellus is used to nourish the developing female gametophyte plant. Also, female gametophyte tissue may be depleted in nourishing the developing embryo or it may remain in the seed. In addition to these tissues, the seed of a flowering plant often contains a specialized nutritive tissue which will be discussed later.

As the ovule matures on the plant, microspores develop within specialized microsporangia. They are produced from *microspore mother cells* or *microsporocytes* (Fig. 23-4) which divide by meiosis to form the haploid spores. (Figure 2-13 shows meiosis in a lilly microsporophyte.) The nucleus of a microspore divides one or more times to produce the microgametophyte or male gametophyte plant. Similar to the femal gametophyte, the male gametophyte plant is always endosporic. The outer wall of the microspore often becomes modified and highly ornate as it matures. This matured microspore is known as a *pollen grain* (Fig. 23-5). A pollen grain always produces sperm which may be large and motile in primitive seed plants, or small and nonmotile in higher seed plants. In either case, the sperm are delivered to the female gametophyte plant by a specialized *pollen tube* (Fig. 23-6). This tube develops when the pollen grain germinates. It is attracted by hormones toward the megagametophyte plant as it grows. Sperm either swim or are carried into this tube and eventually contact the egg at fertilization.

Let us summarize some of the significant characteristics of seed plants. First, the megasporangium of a seed plant with an outer protective layer is known as an ovule. Second, the ovule matures to form a seed. Third, microspores are produced within a microsporangium and mature to become pollen grains. Fourth, a mature seed may contain parts of three separate plant generations. Thus, any remaining nucellus represents the old sporophyte generation, remaining megagametophyte tissue represent the female gametophyte generation and the embryo represents the new sporophyte generation. However, not all seeds contain all of these tissues.

Seed plants have been on earth since at least Carboniferous and possibly Devonian times when they grew with the large spore producing Lycophyta and Equisetophyta as well as with ferns and others. Seed producing plants became more numerous throughout the Mesozoic Era, and at the end of this era they completely dominated the earth's vegetation. Life as we know it presently on earth, especially for humans and their domestic animals, is greatly dependent on seed plants for a variety of reasons. Seed plants include nearly all of our important plants used for building materials, paper manufacturing, watershed protection, and food.

The origin of the seed habit is difficult to trace and probably occured several times independently. By Middle Carboniferous times, heterosporous Lycophyta and Equi-

Figure 23-3. Ovules mature to become seeds. Three seeds are illustrated in the lower left of this fruit. (Courtesy of Macmillan Science Company—Turtox/Cambosco.)

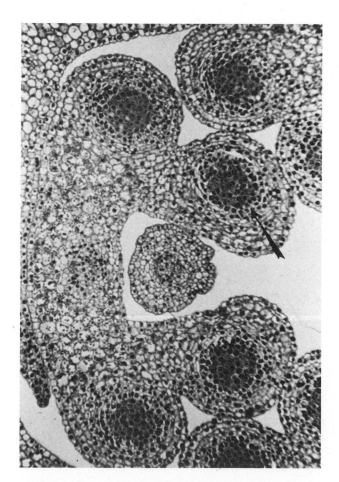

Figure 23-4. This cross section through the male portion of a lily flower shows several microsporangia (arrow) which contain microsporocytes. These microsporocytes will divide by meiosis to produce microspores.

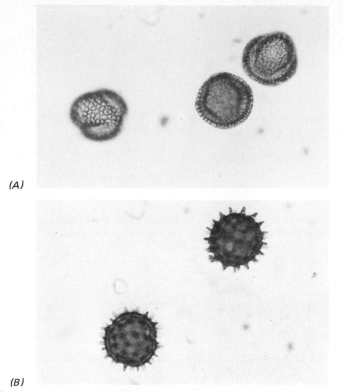

(A)

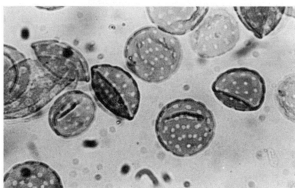

(B)

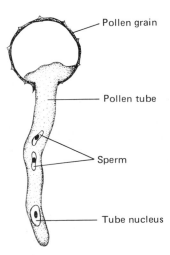

(C)

Figure 23-5. Microspores of seed plants mature to become pollen grains. Pollen grains of three plants are illustrated here. A is lilac, B is arrowroot, and C is pigweed.

setophyta which produced megasporangia much like ovules were known. During the same time, several groups of seed plants were originating as well.

Seed plants are placed into two large groups. *Gymnosperms* are those plants that produce their ovules nakedly on a sporophyll or cone scale. Thus, the ovule of a gymnosperm is not enclosed. *Angiosperms* or flowering plants, on the other hand, produce their seeds entirely enclosed within a specialized protective megasporophyll. Several divisions of plants are gymnospermous including the Cycadophyta, Gnetophyta, Ginkgophyta, and Coniferophyta. Only one plant division, the Anthophyta, is angiospermous. Each of these divisions will be discussed in the next chapters.

FURTHER READING

ANDREWS, H. N., "Early Seed Plants," *Science,* 142:925–931, 1963.

BANKS, HARLAN, *Evolution and Plants of the Past,* Wadsworth Publishing Co., Belmont, Calif., 1970.

PETTITT, J. M., "Heterospory and the Origin of the Seed Habit," *Biological Review,* 45:401–415, 1970.

SMITH, DAVID, "The Evolution of the Ovule," *Biological Review,* 39:137–159, 1963.

THOMSON, R. B., "Evolution of the Seed Habit in Plants," *Proceedings and Transactions of the Royal Society of Canada,* 21:229–272, 1927.

Also see the list of general references on vascular plants at the end of Chapter 19.

Pollen grain

Pollen tube

Sperm

Tube nucleus

Figure 23-6. Pollen grains germinate to produce pollen tubes as shown in this drawing. Sperm are carried down the tube and delivered to the female gametophyte.

24 Division Cycadophyta: the Cycads

Megasporangiate cone of *Zamia*. (Courtesy of Walter Hodge.)

(A)

(B)

Figure 24-1. Two cycads, *Macrozamia* (A) and *Cycas* (B) are shown in these photos. (Courtesy of Macmillan Science Company—Turtox/Cambosco.)

Figure 24-2. This photograph of *Microcycas* shows the stout leaf bases on the stem which form a protective armor. (Courtesy of Macmillan Science Company—Turtox/Cambosco.)

Cycads are gymnosperms with robust stems and leaves that resemble palm fronds. Presently the Cycadophyta contains nine genera, all of which are tropical or subtropical in distribution.

GENERAL CHARACTERISTICS

The cycad sporophyte is large and conspicuous with true roots, stems, and leaves. Most cycads have short, thick stems with large pinnate leaves attached at the crown (Fig. 24-1). Generally this stem is aerial and erect, although in some species it may be nearly entirely buried. The stem is usually unbranched or only sparsely branched and less than 10 ft tall. However, a few are tree-like (Fig. 24-2), and one species ranges up to 60 ft tall.

The outer layer of the stem is an epidermis. However, this tissue is often obscured by a layer of matted epidermal hairs and the hard, woody bases of old leaves remaining after the terminal portion becomes detached. *Leaf bases* are stout or even thorn-like, forming a very efficient armor (Fig. 24-2). In fact, collecting cycads for study often is quite a task since it is difficult to grasp the stem even after the plant has been loosened from the soil.

A wide cortex occurs inside the epidermis (Fig. 24-3). This tissue is composed of large parenchyma cells that are often filled with starch granules. Leaf traces cross this cortex as they pass from the stele to the leaf. The central stele has an exceptionally large pith and a limited amount of vascular tissue. The conductive tissue is arranged in individual vascular bundles, with the xylem interior to the phloem. Such a stele is similar to a siphonostele, although the origin of the vascular tissues is somewhat different. Such a stele common in higher plants is known as an *eustele*. Tracheids are the only conductive cells of the xylem as are sieve cells in the phloem.

Cycads are often grown as ornamental plants because of their beautiful leaves which may range up to about 10 ft long (Fig. 24-1). The petiole of the leaf is large and strong, and the lamina is pinnately divided into strap-shaped or linear segments (Fig. 24-4). Leaves are usually deep green and waxy in appearance and are often used in floral displays.

A typical thin section through the lamina of a cycad leaf would contain an upper epidermis with a thick cuticle. A *palisade mesophyll* layer usually occurs beneath this epidermis (Fig. 24-4). This layer is composed of elongate, tightly compressed cells containing many chloroplasts. These cells are aligned perpendicularly to the epidermis. Most of the photosynthesis of the leaf occurs in this palisade region. Vascular bundles or veins occur between the palisade mesophyll and the *spongy mesophyll* which is directly beneath. This spongy tissue is composed of large, loosely arranged parenchyma cells with prominent intercellular spaces. These cells are photosynthetic although less so than those of the palisade layer. The spongy mesophyll is responsible for much of the exchange of gases between the leaf and the atmosphere. Gaseous exchange is essential for photosynthesis and occurs through the lower epidermis immediately beneath the spongy layer. The lower epidermis is highly cutinized, and the stomata are sunken beneath the layer of the other epidermal cells.

The cycad root system is a large simple or branched *tap root*. This root is nearly as large as the stem in some species, and it is sometimes difficult to determine where the stem begins and the root ends. Smaller roots often originate from the base of the stem or as branch roots from the tap root. These roots often grow on the surface of the ground forming a tangled mass. A zone within the cortex of these aerial roots often becomes filled with cells of the blue-green alga *Anabaena*. The relationship between the alga and the cycad is thought to be *mutualistic*. The cycad provide a protected habitat for the alga which in turn fixes nitrogen for use in the metabolic activities of the cycad.

LIFE HISTORY

Like all seed plants, cycads are heterosporous. Microsporangia and megasporangia are produced on sporophylls which are arranged into strobili or cones. Each strobilus is unisexual, and each plant produces only microsporangiate or megasporangiate cones.

Ovules are produced on leaf-like or highly modified, scale-like megasporophylls (Fig. 24-5). Two or more ovules are generally produced on each sporophyll. Depending on the nature of the megasporophyll, the female cone may resemble a group of tightly compacted, small leaves or a woody cone similar to that well known among conifers (Figs. 24-6 and 24-7).

The cycad ovule is characterized by a single thick integument with a small opening in its apex. This opening, or *micropyle,* allows pollination of the ovule (Fig. 24-8). The nucellus within the integument is a large, many celled tissue. A small chamber develops at the apex of the nucellus below the micropyle. It is the *pollen chamber* which is the site where pollen grains are ultimately deposited and germinate to produce a pollen tube.

A single megasporocyte develops within the nucellus. It divides meiotically to produce four megaspores. Three of these abort and the fourth divides rapidly to produce the megagametophyte plant. The initial divisions of this functional megaspore are so rapid that cell wall formation does not keep pace with nuclear division, resulting in the production of a mass of free haploid nuclei near the center of the nucellus (Figure 24-8). As this process continues, much of the nucellar tissue disintegrates to provide nourishment for the developing gametophyte plant. Cell wall

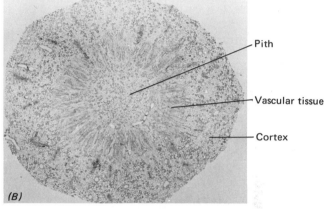

Figure 24-3. *Zamia* stems are illustrated in surface view and cross section. *A* shows a large branched stem and *B* is a cross section through a stem. Note the broad cortex and pith and the relatively small amount of vascular tissue in the stem at *B*.

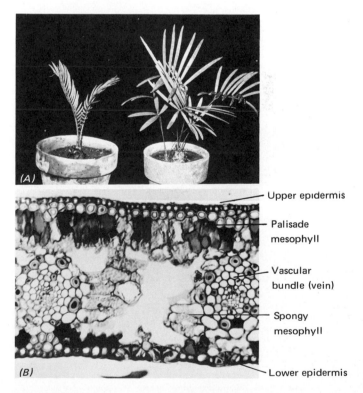

Figure 24-4. *Zamia* plants and leaf detail are shown in these two young plants (*A*) and cross section through a leaf (*B*).

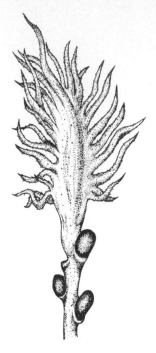

Figure 24-5. Megasporophylls of *Cycas* are more leaf-like than other cycads. The attached ovules are evident here.

    (A)   (B)

Figure 24-6. Microsporangiate cones of *Zamia* are long and slender and bear numerous microsporangia. *A* shows a section through a cone and *B* is a face view. Microsporangia are clearly evident in *A*.

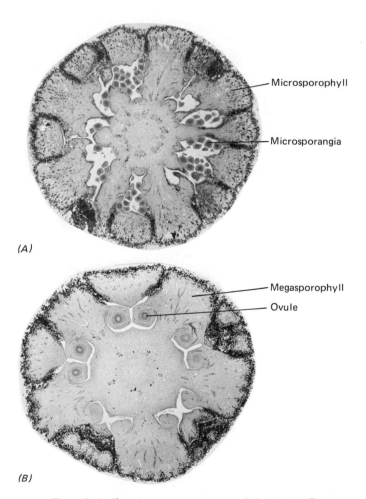

(A)

Microsporophyll

Microsporangia

(B)

Megasporophyll

Ovule

Figure 24-7. The microsporangiate cone of *Zamia* (A) illustrates microsporophylls with attached microsporangia. The megasporangiate cone (B) demonstrates megasporophylls with attached ovules.

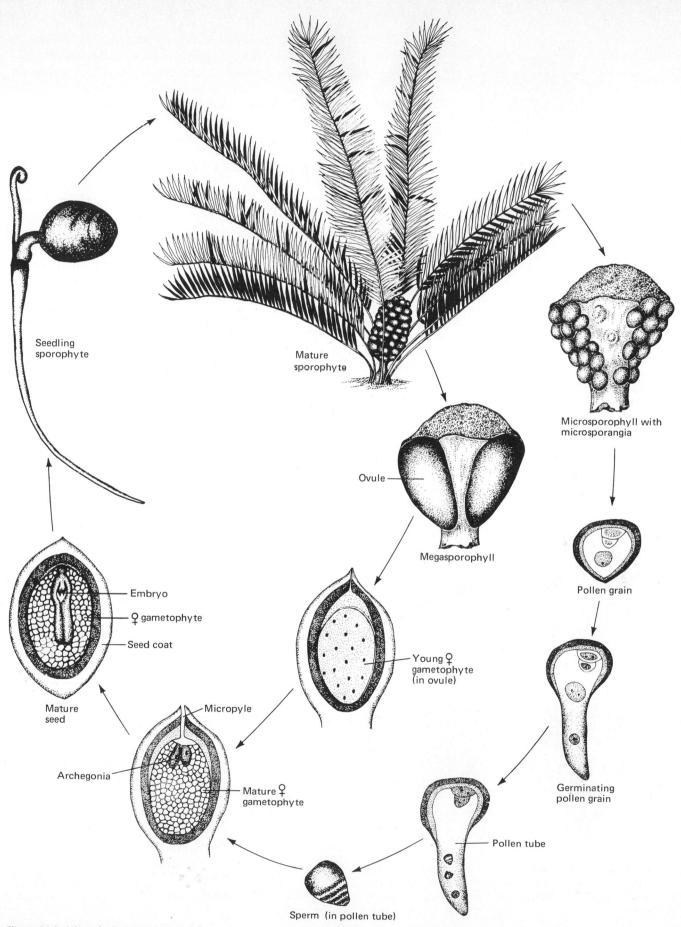

Seedling
sporophyte

Mature
sporophyte

Microsporophyll with
microsporangia

Ovule

Megasporophyll

Pollen grain

Embryo

♀ gametophyte

Seed coat

Young ♀
gametophyte
(in ovule)

Germinating
pollen grain

Mature
seed

Micropyle

Archegonia

Mature ♀
gametophyte

Pollen tube

Sperm (in pollen tube)

Figure 24-8. Life cycle diagram of a cycad.

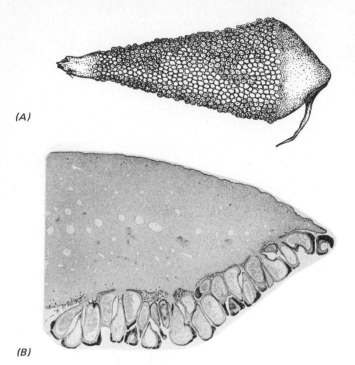

Figure 24-9. Microsporophylls of cycas are large and covered on their undersurfaces with microsporangia. A is a drawing of the undersurface of a sporophyll. B is a photograph of a cross section through a sporophyll illustrating the many microsporangia.

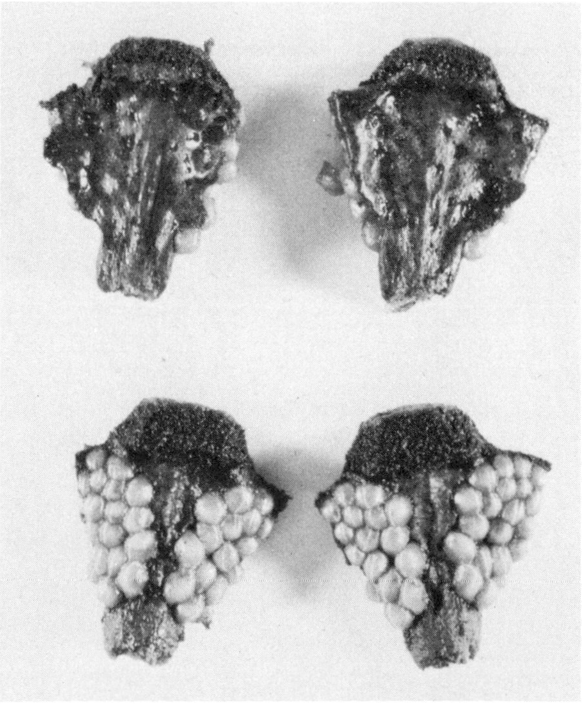

Figure 24-10. *Zamia* microsporophylls are smaller and more scale-like than those of *Cycas*. A shows a top view of these micro-sporophylls and B shows their undersurface.

formation eventually catches up with nuclear division to produce the mature megagametophyte plant that is composed of hundreds of cells. From two to several archegonia are produced at the apex of the megagametophyte plant nearest the pollen chamber. A large egg is produced in each archegonium when the female gametophyte plant is completely mature and ready for fertilization.

Microsporangia are produced on the under surface of microsporophylls (Figs. 24-9 and 24-10). They are smaller than ovules, and are produced in large numbers, often completely covering the lower sporophyll surface. Several microsporocytes are produced within each microsporangium. Each of these cells divides meiotically to produce four microspores. Each microspore remains viable and matures to become a pollen grain (Fig. 24-8). Pollen grains are produced in profusion by cycads, and are shed at maturity when the microsporangium ruptures. As the microsporangia open, the axis of the male cone elongates greatly to separate the sporangia so that pollen grains may be exposed to wind currents. This elongation is surprisingly rapid since the male cone may double its length in a day.

At the time the pollen grains are shed, a sticky droplet of a liquid known as a pollen droplet is extruded from the micropyle of the ovule. Since this droplet is extruded at the time of pollen release, many pollen grains are often trapped in it. After a few days or less, the pollen droplet dehydrates significantly and withdraws into the pollen chamber of the ovule. Everything adhering to this droplet is pulled into this chamber. In this manner pollen grains are introduced into the ovule. Pollen grains continue their development within the pollen chamber to become mature male gametophyte plants.

Often fungi take advantage of the cycad pollination procedure. Thus, fungal spores occasionally become attached to the pollen droplet and are pulled into the pollen chamber. These spores then germinate and attack the succulent tissues of the ovule, often destroying it.

A pollen tube develops as the pollen grain continues to mature. This tube grows through the remaining nucellus toward the megagametophyte plant. After a period of up to several months, the pollen tube eventually reaches the archegonium. Two large, multiflagellate sperm are then produced within the pollen grain. These sperm swim down the pollen tube to the archegonium where one of them fertilizes the egg. The resultant zygote develops in place within the megagametophyte plant to become an embryo (Figure 24-11). The integument matures concurrently to become a seed coat (Figure 24-12). Eventually the seed is shed from the female cone. They often have a bright-colored outer fleshy layer and are gathered and eaten by birds or rodents. When a seed reaches the proper environment, it germinates and the embryo continues development to become a new sporophyte plant.

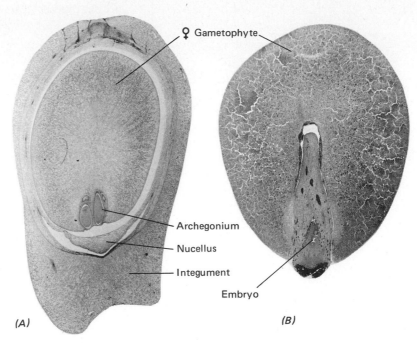

Figure 24-11. *Zamia* ovules in two stages of maturation are shown in these photographs. The ovule at *A* contains two archegonia and is ready for fertilization. The ovule at *B* has been fertilized and contains an embryo. The seed coat is absent at *B*.

♀ Gametophyte

Archegonium

Nucellus

Integument

Embryo

(A)

(B)

ECOLOGY

Cycads occur in tropical or subtropical regions of the earth. They may be prevalent in certain local areas but often are scattered among more prevalent flowering plants and ferns.

Even though cycads occur in tropical regions, they generally inhabit well-drained areas exposed to periodic dry periods. Thus, these plants are adapted to dry conditions by having thick, well-protected leaves with sunken stomata. Cycads are often slow growing and plants encountered in the field are often very old, even though they may be small.

OCCURRENCE AND ECOLOGY

The chief economic importance of cycads is as house and landscape plants. These plants are relatively important in nursery and floral businesses, especially in warmer parts of the world. Cycad leaves are often used in floral displays because of their large size and beautiful shape and color.

Most or all cycads are poisonous to humans and livestock if eaten directly, although the prospect of someone eating a cycad leaf is relatively poor. However, with special preparation, an edible, starchy product is prepared from the stems of some species. This is generally baked into small loaves and eaten. This food was commonly used by Seminole Indians and early white settlers in Florida and is often known as *Seminole bread*. Cycads are still used to a limited extent for this purpose, and Seminole bread may be purchased in some areas in Florida.

Some cycads were involved in coal formation during Mesozoic times. However, the relative importance of these plants in this role is probably much less than conifers.

Figure 24-12. The seed coat of a *Cycas* seed has an outer, brightly colored fleshy layer that attracts animals that aid in seed dispersal.

(A)

(B)

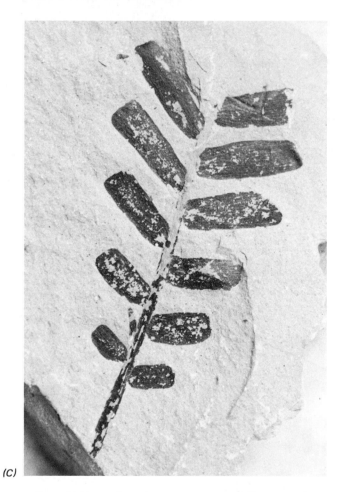

(C)

Figure 24-13. Fossil cycads are common in some Mesozoic rocks. A is a cross section through a petrified cycad-like stem. The xylem is evident at the left and a cone is shown at right center. B is a petrified cycad trunk collected from Maryland. C is a cycad leaf impression. (A is courtesy of Blaine Furness, Brigham Young University.)

SUMMARY

Cycads are gymnospermous seed plants. Ovules of these plants are large, with a single thick integument surrounding a massive nucellus. Pollen grains are drawn into the ovule at pollination by the dehydration of a pollen droplet. Sperm produced by the male gametophyte plant are large and multiflagellate. Fertilization occurs within the ovule, and the zygote divides to form an embryo that is dependent on the female gametophyte plant for its early nourishment.

Cycads inhabit well-drained, often dry habitats of the tropics or subtropics. They are important in nursery and floral businesses, and are responsible for the occasional poisoning of livestock.

FURTHER READING

ARNOLD, C. A., "Origin and Relationships of the Cycads," *Phytomorphology,* 3:51–65, 1953.

BECK, CHARLES, "The Appearance of Gymnospermous Structure," *Biological Reviews,* 45:379–400, 1970.

CHAMBERLAIN, C. J., *The Living Cycads,* University of Chicago Press, Chicago, 1919.

CHAMBERLAIN, C. J., *Gymnosperms: Structure and Evolution,* University of Chicago Press, Chicago, 1935. Also Dover Publishing Co., New York, 1966.

Also see the list of general references on vascular plants at the end of Chapter 19.

25 Division Ginkgophyta

Single leaf of *Ginkgo biloba* showing the characteristic pattern of veins.

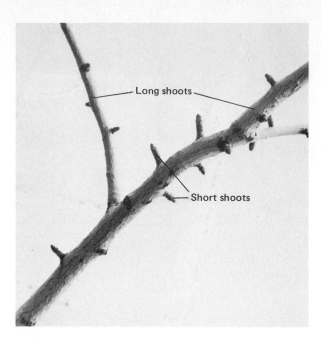

Figure 25-1. This branch of *Ginkgo biloba* shows long shoots and short shoots.

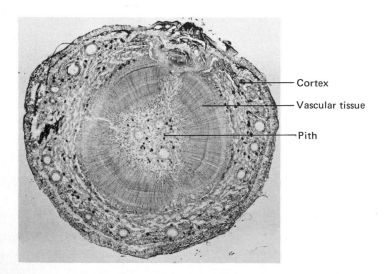

Figure 25-2. This cross section through a short shoot of *Ginkgo biloba* shows the arrangement of tissues.

The Ginkgophyta is an ancient plant division that has been in existence since Late Carboniferous times. These plants were most prevalent during the Mesozoic Era when they were common in many areas of the world. The genus *Ginkgo* originated during early Mesozoic times. This genus still exists as *Ginkgo biloba,* which is the sole living representative of the entire division.

Many Ginkgophyta were well known from the fossil record before the discovery of living *Ginkgo biloba* in China. This species probably would have become extinct as did all of its relatives except that it is a beautiful ornamental tree which was careful cultivated by the Chinese for thousands of years.

GENERAL CHARACTERISTICS

Ginkgo biloba is a medium-sized tree which grows to a maximum of 100 ft tall. It consists of a central stem or trunk with lateral or side branches. These branches originate from buds in the axils of leaves and are of two types. *Long shoots* are relatively long leaf-bearing branches (Fig. 25-1). *Short shoots* or *spur shoots* (Figs. 25-1 and 25-2) are much shorter than long shoots since they have greatly shortened internodes. They produce numerous leaves and reproductive structures and develop along the long shoots.

Leaves are basically fan-shaped with long petioles (Fig. 25-3), although they differ somewhat in shape according to where they occur. Those produced on long shoots are deeply lobed, while those on short shoots are nearly entirely or only slightly lobed. Leaf veins are prominent and dichotomously branched (Fig. 25-3). Their prominence in the lamina makes the leaf very beautiful and gives the plant the common name "maidenhair" tree.

Leaves of fossil *Ginkgophyta* are common from Mesozoic strata. These fossil leaves are more deeply dissected than those of the modern *Ginkgo biloba*. A direct evolutionary line apparently exists from species with long, thin leaf divisions to those with fewer and shorter leaf divisions and then to the modern *Ginkgo biloba* with nearly fan-shaped leaves.

The root system of *Ginkgo biloba* is large and well developed. It is similar to the stem in external form since it is composed of a central axis with many lateral branches. These branch roots range from very large to very small. The smallest are produced in profusion and absorb most of the water and mineral nutrients used by the plant.

LIFE HISTORY

Ovules are produced in pairs at the tip of a short stalk which arises from a spur shoot of a female tree (Fig. 25-4). Each ovule is large, with a single, thick integument surrounding a massive nucellus (Fig. 25-5). A micropyle occurs at the tip of the ovule, and a pollen chamber is formed in the nucellus immediately inside of the micro-

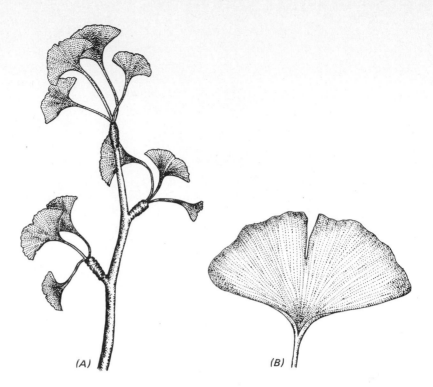

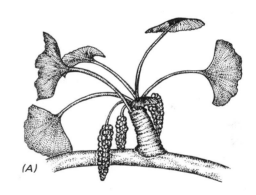

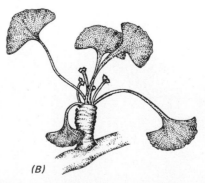

Figure 25-3. A long shoot of *Ginkgo biloba* with attached short shoots is shown at *A*. Note that most leaves occur on the short shoots. *B* is a single leaf showing the characteristic vein pattern.

(A)

(B)

pyle. A single megaspore mother cell differentiates in the nucellus and divides to produce four megaspores. Three of these abort, and the fourth divides to produce the mega-gametophyte plant. This development is similar to that among cycads, in that it is free nuclear at first and then cellular. The megagametophyte plant is large and well developed and generally produces two archegonia at its apex. The female gametophyte plant continues development after pollination occurs but prior to fertilization. Microspores are produced prior to the maturation of the megagametophyte plant. They are produced in micro-sporangia aggregated into strobili which develop on short shoots of male trees (Fig. 25-4). Each microstrobilus has a central axis with sporangia devolping on attached sporangiophores (Fig. 25-6). Microspores mature within the sporangium to become pollen grains. These are shed in the early spring prior to the full development of leaves. Pollen grains are carried by wind currents until one or more of them eventually is carried to the ovule and introduced into the pollen chamber. The pollen grain germinates to produce a pollen tube that grows through the nucellus toward the archegonium. Eventually two large, multiflagellate sperm are produced and the egg is fertilized.

Early development of the embryo proceeds by free nuclear division followed by simultaneous cell wall formation (Fig. 25-7). The embryo is dependent on the female gametophyte tissue for its early nourishment.

The seed coat is divided into two distinct layers (Fig. 25-8). The outer layer is soft and fleshy while the inner layer is hard and resistant, functioning well in protecting the embryo. The fleshy layer contains a high concentration of lactic acid which creates a favorable growth substrate

Figure 25-4. Short shoots of *Ginkgo biloba* bear the reproductive structures. The shoot at *A* bears mature male strobili and the shoot at *B* bears several pairs of immature ovules.

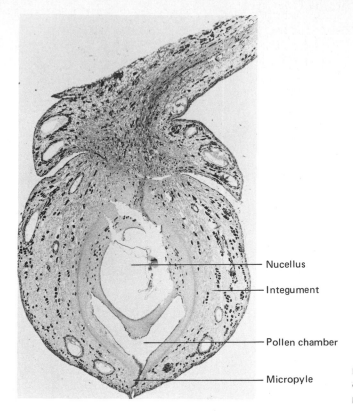

Nucellus

Integument

Pollen chamber

Micropyle

Figure 25-5. This longitudinal section through an ovule of *Ginkgo biloba* demonstrates the principle components of the ovule prior to fertilization.

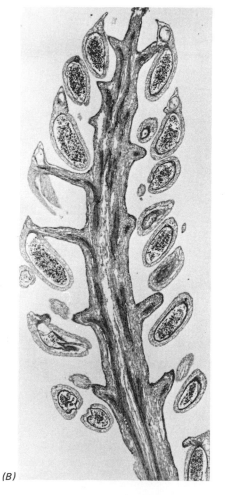

(A)

(B)

Figure 25-6. *Ginkgo biloba* male strobili are illustrated here. *A* shows strobili and immature leaves at the tip of a short shoot. *B* is a longitudinal section through a male strobilus showing the microsporangia attached to sporophylls.

for the many bacteria that soon inhabit it. As the bacteria decompose this fleshy layer, a miserable odor is produced that has been described as resembling sour milk, raw sewage, spoiled meat, or almost any other foul smell. Because of this odor, the female tree is rarely planted, or is often quickly disposed of if planted by mistake. Only a hardy botanist would want a female *Ginkgo* tree in the yard, and even then the neighbors would probably have it removed when the botanist was at the laboratory. It has been postulated that this bacterial action in the outer seed coat has an adaptive value since it raises the temperature of the seed somewhat and thus facilitates the continued development of the embryo.

OCCURRENCE AND ECOLOGY

Ginkgo biloba probably occurred naturally under cool temperate to warm temperate conditions. It is presently cultivated throughout the world in a wide variety of habitats. This plant is hardy to temperatures around −30°F, although the reproductive cycle is seldom completed in colder habitats.

IMPORTANCE

At one time it was hoped that *Ginkgo* would provide a valuable source of lumber. However, this proved impractical because of its slow growth and brittle wood.

Ginkgo is currently grown commercially in the Orient for seeds that provide a locally desirable source of food. *Ginkgo* is also a significant nursery plant since a large amount of money is spent annually to develop new varieties and to purchase these trees for landscape planting.

SUMMARY

The Ginkgophyta is an ancient division of large, gymnospermous trees. This division was more abundant in the geologic past than at present and is now represented only by the single species *Ginkgo biloba*. Even this tree probably would have become extinct if it had not been cultivated for its beauty by Chinese gardeners for several centuries.

Ginkgo trees are unisexual, producing either male or female reproductive structures on short shoots among the leaves. Ovules are not produced within a strobilus, although microsporangia are. The life history of *Ginkgo biloba* is similar in many respects to that of a cycad.

FURTHER READING

Li, Hui-lin, "A Horticultural and Botanical History of *Ginkgo,*" *Bulletin of the Morris Arboretum,* 7:3–12, 1956.

Seward, A. C., "The Story of the Maidenhair Tree," *Science Progress,* 32:420–440, 1938.

Also see the list of general references on vascular plants at the end of Chapter 19.

Figure 25-7. This longitudinal section through the megagametophyte of *Ginkgo* illustrates an imbedded embryo.

(A)   (B)

Figure 25-8. Two mature seeds of *Ginkgo* are shown here. The seed at A has the outer fleshy layer of the seed coat intact. This layer has been removed from the seed at B.

26 Division Gnetophyta

Closeup of several stems of *Ephedra* showing the tiny leaves.

The division Gnetophyta is one of the most interesting but problematical groups of living gymnosperms. These plants are considered to be only distantly related to any other plants presently living on earth, and several characteristics of Gnetophyta are unique.

Only three genera of plants, *Gnetum, Ephedra* and *Welwitschia,* are presently placed in this division. These genera are not closely related to each other and so are placed in separate orders. However, they do exhibit several similar characteristics that align them together in the same division.

GENERAL CHARACTERISTICS

Several features of Gnetophyta are unique among all gymnosperms. One of the most interesting of these is that all three genera produce vessel elements in their xylem (Fig. 26-1). Most flowering plants contain vessel elements as well, and this has led some botanists to suggest that *Gnetophyta* might have been ancestral to the angiosperms.

Another feature common to both Gnetophyta and angiosperms is the production of an ovule with two integuments rather than one. This is also characteristic of a few other gymnosperms, but together with the vessel elements in the xylem, has reinforced the idea of some botanists that Gnetophyta gave rise to the flowering plants. However, Gnetophyta differ from angiosperms in so many other important characteristics, particularly in reproductive morphology, that this theory does not seem plausible and is rejected by nearly all modern botanists.

The three genera of Gnetophyta are very different in appearance. *Ephedra* is a highly branched shrub (Fig. 26-2) which superficially resembles a shrubby *Equisetum* since it has highly reduced leaves and jointed stems. *Gnetum,* on the other hand, is a vine or small tree resembling a flowering plant in outward appearance. *Welwitschia* does not resemble any other plant on earth. It has a short, thick, disk-shaped stem with two long leaves attached at its edge (Fig. 26-3). This stem efficiently stores water which is scarce in the African desert where *Welwitschia* occurs. The reproductive structures of *Welwitschia* occur along the edge of the stem with the leaves. The total plant is certainly odd in appearance.

The leaves of the three genera of Gnetophyta differ widely from each other. However, all are produced opposite each other along the stem (opposite arrangement). This is uncommon among gymnosperms, but is not restricted to Gnetophyta. *Gnetum* produces rather large, oval-shaped leaves with reticulate venation much like those of some flowering plants. *Ephedra* has very small, inconspicuous, scale-like leaves. These are produced along green stems that carry on the majority of photosynthesis of the plant. *Welwitschia* has two large, strap-shaped

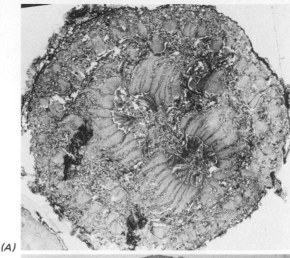

(A)

(B)

Figure 26-1. Stems of Gnetophyta illustrate unusual features in their anatomy. A young stem of *Welwitschia* is illustrated at *A* and *Ephedra* is illustrated at *B*. Note the conspicuous vessel elements in the *Ephedra* wood which are missing from other gymnosperms.

Figure 26-2. This *Ephedra* bush was photographed growing on the southeastern Utah desert.

Figure 26-3. *Welwitshia* is an unusual plant that grows only in the desert along the southwestern coast of Africa. (Courtesy of Macmillan Science Company—Turtox/Cambosco.)

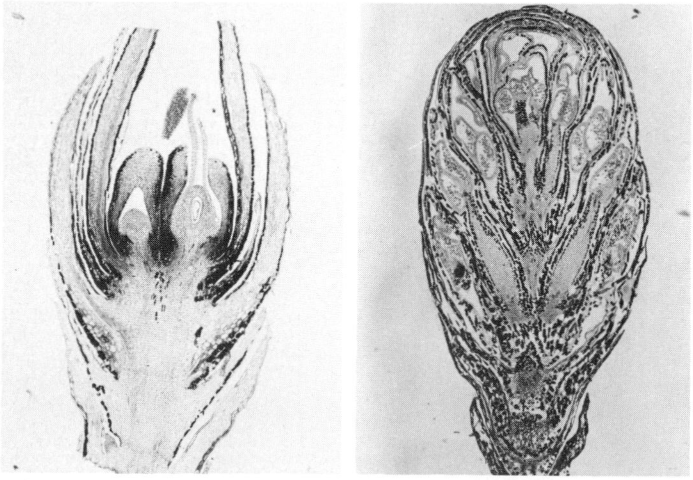

(A)

(B)

Figure 26-4. Longitudinal sections of *Ephedra* cones are shown here. *A* is a female cone exhibiting two ovules. *B* is a male cone demonstrating several microsporangia.

leaves that may reach several feet in length. These leaves originate from the embryo and last throughout the life of the plant which may be over a hundred years.

Gnetophyta produce ovules in which the inner integument forms a long tube at the apex of the ovule. This *micropylar tube* or *pollination tube* extends well beyond the outer integument and sterile protective *bracts* surrounding the ovule (Fig. 26-4). A pollen droplet is formed in the ovule and extruded from this tube. Pollen grains adhere to this droplet and are pulled into the pollen chamber.

The male cone of Gnetophyta is composed of a central axis with attached bracts and sporangiophores. Microsporangia are produced at the tips of the sporangiophores (Figs. 26-5 and 26-6). Since this cone produces bracts as well as sporangiophores, it is a compound cone. Only Gnetophyta among all gymnosperms produce compound male cones. The female cone of Gnetophyta also is compound because it also contains protective bracts. However, this is common among conifers as well.

Another feature unique among gymnosperms occurs in *Gnetum* and *Welwitschia* but not in *Ephedra*. A naked egg is produced by the megagametophyte in the former two genera rather than enclosed within an archegonium as in most gymnosperms. This is another feature that is similar to flowering plants and is thought to be advanced.

The life history of Gnetophyta is similar to that of cycads and *Ginkgo*.

OCCURRENCE AND ECOLOGY

Gnetum is a relatively small genus containing about thirty species. These plants are restricted to the tropics, particularly rain forests. Most species of *Gnetum* are epiphytic vines, although a few are tree-like.

Ephedra also contains about thirty different species. These plants are restricted to dry or arid habitats to which they are well suited with their reduced leaves and photosynthetic stems. Southwestern America is one common site of growth of *Ephedra*. This plant was widely used by the American Indians and early settlers of that area to make a beverage. Thus, *Ephedra* is often referred to as *Mormon tea* because of its use by the early Mormon pioneers.

Welwitschia contains only a single species restricted to southwestern deserts of Africa.

IMPORTANCE

The division Gnetophyta is relatively unimportant to humans although several uses of these plants have been made. *Gnetum* is occasionally cultivated as an ornamental plant, especially those species that are small trees. *Ephedra* is still used to some extent as a mild beverage, and the drug ephedrine is extracted from this plant as well. This drug is used in nose drops and inhalants for the treatment of hay fever and colds since it causes the shrinkage of mucous membranes in nasal passages.

SUMMARY

The division Gnetophyta is composed of the three genera *Gnetum, Welwitschia,* and *Ephedra.* These three genera are very different in outward appearance but are similar in many important respects. Thus, the ovule has two integuments rather than one, and the inner one forms a micropylar tube that aids in pollination. In addition, vessel elements occur in the xylem of these plants. Both male and female cones are compound since they produce sterile protective structures or bracts. The female gametophyte plant of *Gnetum* and *Welwitschia* produces an egg directly rather than in an archegonium.

FURTHER READING

CHAMBERLAIN, C. J., *Gymnosperms: Structure and Evolution,* University of Chicago Press, Chicago, 1935. Also Dover Publishing Co., New York, 1966.

EAMES, ARTHUR, "Relationships of the Ephedrales," *Phytomorphology,* 2:79–100, 1952.

Also see the list of general references on vascular plants at the end of Chapter 19.

(A)

(B)

Figure 26-5. Male and female cones of *Ephedra* are unusual in construction. *A* is a stem with four attached female cones. *B* is a stem with several male cones attached.

Figure 26-6. Female cones of *Gnetum* shown here are produced in whorls along the stem.

27 Division Coniferophyta: the Conifers

Spruce trees near the timber line in the Wasatch Mountains of Utah.

The Coniferophyta is the largest and most economically important division of gymnosperms. Conifers are familiar to most people since they include the redwoods, pines, firs, spruces, Douglas fir, larch, junipers and many other common trees (Fig. 27-1).

GENERAL CHARACTERISTICS

The concept of elongation of a plant by the addition of new tissues from an apical meristem has been discussed in earlier chapters. All higher plants exhibit this type of growth when stems and roots elongate, when branches develop, and when new leaves are formed. Growth of this type is known as *primary growth,* and the tissues formed by primary growth are primary tissues. In addition to primary growth, a few of the lower vascular plants, most gymnosperms including all conifers, and many flowering plants demonstrate *secondary growth.* This type of growth causes an increase in diameter of the stem or root and is responsible for the tremendous girth of large trees. Characteristics of primary stems and roots are quite different from those of secondary stems and roots.

The primary stem of a conifer has an outer epidermis with scattered stomata. It develops from and is continuous with the outermost cells of the apical meristem. This tissue has an outer cuticle that aids in protecting the plant against water loss. The cortex is immediately inside the epidermis. This tissue is composed of large parenchyma cells that are often photosynthetic. Some of the cells within the cortex may become filled with resins, and *resin ducts* are often found in this tissue as well. Resin ducts are hollow tubes lined by specialized cells which secrete resins into the duct. The function of these resins is not entirely known, although they probably serve as a protective mechanism to the plant by sealing wounds and discouraging insect attack.

The eustele of a conifer is composed of a ring of separate vascular bundles surrounding a central pith. Xylem occurs toward the center of the stem in each bundle and phloem occurs outwardly. A region of meristematic cells occurs between the xylem and phloem in these bundles. Those cells form the *vascular cambium* that is responsible for producing secondary vascular tissues. The pith is composed of large, rounded parenchyma cells similar to those of the cortex.

Primary growth of a stem or root is responsible for elongation of that organ. Thus, for any given point on a stem or root, primary tissues are deposited by the apical meristem only during the first year, or during part of the first year of growth. All further development of the organ at that point occurs by secondary growth. Therefore, if you were to pound a nail into a young tree trunk, the nail would always stay at the same height because elongation of the stem occurs only during the first year.

Secondary growth begins when the vascular cambium becomes active and begins to produce new xylem cells to-

Figure 27-1. Conifers are mostly large and tree-like, though they differ in appearance. *A* is blue spruce (*Picea pungens*); *B* is black pine (*Pinus nigra*); *C* is giant redwood (*Sequoiadendron giganteum*); and *D* is bald cyprus (*Taxodium distichum*). (*C* and *D* are courtesy of Walter Hodge.)

ward the inside and phloem cells toward the outside of the stem. As this occurs, some of the parenchyma cells between the separate vascular bundles become meristematic and also produce secondary xylem and phloem. These cells connect the vascular cambia of the separate vascular bundles so that secondary xylem and phloem are produced in a ring throughout the entire circumference of the stem rather than only in the individual bundles (Fig. 27-2). The result of such secondary growth is that the primary eustele is soon obliterated and the vascular tissue of the plant occupies essentially the entire interior of the stem (Fig. 27-2).

Xylem and phloem are produced by the vascular cambium at approximately the same rate. However, the cells of the xylem are much more resistant to damage or compression than are the cells of the phloem. Consequently, phloem cells are continually compressed and destroyed after a relatively short functional period, and only a thin layer of functional phloem surrounds the cambium. Since the cells of the xylem are thick walled and resistant, they are not compressed. Thus, xylem produced by the secondary growth of one year is added to that formed the previous year so that the stem continues to increase in diameter by yearly increments.

Secondary xylem cells are normally larger and are produced more rapidly during the spring of the growing season. Conversely, xylem cells produced during the summer months are smaller in diameter and have relatively thicker walls. Because of this, when a cross section of a conifer stem is examined carefully, these regions of thick and thin walled cells are evident as *annual rings* (Figs. 27-3 and 27-4). The age of a conifer may be determined by counting each light and dark colored band of xylem and dividing by two since one light colored ring of large, thin walled cells and one dark colored ring of smaller, thick walled cells are produced during a single growing season.

It is striking to examine the stump of an ancient redwood tree or other large conifer and study the annual rings. This can give a person an interesting historical perspective since, with a little counting, the actual annual ring can be found that was formed during the year when the American Civil War began, or the Revolutionary War ended, or Columbus discovered America, and so on even far back into the Middle Ages. This simple exercise seems to stimulate a new reverence for these giant trees, and to emphasize the short life span of human beings and our great responsibility for conserving our irreplaceable natural resources. It is also possible by studying the annual rings in detail to learn a great deal about the environmental conditions present during the time the tree was growing (Fig. 27-5). This is particularly helpful in reconstructing the climate of the past by studying tree rings of fossil trees of known age.

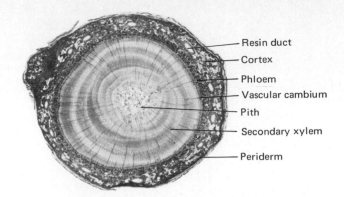

Resin duct
Cortex
Phloem
Vascular cambium
Pith
Secondary xylem
Periderm

Figure 27-2. Secondary growth is the process that increases the diameter of plants. A cross section of a young *Pinus* stem illustrating early secondary growth is illustrated here.

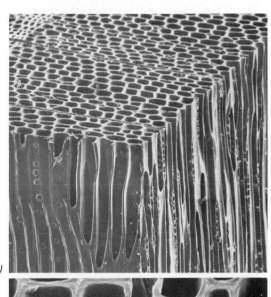

(A)

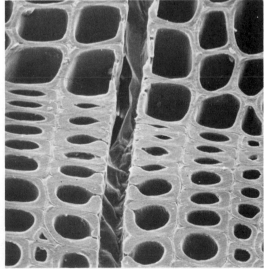

(B)

Figure 27-3. These scanning electron micrographs show the wood of *Pinus*. A three dimensional view is presented at A. Note the pits on the sides of the tracheids. B is a closeup view of an annual growth ring. Note the thick walls of the summer tracheids and the relatively thin walls of the spring tracheids. (Photographs courtesy of B. A. Meylan and B. G. Butterfield from *Three Dimensional Structure of Wood*, Syracuse University Press, Syracuse, N. Y., 1972. Used by permission of Syracuse University Press.)

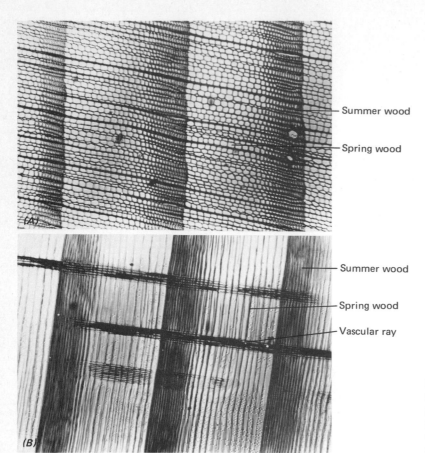

Summer wood

Spring wood

Summer wood

Spring wood

Vascular ray

(A)

(B)

Figure 27-4. The wood of *Pinus* is shown in these photographs taken with the light microscope. *A* is a cross section through a *Pinus* stem. Note the bands of spring and summer wood. *B* is a longitudinal section through a *Pinus* stem. Note the pits on the walls of the tracheids. Also note the vascular rays that function in lateral conduction in the stem.

As the stem increases in diameter with the addition of new vascular tissues, the epidermis and cortex do not keep pace with this expansion and are sloughed off. A second cambium develops in the cortex or phloem that produces a secondary protective tissue which replaces these primary tissues. This *periderm* tissue is composed largely of cork cells that help to protect the plant against the infestation of insects and fungi and against water loss (Fig. 27-6). The periderm together with the secondary phloem which becomes crushed each year form the *bark* of the tree. Periderm is often produced in unequal amounts around the stem resulting in bark with a roughened texture. On the other hand, some trees produce a relatively smooth bark. The characteristic pattern and color of this tissue is often helpful in the identification of woody plants (Fig. 27-7).

Conifer roots are protostelic throughout their development. They grow by primary and secondary growth similar to the stem and are often quite large. A periderm also is produced by the root to replace the cortex and epidermis as it increases in diameter. It is often superficially difficult to differentiate a secondary stem from a secondary root. The roots that absorb water and mineral nutrients from the soil are the small, often microscopic rootlets occurring at the tips of the branching root system. These roots are

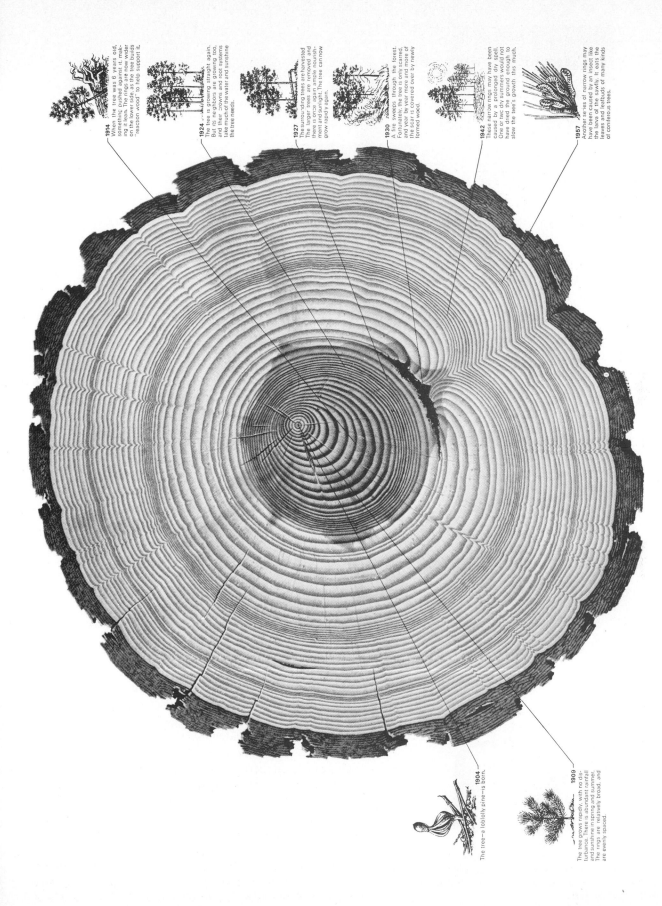

1914 When the tree was 6 years old, something pushed against it, making it lean. The rings are now wider on the lower side, as the tree builds "reaction wood" to help support it.

1924 The tree is growing straight again. But its neighbors are growing too, and their crowns and root systems take much of the water and sunshine the tree needs.

1927 The surrounding trees are harvested. The larger trees are removed and there is once again ample nourishment and sunlight. The tree can now grow rapidly again.

1930 A fire sweeps through the forest. Fortunately, the tree is only scarred, and year by year more and more of the scar is covered over by newly formed wood.

1942 These narrow rings may have been caused by a prolonged dry spell. One or two dry summers would not have dried the ground enough to slow the tree's growth this much.

1957 Another series of narrow rings may have been caused by an insect like the larva of the sawfly. It eats the leaves and leafbuds of many kinds of coniferous trees.

1904 The tree—a loblolly pine—is born.

1909 The tree grows rapidly, with no disturbance. There is abundant rainfall and sunshine in spring and summer. The rings are relatively broad, and are evenly spaced.

Figure 27-5. The study of the yearly growth rings of a tree can yield valuable information concerning its environment. (Courtesy of St. Regis Paper Company.)

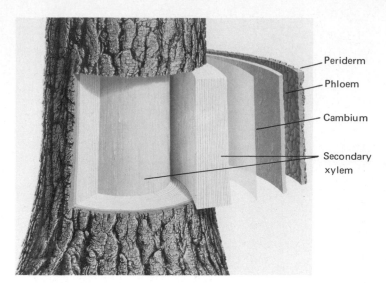

Figure 27-6. This diagram of the tissues in a conifer stem shows the arrangement of tissues following secondary growth. (Courtesy of St. Regis Paper Company.)

The labels on the figure read:
- Periderm
- Phloem
- Cambium
- Secondary xylem

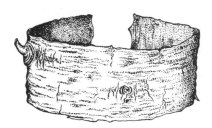

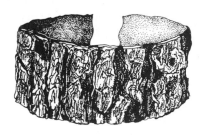

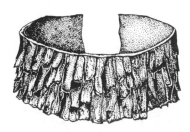

Figure 27-7. These drawings show the bark patterns of three conifers. This bark pattern is often so characteristic that it is helpful in conifer identificaton.

often mycorrhizal, and the fungi associated with conifer roots are often very host specific.

Conifer leaves are generally quite small and needle- or awl-shaped, although some are broad and flattened (Fig. 27-8). These leaves may or may not be produced on specialized short shoots or branches along the stem. They are produced singly in several species including firs and spruces or in groups of two to five in most species of *Pinus*. The number of leaves per group is an important characteristic for conifer identification.

The internal structure of some conifer leaves is somewhat similar to that of a cycad. The outermost layer is the epidermis (Fig. 27-9). This layer generally has a thick cuticle and contains stomata that are sunken beneath the level of the other epidermal cells. A region of thick-walled cells known as the *hypodermis* often occurs immediately beneath the epidermis (Fig. 27-9). The hypodermis lends strength and rigidity to the leaf. A mesophyll composed of thin-walled parenchyma cells containing many chloroplasts is present inside the hypodermis. The cells of the mesophyll may be tightly arranged or rather loose and irregularly placed. The bulk of photosynthesis for the entire plant occurs in this leaf region. One or two leaf bundles or veins occurs near the center of the leaf. These are composed of a cap of xylem tissue oriented toward the upper leaf surface with a region of phloem beneath. These vascular bundles are imbedded in a tissue of thin-walled parenchyma cells that is in turn surrounded by a cylinder of thick-walled endodermal cells.

Many of the characteristics of the conifer leaf are designed to aid the plant against excess water loss, and are even similar to some characteristics of the leaves of desert plants. This perhaps seems odd since many conifers grow in regions where there is apparently abundant moisture. However, most conifers bear their leaves throughout the year, and of course water is not available to the plant during winter months in cold regions since it freezes in the

(A) (B)

Figure 27-8. Most conifers produce needle-shaped leaves, although there are several exceptions to this rule as shown in these photographs. *A* is *Podocarpus,* which has strap-shaped leaves, and *B* is *Auricaria* with awl-shaped leaves.

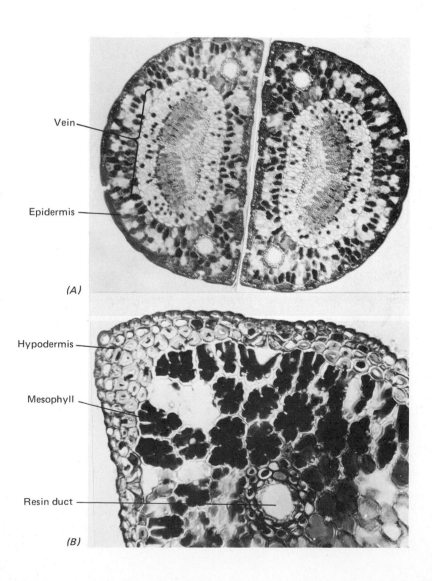

Vein

Epidermis

(A)

Hypodermis

Mesophyll

Resin duct

(B)

Figure 27-9. The structure of a typical needle-shaped conifer leaf is shown in these photographs of *Pinus* leaf. *A* shows a low power magnification of two leaves, and *B* is a closeup of the edge of a single leaf illustrating the hypodermis.

(A)

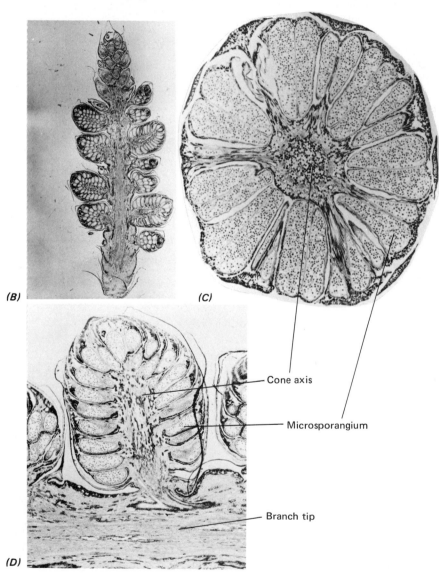

(B)

(C)

(D)

Cone axis

Microsporangium

Branch tip

Figure 27-10. Male cones of *Pinus* are shown in these photographs. *A* is a branch tip with several male cones. *B* is a longitudinal section through a similar branch tip. *C* is a longitudinal section through a single male cone. *D* is a cross section through a single cone.

ground. The drought-resistant characteristics of conifer leaves are necessary to protect the plant during such conditions and represent an important adaptation to the environment.

THE CONIFER LIFE HISTORY

Conifer strobili are unisexual in all species, and are formed at the tips of short lateral shoots. Male and female cones often are produced on the same plant, although in some they develop on separate plants. When both sexes are produced on the same plant, the species is *monoecious*. When they are produced on separate plants, the species is *dioecious*. In many monoecious conifers, the male and female cones are produced on different parts of the plant. Male and female cones on the same plant often mature at different times to encourage sexual reproduction between separate parents so that maximum genetic recombination is encouraged.

The male cone is simple in construction. It is composed of a central axis with microsporophylls attached in a tight spiral arrangement (Fig. 27-10). Microsporangia are attached to the lower surface of these sporophylls. Each microsporangium may produce hundreds of microspores. One or two air bladders are often formed on these microspores as they mature to become pollen grains (Fig. 27-11). These bladders give the pollen grain buoyancy in the air which is essential since conifers are wind pollinated. Conifer pollen often travels hundreds of miles on air currents because of the efficiency of these bladders. It is not uncommon to walk in the forest during the spring and find the entire ground yellow and the air containing a fine yellow powder. This occurs when conifer pollen is shed from the male cones. During this time, if you shake or brush against one of these gymnosperms, great clouds of pollen may be released, coloring your clothing and hair.

Immature female cones are often slightly larger than male cones. Unlike male cones, these megastrobili are compound structures. A central axis is present similar to that of the male cone. Ovule-bearing structures are often called *cone scales* or *ovuliferous scales* (Fig. 27-12). These structures are usually large and conspicuous, comprising the bulk of the tissue of the cone. They differ widely in shape and size among different conifer species (Fig. 27-13). Bracts occur between successive cone scales. These are generally small and inconspicuous and do not seem to have a function. They are probably evolutionary remnants of larger, much more conspicuous bracts that occurred in some of the early conifers known only from the fossil record.

The arrangement of cone scales and bracts is used in the identification of conifers. For instance, *Pseudotsuga*, or Douglas fir, exhibits a three-pointed bract that extends beyond the cone scales and is very evident when the cone

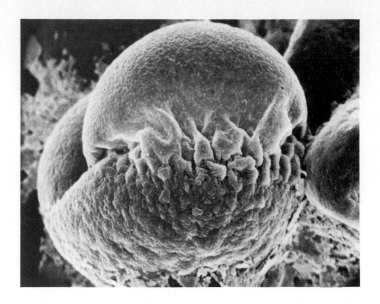

Figure 27-11. Pollen grains of *Pinus* have two air bladders shown here oriented toward the top of the photograph. These are hollow, allowing the grain to be buoyant on air currents and thus aiding in wind dissemination. (Scanning electron micrograph courtesy of James V. Allen, Brigham Young University.)

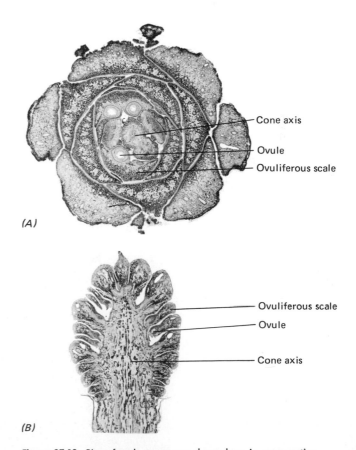

(A)

(B)

Figure 27-12. *Pinus* female cones are shown here in cross section (A) and longitudinal section (B).

(A) Digger pine
(*Pinus sabiniana*)

(B) Sugar pine
(*Pinus lambertiana*)

(C) Larch (*Larix* sp.)

(D) Pinyon pine
(*Pinus edulis*)

(E) Southern hemisphere pine
(*Auracaria* sp.)

(F) Douglas fir
(*Pseudotsuga taxifola*)

Figure 27-13. Cones of conifers are distinctive enough to be used in classification. Several cones are shown in these photographs.

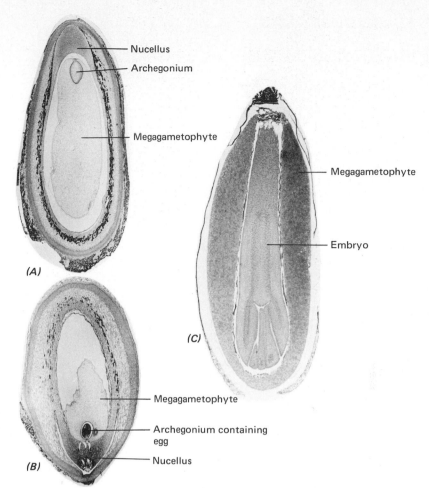

Nucellus

Archegonium

Megagametophyte

(A)

Megagametophyte

Embryo

(C)

Megagametophyte

Archegonium containing egg

Nucellus

(B)

Figure 27-14. These longitudinal sections through the ovules of *Pinus* show various stages in maturation. *A* shows an ovule containing a mature female gameto-phyte (megagametophyte) with an apical archegonium. The ovule at *B* shows fertilization. *C* is the internal por-tion of a mature ovule (seed) showing the embryo im-bedded in the megagametophyte tissue.

is examined (Fig. 27-13*F*). This is the only conifer with such an arrangement of bracts, and so may be easily iden-tified. In addition, several other conifers show characteris-tic arrangement of the bracts.

The typical conifer ovule is similar in many respects to that of a cycad (Fig. 27-14). A single large integument surrounds a many-celled nucellus. A single megasporocyte develops within the nucellus and divides to produce a tetrad of megaspores. Three of these abort and the fourth divides rapidly to produce a free nuclear megagametophyte. The formation of cell walls soon follows and the female gametophyte matures with the formation of one or more archegonia (Fig. 27-14). These archegonia each contain a single large egg.

Pollen grains are released at maturity and are blown about passively. Some eventually contact a female cone where they may be sifted down to the ovules. Pollen grains continue development in the ovule by the formation of pollen tubes that grow through the remainder of the nucel-lus toward the female gametophyte (Fig. 27-15). A pollen tube may take from a few weeks to over a year to reach the female gametophyte among the different conifer species. When the tip of the tube reaches the megagametophyte, two *nonmotile* sperm are released and one fertilizes the

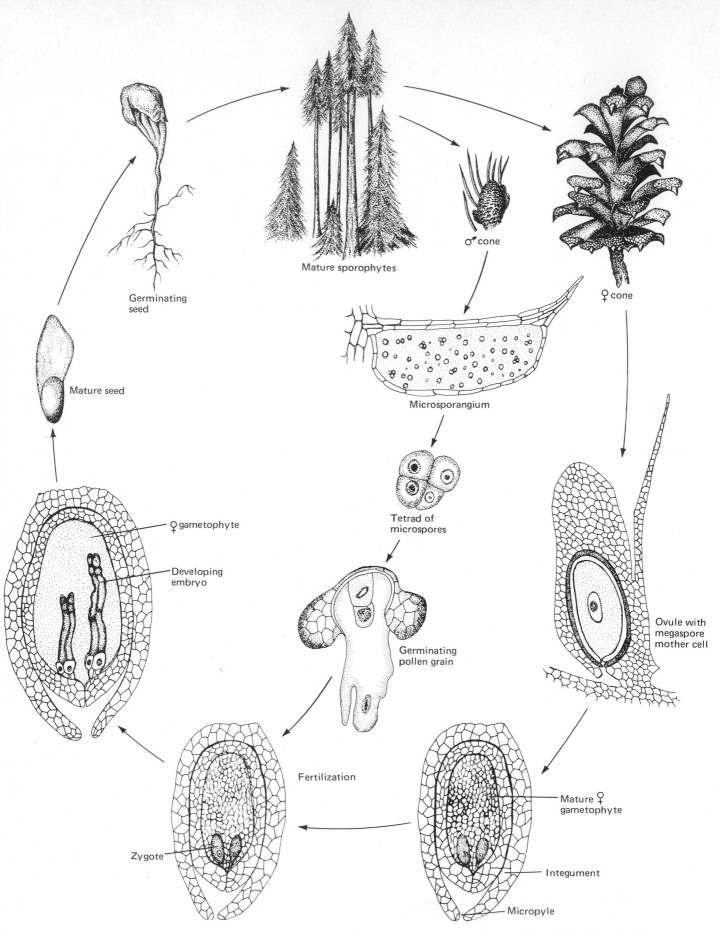

Germinating seed

Mature sporophytes

♂ cone

♀ cone

Microsporangium

Tetrad of microspores

Mature seed

Germinating pollen grain

Ovule with megaspore mother cell

♀ gametophyte

Developing embryo

Fertilization

Zygote

Mature ♀ gametophyte

Integument

Micropyle

Figure 27-15. Life cycle diagram of *Pinus*.

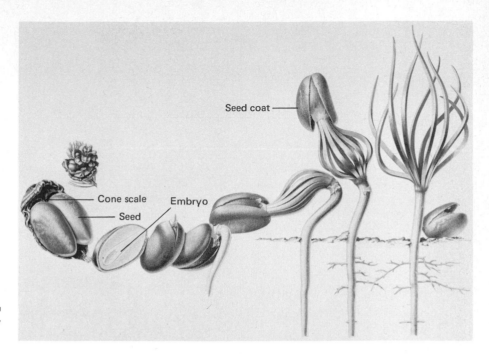

Figure 27-16. This diagram shows germination of a pine seed to produce a young sporophyte. (Courtesy of St. Regis Paper Company.)

Seed coat

Cone scale

Seed

Embryo

egg. This process may occur more than once in each ovule resulting in more than one egg becoming fertilized. However, as the zygotes develop into embryos, generally all but one die so that each seed usually contains only one mature embryo. As this process occurs, the female gametophyte tissue is used to nourish the embryo. The integument matures concurrently to form the seed coat. This seed coat differs among the various conifer seeds according to hardness and structure and is often characteristic.

The female cone of some species disintegrates as the seeds mature or during the following winter to disseminate the seeds. Other conifers produce resistant cones that may take several years to decay to release their seeds. Still others do not open until they are heated to a rather high temperature. Thus, in nature, the seeds of these species are not released until the tree is exposed to a forest fire which insures reseeding. A germinating seed is shown in Figure 27-16.

OCCURRENCE AND ECOLOGY

Conifers occur in most habitats on earth. They are especially prevalent in temperate regions of both the Northern and Southern Hemispheres. The number of individual conifers in some habitats is extremely large. This becomes particularly evident when flying over parts of North America or Europe where thousands of square miles are covered by nearly pure stands of conifers (Fig. 27-17).

Some conifers are extremely restricted in their present distribution. For instance, coast redwoods are restricted to northern California and southern Oregon coastlines. Sierra redwoods are restricted to some parts of the Sierra Nevada Mountains of California. Similar limited distribution patterns are known for some Southern Hemisphere

Figure 27-17. Hundreds of thousands of square miles of the earth's surface are covered by dense stands of conifers such as those in the Uinta Mountains of Utah shown here.

Figure 27-18. One economically important use of conifers is in the manufacture of turpentine. This apparatus is used for steam distillation in the purification process during turpentine production. (Courtesy of U. S. Department of Agriculture.)

conifers as well. The reason for these narrow distributions is problematical since the fossil record indicates that many of these species enjoyed a much wider distribution in the recent fossil past. Glaciation during the Pleistocene Epoch probably is responsible for at least some of these distribution patterns.

IMPORTANCE

The division Coniferophyta is one of the most important plant divisions on earth both in natural occurrence and in its use by humans. Many conifers are important landscaping plants and also support an economically significant Christmas tree business. More important is the use of conifer wood for building materials and in the manufacture of paper and other products (Fig. 27-18).

Perhaps the most important role of conifers is in the ecosystem of the earth. Conifer forests provide recreation sites for people as well as habitat for hundreds of different animals. Coniferous trees are extremely important in soil conservation and watershed protection of literally millions of acres of the earth's surface. Conifers were very prevalent during the geologic past and comprise most of the Cretaceous coals of the world.

FOSSIL CONIFERS

Fossil conifers were first known from the Carboniferous Period (Fig. 27-19). They rapidly became more numerous in the earth's floras until they were dominant during the

Figure 27-19. This figure shows a reconstruction of the ancient conifer-like fossil *Cordaites* which was an important Carboniferous plant.

(A) Small branch of *Sequoia*

(B) Two small branches of *Taxodium*

(C) Single leaf of *Podocarpus*

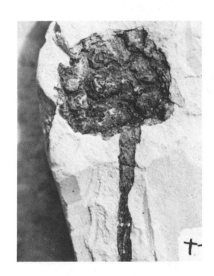

(D) Compressed cone of *Metasequoia*

Figure 27-20. Conifer fossils may be collected from many localities throughout the world. Those illustrated here were collected from the western United States.

(E) Petrified cone of *Pinus*

(F) Petrified conifer wood

early Mesozoic Era. They apparently became abundant because of a worldwide shift in climate which selected against the large lycopods and horsetails in favor of the conifers. Nearly all fossil floras from the Mesozoic contain several conifer fossils that appear to be closely related to modern conifers (Fig. 27-20). Following the development of the angiosperms later in the Mesozoic, conifers became relatively less important, although they are still very abundant on the earth at the present time.

SUMMARY

Conifers are the most important of all gymnospermous plants. They are widely used in lumber, paper, landscaping, and other industries and are indispensable in many watersheds.

Primary conifer stems are eustelic. All conifers exhibit secondary growth through the activity of a vascular cambium that produces new vascular tissues and a cork cambium that produces the periderm. Some conifers reach a great diameter by secondary growth.

Many conifers are adapted to living under dry conditions even though they have abundant moisture during much of the year. Thus, conifer leaves are often small and needle- or awl-shaped with the epidermis having a thick cuticle and sunken stomata. In addition, the leaves of many conifers contain a hypodermis and an endodermis.

Conifer cones are always unisexual and may occur on the same or separate plants. The male cone is a simple structure which produces large amounts of pollen. The female cone is compound since it contains bracts as well as ovuliferous scales.

Conifer sperm are nonmotile and are carried to the archegonia by cytoplasmic streaming in the pollen tube. Fertilization occurs within the ovule and a mature seed develops which contains the embryo. Seeds may be shed soon from the female cones or may be retained for long periods of time. Each seed has the capacity to germinate to produce a new sporophyte plant.

FURTHER READING

CHAMBERLAIN, C. J., *Gymnosperms: Structure and Evolution,* University of Chicago Press, Chicago, 1935. Also Dover Publishing Co., New York, 1966.

FERGUSON, C. W., "Bristlecone Pine: Science and Esthetics," *Science,* 159:839–846, 1968.

LAMBERT, DARWIN, "Timberline Ancients are Rewriting History," *Natural History,* 82(1):44–47, 1973.

SCHULMAN, E., "Bristlecone Pine, Oldest Known Living Thing," *National Geographic,* 113:355–372, 1958.

Also see the list of general references on vascular plants at the end of Chapter 19.

28 Division Anthophyta: the Flowering Plants

Peach blossom. (Courtesy of U. S. Department of Agriculture.)

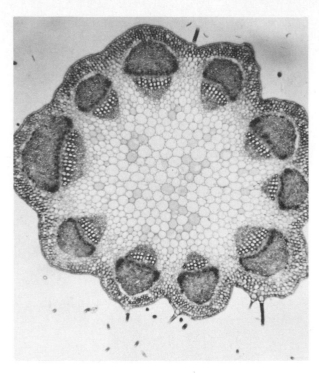

Figure 28-1. The dicot stem contains an eustele where a ring of vascular bundles surround a central pith. This stem cross section shows clover.

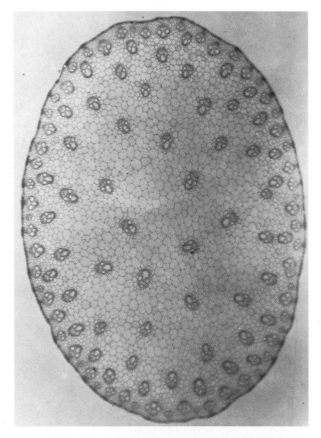

Figure 28-2. Monocot stems are characterized by atactosteles with scattered vascular bundles as seen in this cross section through corn.

Anthophyta or flowering plants are the most highly evolved and successful of all plants presently occurring on earth. Flowering plants are often known as angiosperms (meaning enclosed seed) since they produce seeds enclosed within a protective covering rather than exposed on a megasporophyll or cone scale as among gymnosperms. Angiosperms are so successful that there are presently more than twice as many species of flowering plants on earth as all other plant species combined.

GENERAL CHARACTERISTICS

Angiosperms are divided into two groups or classes. These are the Monocotyledonae and the Dicotyledonae, which are commonly referred to as monocots and dicots. Dicots outnumber monocots by a sizeable majority, although monocots include some of the most important plants on earth. For instance, the monocot family Graminae contains the cereal grains that are so essential as food plants for humans and our livestock.

The dicot stem contains an eustele (Fig. 28-1). The vascular tissue of this stem is divided into several individual vascular bundles surrounding a central pith. These bundles have a region of phloem external to a region of xylem similar to those of a conifer. A vascular cambium often occurs between the two and occasionally a second region of phloem develops internal to the xylem.

Monocots have a unique, characteristic stele known as an *atactostele* (Fig. 28-2). This is composed of individual vascular bundles similar in many respects to those of a dicot. However, monocot vascular bundles lack a vascular cambium and are scattered throughout the stem rather than arranged in a ring. Thus, the atactostele lacks a definite pith and cortex.

Secondary growth occurs in many angiosperms and is particularly evident in long lived or *perennial* dicots (Figs. 28-3, 28-4, and 28-5). The pattern of secondary growth in a dicot is very similar to that of a conifer. The vascular cambium becomes active to initiate the process, producing xylem toward the inside of the stem and phloem to the outside. Vessel elements and tracheids produced in the xylem are not readily compressable, and so the stem increases in diameter by the addition of secondary xylem. Conversely, sieve tube elements and companion cells of the phloem are easily compressed and are sloughed off after a period of activity when they are replaced by newly formed cells. The epidermis is usually lost as the stem enlarges and is replaced by periderm produced by a cork cambium. Just as in conifers, the pattern of outer bark may be used in angiosperm identification (Fig. 28-6).

Secondary growth is absent in most monocots. However, a few perennial monocots demonstrate a peculiar, characteristic type of secondary growth. In this type of growth, a vascular cambium develops from parenchyma tissue near the periphery of the stem. This cambium is not associated with vascular tissues as it differentiates, although it soon begins to produce secondary xylem and phloem. These tissues are produced in an unusual manner since they all develop toward the inside of the stem rather than xylem toward the inside and phloem to the outside. Thus, a monocot vascular cambium at first produces a group of xylem cells and then a group of phloem cells to produce discrete secondary vascular bundles. These secondary bundles are often smaller, rounder, and more compact than those of the primary stem, and in many species they are composed of a ring of xylem cells surrounding a core of phloem cells. Neither primary nor secondary vascular bundles contain a vascular cambium. This unusual type of secondary growth accounts for the stringy nature of the wood of certain monocots which is composed almost entirely of secondary vascular bundles.

A wide variation in leaf shape, size, and anatomy occurs among angiosperms. Some species produce very small, inconspicuous leaves while others may produce leaves several feet long (Fig. 28-7). Some leaves, such as certain cactus needles, are highly modified to protect against predators and water loss. Others are highly dissected or divided compound leaves, simple rounded leaves, or long strap-shaped leaves. Literally thousands of different leaf variations are known among different angiosperm species (Figs. 28-8 and 28-9). Regardless of this great variation, all angiosperm leaves are megaphylls since they exhibit one or more leaf gap and generally several to many leaf veins. Most angiosperm leaves are divided into a more or less evident petiole and a broadened blade or lamina.

The lamina is similar in construction among many flowering plants. It has an upper epidermis with a cuticle of varying thickness depending on environmental factors. Stomata are usually either infrequent in this upper epidermis or absent entirely. A layer of perpendicular palisade parenchyma with very few intercellular spaces develops

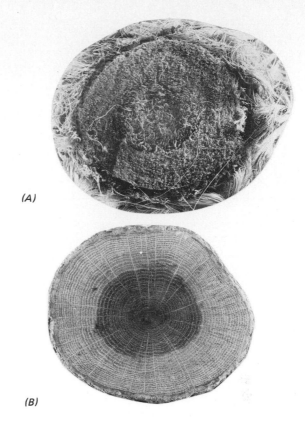

(A)

(B)

Figure 28-3. Cross sections through the stems of a "woody" monocot (A) and dicot (B) are shown here. The monocot stem appears fibrous because of the scattered individual vascular bundles. The dicot stem shows secondary growth and is very dense and hard because of the compact xylem cells produced toward the center of the stem. Note the distinct annual rings in the secondary xylem.

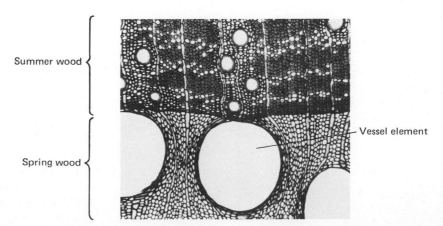

Summer wood

Spring wood

Vessel element

Figure 28-4. This cross section through the secondary xylem (wood) of an oak shows very large vessel elements in the spring wood.

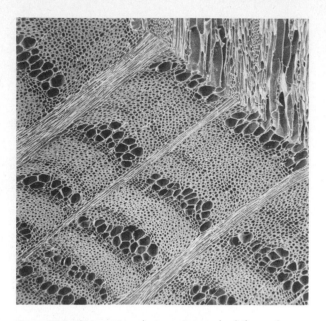

Figure 28-5. This scanning electron micrograph of the angiosperm *Knightia excelsa* shows that the vessel elements in this plant are all produced in the early spring wood. (Courtesy of B. A. Meylan and B. G. Butterfield from *Three Dimensional Structure of Wood*, Syracuse University Press, Syracuse, N. Y. 1972. Used by permission of Syracuse University Press.)

beneath the epidermis (Fig. 28-10). This tissue represents the efficient photosynthetic region of the leaf. The spongy mesophyll is generally present below the palisade layer. This tissue is photosynthetic but the cells are loosely arranged with large intercellular spaces between them. Gases necessary for photosynthesis accumulate in these spaces until they are used. In addition, gases released by respiration and as a byproduct of photosynthesis accumulate in these until they are either used by the plant or released to the outside atmosphere. Thus, the spongy mesophyll is often referred to as a region of gaseous exchange. Vascular bundles or veins occur in the zone between the palisade and spongy layers (Figure 28-10). These veins are surrounded by thin-walled parenchyma cells that transport water and mineral nutrients from the xylem to the photosynthetic cells, and sugars from the photosynthetic cells to the phloem for distribution to other parts of the plant. The lower epidermis is also cutinized and generally has a higher number of stomata than the upper epidermis since it borders the spongy mesophyll.

Both the upper and more commonly the lower leaf epidermis may contain hairs or scales (Fig. 28-10). These structures act as a protective mechanism against water loss since they lower the rate of air movement across the leaf surface and thus lower the rate of diffusion of water vapor through the stomata. Hairs and scales also protect against the attack of fungal parasites and insect predators. They may be either unicellular or multicellular and are often characteristic in shape and size.

The arrangement of veins in the leaf is often characteristic. Most monocots exhibit parallel veins, and most dicots have reticulate or netted veins. Furthermore, many variations occur in both the parallel and reticulate patterns. A particular venation pattern is usually characteristic of an individual family or genus.

The arrangement of leaves on the stem represents another characteristic used in classifying angiosperms. Thus, some species have alternate leaves while others have opposite or whorled leaves.

The angiosperm root system is variable. Monocots produce only adventitious, fibrous roots. Dicots may produce either fibrous or tap root systems. Some dicot tap roots are used as food. Carrots, beets, sugar beets, and turnips are angiosperms grown commercially for their roots.

A typical dicot root is protostelic with the xylem forming a star or *X*-shaped central cylinder (Fig. 28-11). The phloem usually occurs in strands between the xylem arms. A thin pericycle surrounds the xylem and phloem. This tissue is responsible for the formation of branch roots. An endodermis surrounds the pericycle. It is the innermost layer of a rather wide cortex which usually functions in storing starch or other products. An epidermis is the outermost layer of root tissue. Certain epidermal cells produce

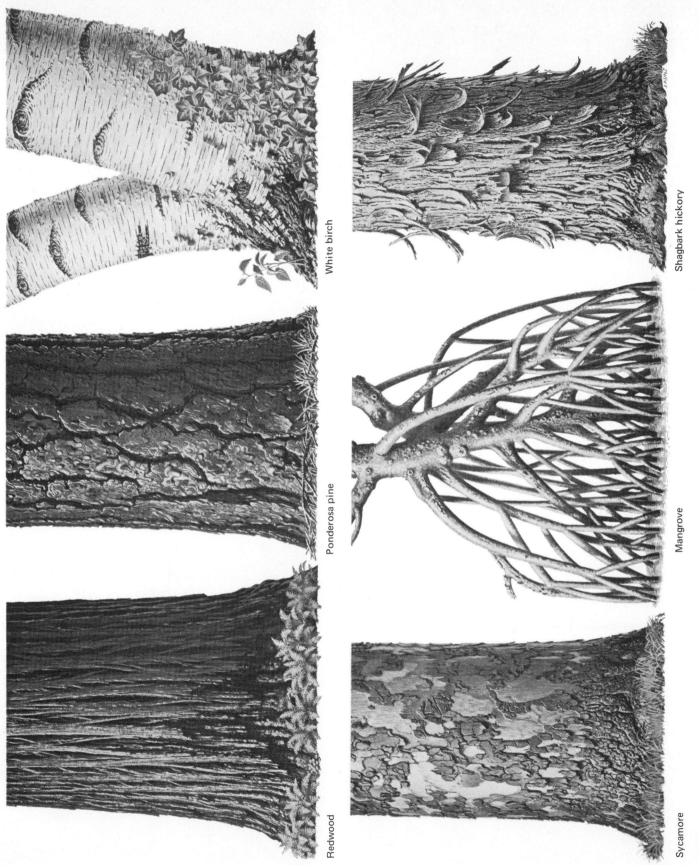

White birch

Shagbark hickory

Ponderosa pine

Mangrove

Redwood

Sycamore

Figure 28-6. Bark patterns of representative conifers and angiosperms are often characteristic and may aid in identification. (Courtesy of St. Regis Paper Company.)

(A)

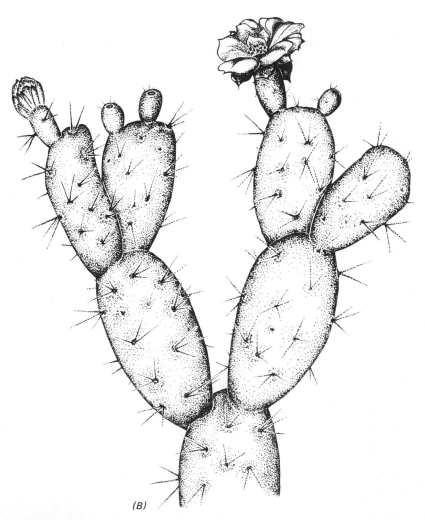

(B)

Figure 28-7. Leaves of angiosperms differ widely in form and function according to several factors, including their environment. The leaves of a water lily (A), for example, may reach more than 2 ft in diameter, while cactus leaves (B) are often small and needle-like.

(A)

(B)

Figure 28-8. This figure shows the modified leaves of two insectivorous plants. The pitcher plant (A) has funnel-shaped leaves with epidermal hairs directed toward the base of the funnel. Insects are attracted to the leaves and directed into the funnel where they are digested. The venus fly trap leaf (B) closes rapidly, entrapping insects which are then digested.

(A)

Figure 28-9. Many cacti (including *Notocactus* (A) and *Carnegia* (B) shown here) show thick, succulent stems, and highly modified, spine-like leaves. (B courtesy of Walter Hodge.)

(B)

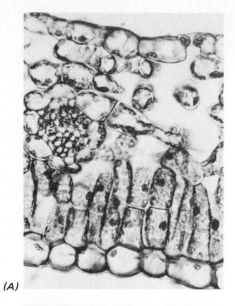

(A)

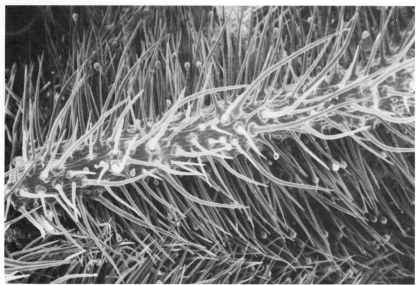

(B)

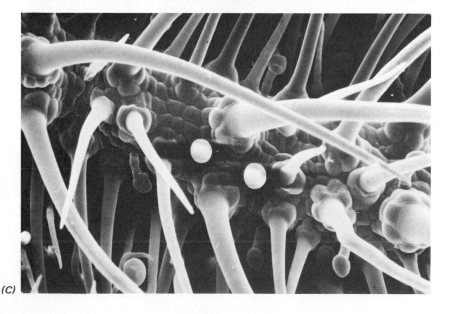

(C)

Figure 28-10. Two angiosperm leaves are shown in this figure. A cross section through a privet leaf illustrating the various internal leaf tissues is shown at A. B and C show scanning electron micrographs of a geranium leaf illustrating prominent epidermal hairs. (Courtesy of James V. Allen, Brigham Young University.)

projections known as root hairs that greatly increase the absorptive area of the root (Fig. 28-12).

Some dicot roots demonstrate secondary growth. This process follows a similar pattern as the same process in a stem. A vascular cambium is present between the xylem and phloem. It becomes active and produces xylem toward the inside of the root and phloem to the outside. The arrangement of secondary vascular tissues in a root is variable and is often characteristic. Secondary xylem may be produced in an entire central cylinder or as separate radiating arms. The amount and duration of secondary growth is likewise variable. A cork cambium often develops as secondary growth proceeds. This cambium produces a periderm that replaces the epidermis and cortex.

Monocot roots are just as distinct in their stelar type as are monocot stems. They are always adventitious, and the entire root system is fibrous. A tap root similar to that of a carrot is never produced by a monocot. The placement of vascular tissues in the stele of a monocot root is unusual since the xylem and phloem are not in close contact. Discrete bundles of xylem alternate with bundles of phloem and both surround a central pith (Fig. 28-13). The vascular tissues are surrounded by a pericycle which is in turn surrounded by an endodermis. A cortex is external to the endodermis, and the epidermis is the outermost tissue. Secondary growth never occurs in monocot roots.

The angiosperm flower is unique in the plant kingdom. Flowers of all shapes and sizes are produced by different species of angiosperms (Fig. 28-14). Some of these are perfectly regular in symmetry while others are contorted into various unusual shapes (Fig. 28-15). Almost any color imaginable is likely to be found among different flowers.

One of the most fascinating studies in biology is that of the relationship between flowers and their pollinators. Floral shape and color have evolved to attract pollinating organisms such as beetles, honey bees, moths, wasps, flies, bumble bees, birds, and bats (Fig. 28-16). Also, pollinating organisms have evolved to become very efficient in obtaining and using nectar and pollen from these plants. Several elaborate experiments have shown that many pollinating insects have the ability to discern and remember different colors and numbers of flower parts as well as their size and shape. These insects can remember which flowers provide desirable nectar or pollen, and they will visit that type of flower repeatedly. Such repeated visits by pollinators generally assure adequate pollination for large numbers of flowers. In general, insects are the most effective pollinators on earth. They have played a very important role in the remarkable success of flowering plants since their development in the Mesozoic Era.

The outermost floral parts of most flowers are often green and leaf-like though variously colored in some flowers. These are the *sepals* which are collectively termed the

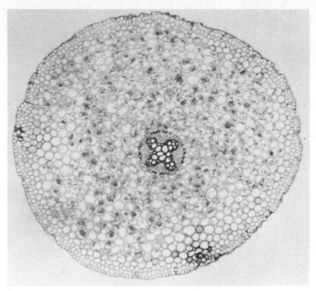

(A)

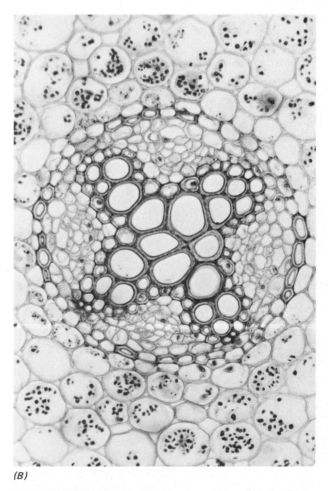

(B)

Figure 28-11. These *Ranunculus* root cross sections show the characteristic dicot root protostele. A is a low power view of the root, and B is a closeup of the stele. Note the bundles of phloem cells alternating with the "arms" of xylem.

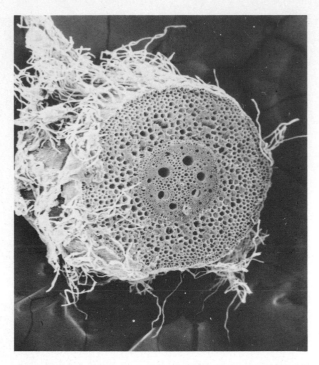

Figure 28-12. A young root of *Zea mays* (corn) is shown in this nicrograph. Note especially the central siphonostele and the external root hairs. (Courtesy of James V. Allen, Brigham Young University.)

calyx. They serve to protect the floral bud. The calyx may be composed of few to many sepals (Fig. 28-17). *Petals* occur inside of the calyx. The petals collectively are known as the *corolla,* which may be composed of few to many petals. The corolla is often brightly colored to attract pollinators, and the petals may be regular and uniform or variously twisted or contorted forming unusual shapes. Either or both the calyx or corolla may be absent from the flowers of some species.

The male reproductive structures of the flower are the *stamens* which collectively comprise the *androecium.* A stamen usually is composed of an elongate *filament* and a terminal *anther* (Figs. 28-17 and 28-18). Microspores are produced within the anther. Much variation occurs among different angiosperms in the shape, size, and number of stamens present in the flower.

The innermost floral parts are the female reproductive structures which comprise the *gynoecium.* This gynoecium is highly variable but is often composed of a single *pistil* (Fig. 28-17). Most pistils have a basal *ovary* and an elongate *style* with a terminal *stigma.* From one to several pistils may be present in the flower to comprise the gynoecium. Variations in the structure of the pistil and in the number of pistils comprising the gynoecium are common. The style may be shortened or absent, or long and well-developed. The stigma may be large and evident with an abundance of epidermal hairs (Fig. 28-19) or rarely it may be absent. The ovary is formed from one to several leaf-like megasporophylls or *carpels* which fuse along one edge to produce a chamber (Fig. 28-20). Ovules are produced within the chamber. All floral parts are produced from a receptacle.

Since all of the structural components of a flower vary widely in number and form, literally hundreds of thousands

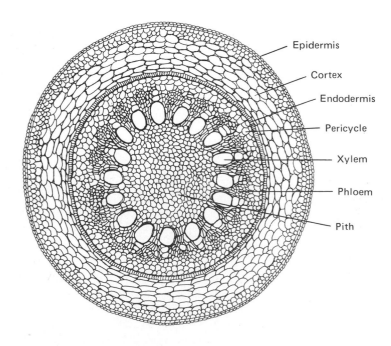

Epidermis

Cortex

Endodermis

Pericycle

Xylem

Phloem

Pith

Figure 28-13. This cross section through a typical monocot root (*Smilax*) illustrates the unusual stele typical of the roots of many monocots.

Figure 28-14. Several different angiosperm flowers are shown here including cactus (A), *Microspermia* (B), *Hibiscus* (C), evening primrose (D), *Fuschia* (E) and hollyhock (F).

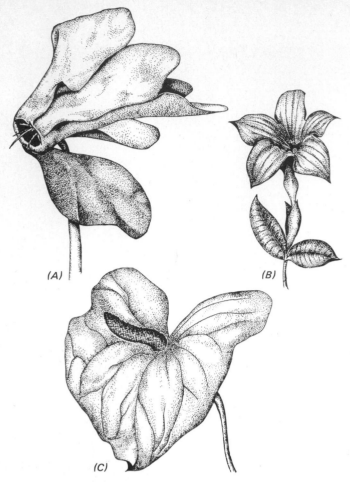

Figure 28-15. The angiosperm flowers illustrated here show wide differences in shape and size. A is *Cyclaman*, B is flax, and C is *Anthurium*.

of variations among angiosperm flowers are possible. If the calyx, corolla, androecium, and gynoecium are all present, the flower is *complete*. If one or more of these floral units is absent, the flower is *incomplete*. In addition, if a flower lacks either the androecium or gynoecium it is *imperfect*. Such a flower is said to be *staminate* if only stamens are present or *pistillate* if only the gynoecium is present.

LIFE HISTORY

From one to thousands of ovules are produced within the ovary among different flowering plants (Fig. 28-20). These ovules are almost always smaller than those of gymnosperms and are protected by two integuments rather than one. The nucellus within the integument is also small, often composed of just a few cells. As the ovule develops, a megaspore mother cell differentiates within the nucellus (Fig. 28-21). This cell is large and conspicuous, and soon divides meiotically to form a linear tetrad of megaspores. In a typical flower, three of these disintegrate, and the fourth, functional megaspore enlarges greatly (Fig. 28-21).

Further development of the female gametophyte is endosporic. This megagametophyte is known as the *embryo sac*. It is extremely reduced, normally being composed of but eight nuclei arranged in seven cells. An egg differentiates nearest the micropyle of the ovule. This egg is flanked on each side by a single cell known as a *synergid*. The function of these synergids is only poorly known and it is probable that they represent a greatly reduced archegon-

(A)   (B)   (C)

Figure 28-16. One of the most important reasons for the great success of flowering plants is their employment of insect pollinators. Three different pollinators are shown in this figure. (Courtesy of Armand Whitehead, Brigham Young University.)

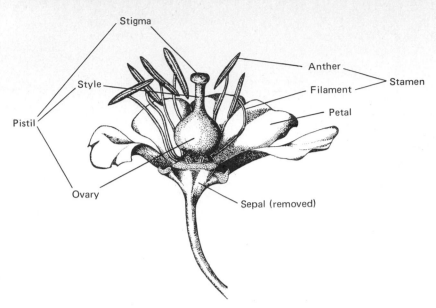

Figure 28-17. This cut away drawing of an angiosperm flower shows the floral tissues.

ium. Recent investigation of some angiosperms has shown that one synergid accepts sperm that are released from the pollen tube. These sperm then pass through the synergid to accomplish fertilization. Many aspects of this process await further study for clarification. Two *polar nuclei* occur near the center of the embryo sac. These nuclei are haploid and function to form a specialized nutritive tissue which feeds the developing embryo. Three *antipodal* cells differentiate at the end of the embryo sac opposite the micropyle (Fig. 28-21). The role of these cells is also unclear, and they likely represent the only vegetative cells of the reduced female gametophyte. Thus, the angiosperm megagametophyte represents the culmination of the evolutionary tendency among higher plants toward the reduction of the gametophyte generation.

As the embryo sac matures, microspores are formed within the anther of the stamen. Several microspore mother cells differentiate within the anther and each of these divides by meiosis to produce a tetrad of microspores. These tetrads are often spherical, but the individual microspores usually separate as they mature to become pollen grains (Figs. 28-22 and 28-23). However, some angiosperms produce large adherent masses of pollen which remain intact and are specialized for pollination by certain insects.

The angiosperm pollen grain contains the most reduced male gametophyte of all vascular plants. This plant is composed of but two cells, a *generative cell* and a *tube cell* (Fig. 28-24).

As the pollen grains mature, they are released from the anther and are carried to the stigma of the same or another flower. The stigma of a flower may accumulate the pollen grains of several different plant species. However, the physiological characteristics of the cells of the stigma and style often preclude the germination of the pollen grains of other plant species. Conversely, pollen grains of the same species as the pollen receptor plant germinate and

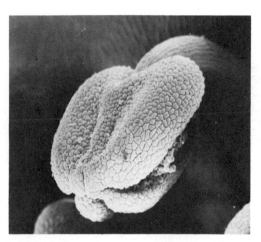

Figure 28-18. This scanning electron micrograph of the anther of an angiosperm (*Lobularia*) shows that the anther has ruptured and is spilling a few pollen grains at the lower right. (Courtesy of James V. Allen, Brigham Young University.)

Figure 28-19. This micrograph shows the surface of the stigma of an angiosperm pistil. Pollen grains land on this stigma surface and germinate to produce a pollen tube. (Courtesy of James V. Allen, Brigham Young University.)

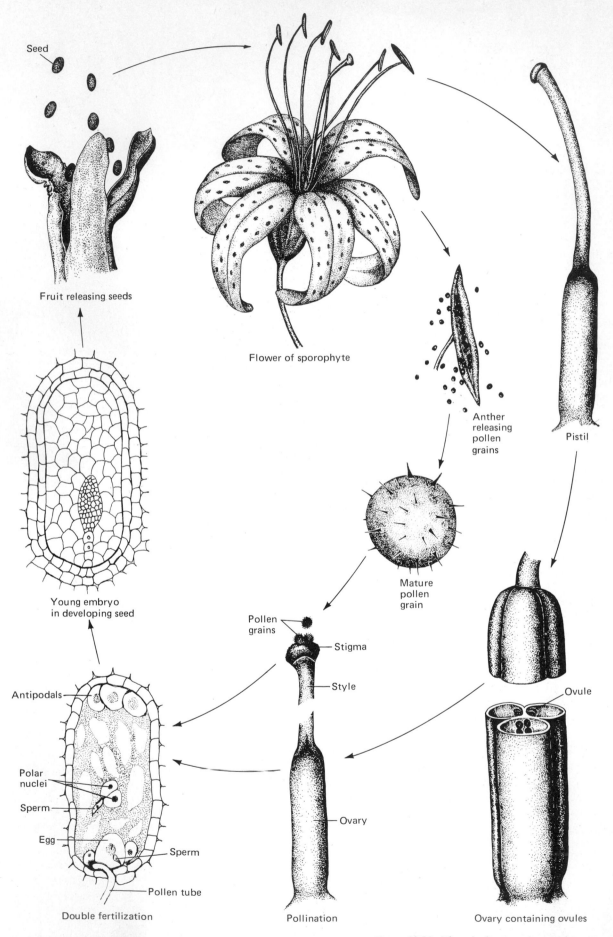

Seed

Fruit releasing seeds

Flower of sporophyte

Anther releasing pollen grains

Pistil

Young embryo in developing seed

Mature pollen grain

Pollen grains

Stigma

Style

Ovary

Ovule

Antipodals

Polar nuclei

Sperm

Egg

Sperm

Pollen tube

Double fertilization

Pollination

Ovary containing ovules

Figure 28-20. Life cycle diagram of an angiosperm.

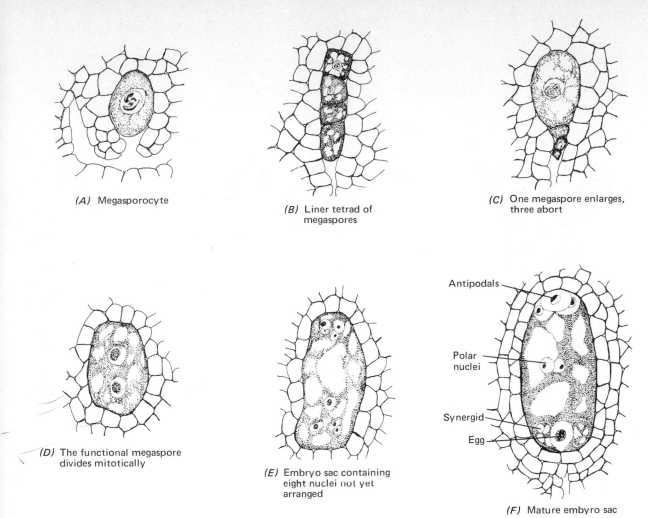

(A) Megasporocyte

(B) Liner tetrad of megaspores

(C) One megaspore enlarges, three abort

(D) The functional megaspore divides mitotically

(E) Embryo sac containing eight nuclei not yet arranged

Antipodals

Polar nuclei

Synergid

Egg

(F) Mature embyro sac

Figure 28-21. Development of a representative angiosperm embryo sac is shown in this series of drawings.

grow rapidly. The pollen grain germinates to form a pollen tube that grows through the stigma and style until it contacts the embryo sac (Fig. 28-25). As the pollen tube grows, the generative cell divides to form two nonmotile sperm. Eventually, the tip of the pollen tube penetrates the embryo sac, and the two sperm are released into a synergid alongside the egg. One sperm then fertilizes the egg to produce the zygote. The second unites with the two polar nuclei of the megagametophyte to produce a $3N$ nucleus. This nucleus is destined to divide rapidly to produce a highly efficient nutritive *endosperm* tissue which nourishes the developing embryo.

This process of the union of one sperm with an egg and the second with the polar nuclei is termed *double fertilization*. This unique process is perhaps the best characteristic of an angiosperm since it only occurs within the plants of this division. This is especially important since several other groups of plants known from the fossil record produced structures somewhat similar to the angiosperm ovary and some modern angiosperms do not produce entirely enclosed ovules. Nevertheless, none of these fossil plants exhibited double fertilization, whereas all modern

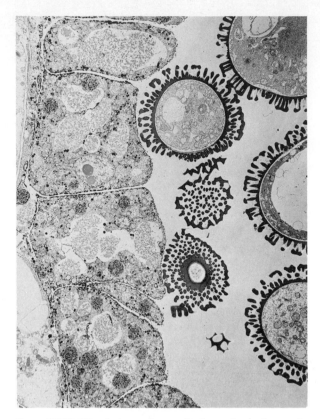

Figure 28-22. This electron micrograph shows developing angiosperm pollen grains. The pollen grains are the spherical objects on the right of the photograph. Note the thickened, uneven walls on the grains. The large cells at the left are the inner cells of the anther which function to nourish the developing grains. (Courtesy of H. G. Dickinson, University of Reading. Used by permission of Cytobios.)

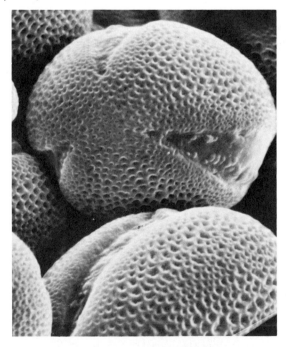

Figure 28-23. This scanning electron micrograph shows mature pollen grains of the angiosperm *Hedysarum*. (Courtesy of James V. Allen and Terry Northstrom, Brigham Young University.)

flowering plants do exhibit this unique process regardless of the structure of their ovary.

It should be mentioned that the processes of megagametophyte development in particular as well as microgametophyte development and fertilization vary according to species and in some are quite different from that described in this chapter. For instance, plants which produce $5N$ or $7N$ endosperm are not uncommon, and the mature embryo sac may vary from four nucleate to sixteen nucleate according to the species.

The embryo develops in place within the embryo sac. As the embryo matures it is nourished by the endosperm. Some angiosperms produce a mature seed that essentially lacks endosperm and is almost entirely filled with a large embryo. Other species produce mature seeds with a large endosperm and a relatively small embryo (Fig. 28-26). The germination of these different types of seeds differs somewhat, although the early nutritive tissue is endosperm in both cases.

As ovules mature to become seeds, the ovary matures simultaneously to become the *fruit* (Fig. 28-27). Hence, a mature fruit contains one to many seeds. The function of the seed is of course to germinate and grow, thereby insuring that the species does not become extinct. The function of the fruit is much the same. Some fruits are large and fleshy and attract animals which use the fruit for their own good and discard the seeds some distance from the parent plant. Other fruits are thorny or prickly and easily become attached to the fur of an animal or the clothing of a man and thus are widely dispersed. Still other fruits are modified so that they are easily disseminated by the wind (Figs. 28-28 and 28-29). Thus, the role of the fruit is generally twofold. First it protects the developing ovules and second, it often aids in the dissemination of mature seeds.

The phenomenal success of angiosperms in the world's floras is noteworthy. This success is especially significant since the Anthophyta is the last major plant division to have evolved. Flowering plants have only existed since Late Jurassic or Early Cretaceous times while many of their competitors have existed much longer.

Flowering plants exhibit a combination of unique factors that account for their unparalleled success. One important factor is the short angiosperm life cycle. Most flowering plants produce a crop of mature seeds in a single year, and many species are able to pass from seed to mature seed in a matter of a few weeks. This has allowed these plants to be more efficient in occupying new territory than gymnosperms, which often take two or more years to complete their life cycle.

The use of insects as pollinating agents is one of the most important factors in the success of angiosperms. These pollinators are much more efficient than the wind and most other pollinating agents and increase the likelihood that an ovule will mature into a seed containing a

(A) Microspore mother cell
(microsporocyte)

(B) Tetrad of microspores

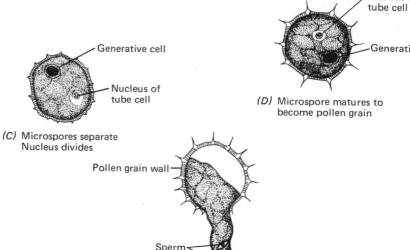

Generative cell

Nucleus of
tube cell

(C) Microspores separate
Nucleus divides

Nucleus of
tube cell

Generative cell

(D) Microspore matures to
become pollen grain

Pollen grain wall

Sperm

Pollen tube

Tube nucleus

(E) Pollen grain germinates
to form pollen tube

Figure 28-24. Pollen grain maturation and germination in an angiosperm are shown in this series of drawings.

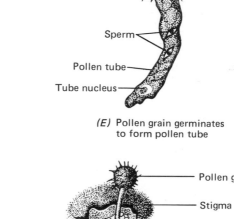

Pollen grain

Stigma

Style

Pollen tube

Egg

Synergid

Polar nuclei

Ovary

Antipodal

Ovule

Figure 28-25. This diagrammatic cutaway drawing of an angiosperm pistil shows the process of fertilization. The pollen grain lands on the stigma where it germinates to produce a pollen tube. This pollen tube grows toward the ovule which it eventually penetrates. The tip of the tube enters a synergid and ruptures to release the sperm. One sperm then fertilizes the egg and the second unites with the polar nuclei.

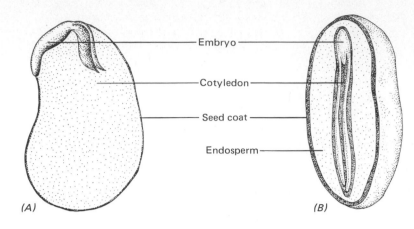

Embryo

Cotyledon

Seed coat

Endosperm

(A)   (B)

Figure 28-26. The longevity of endosperm tissue within the seed differs among different angiosperm species. The seed at A is mature and lacks any endosperm tissue. The endosperm in this seed was used to form the large cotyledons. The embryo in the seed at B has relatively small cotyledons, and endosperm tissue is still present. The endosperm in this seed will be used by the developing embryo at the time of seed germination.

(A)   (B)

Figure 28-27. The angiosperm flower contains an ovary that matures to become a fruit. A is a twig with several peach blossoms, and B is a peach tree with several mature fruits. (Courtesy of U. S. Department of Agriculture.)

living embryo. Insects also act as isolating mechanisms that segment plant populations and allow evolution to proceed independently on the separate segments.

The ovary is also partly responsible for the success of flowering plants. This structure protects the developing ovules and acts as a dissemination agent for mature seeds. When all of these factors are considered together, the angiosperms must be considered to be the most successful division of plants on earth today.

OCCURRENCE AND ECOLOGY

It is difficult to make a summary statement concerning the ecology of flowering plants. This subject is extremely important and most universities throughout the world devote many courses to its study. Angiosperms are found in most habitats on earth. They are most diverse in tropical regions, and many unknown flowering plants still await discovery in the world's tropics.

Anthophyta are also prevalent in temperate regions where they often comprise the dominant vegetation. Even in northern coniferous forests, flowering plants often form much of the understory vegetation. The floras of the coldest parts of the world are dominated by mosses, lichens, and other lower plants. Angiosperms are usually reduced in such environments, but are still present.

Many angiosperms are epiphytes, and some are parasites on other vascular plants. Mistletoe is one such famous parasite. Many species are adapted to living under extremely dry, hot conditions while others are aquatic in both fresh and marine waters.

IMPORTANCE

Several plant divisions, including algae, fungi, and flowering plants, are indispensable for the survival of human beings. Algae are especially essential in their role of primary production and oxygen generation. Fungi are particularly significant as decomposers and in mycorrhizal associations with the roots of higher plants.

Angiosperms are important for a wide variety of reasons. They are perhaps primarily significant as food plants for humans and their livestock. All cereal grains, all fruits, and most leaves and stems used as food by humans are produced by angiosperms (Fig. 28-30). In addition, most of the plant material eaten by cattle, sheep, and hogs is angiospermous.

Some angiosperms are lumber trees used for producing building materials and pulp for the paper industry. Many flowering plants are used in floral, landscaping, and drug industries (Fig. 28-31). In addition, most of the earth's watersheds are protected by flowering plants which stabilize the soil and provide habitat and food for thousands of animal species. Indeed, the Anthophyta is an indispensable

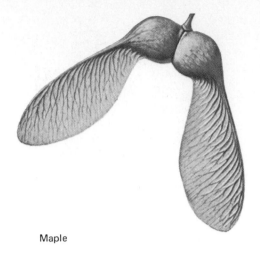

Maple

White pine

Mangrove

Coconut

Apple

Cherry

Willow

Witch hazel

Pecan

Black walnut

Beech

Oak

Figure 28-28. Seeds are disseminated in various ways in different plants. This figure shows the mechanisms of seed dispersal of several representative seed plants (mostly angiosperms). (Courtesy of St. Regis Paper Company.)

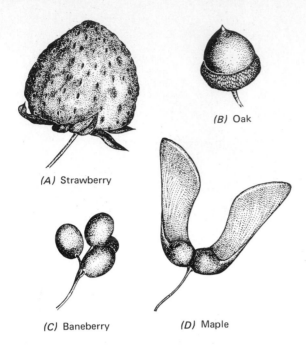

(A) Strawberry

(B) Oak

(C) Baneberry

(D) Maple

Figure 28-29. A few representative angiosperm fruits (fully matured ovaries) are shown here.

division of plants. Civilization as we know it would not be possible without them.

The great importance of flowering plants has been one primary reason for the development of significant conservation methods during the past half century. Civilization has grown to such an extent that only careful planning and thoughtful use of our natural resources will allow its continued development and the improvement of the life style of impoverished peoples throughout the world.

FOSSIL ANGIOSPERMS

The fossil record of angiosperms is relatively good when compared to that of many other plant groups. Even so many problems still remain unsolved. The oldest undoubted flowering plant fossils have been collected from rocks no older than Late Jurassic or Early Cretaceous. Reports of such fossils from earlier rocks have proven to be erroneous or at least are still in question. The problem is that many of the earliest fossil angiosperms appear to be well advanced forms. Many paleobotanists postulate that such plants represent species which had already evolved for long periods of time. These scientists therefore argue that the fossil record cannot be accepted at its face value and that angiosperms must be much older than Early Cretaceous. Other paleobotanists feel that the fossil record is accurate and that during the Cretaceous Period a tremendous period of angiosperm evolution occurred to produce the relatively advanced fossils of the time. This is supported by the fossil record of insects, since the higher pollinator insects began to evolve during Jurassic and Cretaceous times. Such an event could have been responsible for a tremendous period of angiosperm evolution and adaptation responsible for the rapid development of flowering plants during that time. Paleobotanists will continue to search for clues to this problem which Charles Darwin called an "abominable mystery."

Angiosperms are best known from the fossil record as leaf compressions (Fig. 28-32). These are often very well preserved so that cuticle and even internal anatomy may occasionally be studied. A few flowers are known from the Tertiary Period and fossil seeds and fruits from Tertiary rocks are not uncommon.

Petrified woods of both angiosperm stems and roots are also known. These woods are occasionally very well preserved so that thin sections may be cut and examined beneath the microscope. They are similar in composition to modern woods and can generally be placed into a modern family.

SUMMARY

Angiosperms are among the most important of all plants on earth both directly to humans economically and in plant

(A)

(B)

Figure 28-30. Rice, soy bean, and wheat are three of the most important food plants in the world. All three are angiosperms as are most other important crop plants. A is a photograph of soy-bean harvesting, B shows rice harvesting, and C shows wheat **harvest-**ing. (Courtesy of U. S. Department of Agriculture.)

(C)

Figure 28-31. *Papaver somniferum* is the economically important opium poppy. This illustration shows a mature fruit exuding the milky substance from which opium is derived. (Courtesy of Walter Hodge.)

communities throughout the world. More than 250,000 species of angiosperms are presently known, which is more than twice as many as all other plant species combined.

Flowering plants share several common characteristics. First and perhaps more important, all angiosperms produce highly specialized reproductive structures or flowers. Second, angiosperms demonstrate the unusual phenomenon of double fertilization. Third, angiosperm ovules are produced inside of a protective megasporophyll or carpel.

Two classes of Anthophyta are commonly recognized. The first Dicotyledonae, is characterized by eustelic stems, protostelic roots, and reticulate leaf venation. The second, Monocotyledonae, is characterized by atactostelic stems, an unusual root stele and parallel leaf venation. Monocots contain the cereal grains which are the most important food plants on earth.

Angiosperms have been so successful in their inhabitation of the earth for a number of reasons, including a short life cycle, the use of insect pollinators, and the production of a protective ovary.

FURTHER READING

BENSON, L., *Plant Classification*, D. C. Heath and Co., Lexington, Mass., 1957.

(A) *Platanus*

(B) *Platanus*

(C) *Acer*

(D) *Allophylus*

Figure 28-32. Several angiosperm leaf compression fossils are shown in this figure.

Cronquist, A., *The Evolution and Classification of Flowering Plants,* Houghton Mifflin, Boston, 1968.

Davenport, Demorest, "The Esthetics of Orchid Pollination," *Natural History,* 77(4):66–69, 1968.

Dovring, Folke, "Soybeans," *Scientific American,* 230(2): 14–21, 1974.

Eames, A. J., *Morphology of the Angiosperms,* McGraw-Hill Book Co., New York, 1961.

Eames, A. J. and L. H. MacDaniels, *An Introduction to Plant Anatomy,* McGraw-Hill Book Co., New York, 1947.

Esau, K., *Plant Anatomy,* Second edition, John Wiley & Sons, New York, 1965.

Galston, Arthur, "Plants Have a Few Tricks Too," *Natural History,* 81(10):26–29, 1972.

Galston, Arthur, "Turning Plants Off and On," *Natural History,* 82(9):26–34, 1973.

Grant, Verne, "The Fertilization of Flowers," *Scientific American,* 184(6):52–56, 1951.

Heiser, Charles, *Seed to Civilization: The Story of Man's Food,* W. H. Freeman and Co., San Francisco, 1973.

Jensen, William, "Fertilization in Flowering Plants," *Bioscience,* 23(1):21–27, 1973.

MANGELSDORF, PAUL, RICHARD MACNEISH, and W. C. GALINAT, "Domestication of Corn," *Science,* 143:538–545, 1964.

PALMER, JOHN, "The Rhythm of the Flowers," *Natural History,* 80(7):64–73, 1971.

PIRIE, N. W., "Orthodox and Unorthodox Methods of Meeting World Food Needs," *Scientific American,* 216(2):27–35, 1967.

REITZ, R., "New Wheats and Social Progress," *Science,* 169:953–955, 1969.

SPENCER, PATRICIA, "The Turning of the Leaves," *Natural History,* 82(8):56–63, 1973.

Also see the list of general references on vascular plants at the end of Chapter 19.

Glossary

ADVENTITIOUS ROOT A root developing from a stem or occasionally from a leaf. For example, prop roots of a corn plant are adventitious.

AECIOSPORE A dikaryotic spore produced by some rusts (Uredinales) within an aecium.

AECIUM An often cup-shaped, aeciospore-bearing sporangium common in rusts (Uredinales). The aecium ruptures through the epidermis of the host plant to release its aeciospores.

AGAR A substance in the cell walls of some Rhodophyta which is commercially extracted and used in producing medium for culturing many microorganisms.

AKINETE A thick-walled, nonmotile algal spore formed from a single vegetative cell.

ALGIN A substance produced in the cell walls of Phaeophyta. Algin is commercially important for use in several industries.

ALTERNATION OF GENERATIONS The alternation of a haploid, gamete producing plant with a diploid, meiospore producing plant.

ANAPHASE The stage of nuclear division when chromatids or homologous chromosomes separate and migrate toward opposite poles of the mother cell.

ANDROECIUM The term used for the entire complement of stamens in an angiosperm flower.

ANGIOSPERM A plant which produces ovules enclosed within an ovary and in which double fertilization occurs. Angiosperms are commonly known as flowering plants.

ANISOGAMETE A motile gamete which differs in size from the motile gamete with which it unites to produce a zygote.

ANISOGAMY The process of fusion between two motile gametes when one is larger and less motile than the other.

ANNUAL RING An increment of xylem formed in one growing season. Generally, a region of spring wood and a region of summer wood are evident in each annual ring.

ANNULUS (1) A ring or band of cells on a sporangium which aids in rupturing the sporangium thus aiding in spore dispersal. (2) A ring of tissue on the basidiocarp stipe of many Basidiomycetes.

ANTHER The apical, often swollen portion of a stamen containing microsporangia which are responsible for pollen grain formation.

ANTHERIDIUM The male gametangium of all nonseed plants. Antheridia are unicellular among algae and fungi, and are covered by a protective layer of tissue in higher plants.

ANTIPODAL A vegetative cell in the embryo sac (megagametophyte plant) of an angiosperm. These cells usually disintegrate upon fertilization of the egg and polar nuclei.

APICAL CELL A meristematic cell at the tip of an organ which divides to produce the other cells of the organ.

APICAL MERISTEM The apex of an organ (generally a stem or root) composed of several to many meristematic cells which divide to produce the other cells of the organ.

APOTHECIUM An open, cup-shaped or disk-shaped ascomycete fruiting body. An apothecium contains a definite hymenial layer.

ARCHEGONIUM The multicellular female gametangium of higher plants. An archegonium usually consists of a swollen basal venter and an elongate neck.

ASCOCARP The fruiting body of an ascomycete. Sexual reproduction of most Ascomycetes occurs within an ascocarp.

ASCOGENOUS HYPHAE Dikaryotic hyphae originating from plasmogamy of an antheridium and ascogonium. Ascogenous hyphae develop within an ascocarp and produce asci.

ASCOGONIUM The female gametangium of an ascomycete. This structure often produces a trichogyne which contacts an antheridium and thus aids in plasmogamy in sexual reproduction.

ASCOSPORE The meiospore of an ascomycete. This spore is produced within a sac-like ascus.

ASCUS A swollen meiospore containing cell originating from an ascogenous hypha. Karyogamy and meiosis occur within the ascus.

ASEXUAL REPRODUCTION Any process of reproduction wherein gametic union and reduction division do not occur.

AUTOECIOUS The term applied to a rust (Uredinales) which completes its life cycle on a single host.

AUTOTROPHIC The ability to manufacture food by photosynthesis or chemosynthesis. All green plants are autotrophic although some may rely partly on other food sources.

AUXOSPORE A spore produced by many diatoms which is normally silica walled. This spore is often produced by gametic union and the subsequent enlargement of the zygote.

BASIDIUM A club-shaped meiospore bearing cell produced by Basidiomycetes. Basidia are often produced in a definite hymenial layer.

BLOOM A heavy growth of algae which develops under certain environmental circumstances, especially when abundant light and nutrients are present.

BRACT (1) A sterile leaf-like structure produced by many angiosperms. (2) The sterile structure produced in some gymnosperm strobili associated with the sporophylls.

CALYX The collective term for the entire complement of sepals in an angiosperm flower.

CAPILLITIUM The hypha-like strands which develop in the sporangium of Myxomycetes. Spores develop intermingled among the capillitia.

CAPSULE The apical portion of a sporophyte plant of a bryophyte. The capsule is the location of meiospore production.

CARINAL CANAL The large air canals produced in the vascular bundles of *Equisetum*.

CAROTENE A yellowish or orange pigment found in the chromoplasts of all green plants.

CARPEL The ovule-bearing megasporophyll of an angiosperm. A single or several carpels comprise an angiosperm ovary.

CARPOGONIUM The female gametangium of Rhodophyta. This structure is unicellular with an elongate trichogyne. The carpogonium is similar to the ascogonium of Ascomycetes.

CARPOSPORE A spore produced by the carposporophyte generation of advanced Rhodophyta. Carpospores may be haploid or diploid depending upon the species.

CARPOSPOROPHYTE The "extra" generation produced by higher Rhodophyta. The carposporophyte plant may be either haploid or diploid and is produced directly by the division of the zygote.

CAULIDIUM The stem-like structure lacking vascular tissue which is produced by "leafy" liverworts and mosses.

CELL MEMBRANE The differentially permeable membrane surrounding the protoplast of a cell. This membrane is immediately inside the cell wall of most plant cells.

CELL WALL The permeable, often rigid, outermost structure common in most plant cells. The cell wall is often composed of cellulose although this differs among various plant divisions.

CENTRAL BODY The interior, often darkly staining portion of a blue-green algal cell. The central body is thought to function similar to a nucleus although it lacks true chromosomes and a nuclear membrane.

CENTRAL CAVITY The hollow central portion of a stem caused by the expansion of the pith. *Equisetum,* for instance, demonstrates a central cavity.

CENTRAL CELL A cell of the central filament of a polysiphonous thallus of some Rhodophyta. The central cell is surrounded by four or more pericentral cells.

CENTROPLASM The term applied by some botanists to the central body of a blue-green algal cell.

CHLOROPHYLL A green photosynthetic pigment common to all autotrophic plants. Several types of chlorophyll are known which are characteristic of different plant divisions. Chlorophyll occurs in chloroplasts in all green plants except Cyanophyta and photosynthetic bacteria.

CHLOROPLASM The colored chlorophyll-containing cytoplasm of a blue-green algal cell.

CHLOROPLAST A specialized type of chromoplast which contains chlorophyll and is responsible for photosynthesis.

CHROMATID A fundamental unit of the chromosome. Each chromosome is composed of two chromatids which separate at cell division.

CHROMATIN MATERIAL The chemical components of the chromosome prior to their coiling to become recognizable chromosomes. The interphase nucleus is said to contain chromatin material.

CHROMOPLASM The term some botanists use in referring to the chloroplasm of a blue-green algal cell.

CHROMOPLAST A membrane bound structure which contains colored pigments such as chlorophyll and carotenoids.

CIRCINATE VERNATION A characteristic, tightly coiled condition of immature leaves in the bud. This type of vernation is characteristic of ferns and cycads.

CLAMP CONNECTION A protrusion from a basidiomycete hyphal cell which connects two adjoining cells. A nucleus migrates through the clamp connection at cell division to maintain the dikaryotic condition.

CLASS A taxonomic grouping of plants with closely allied characteristics within a division above the rank of order. A class may contain from one to several plant orders.

CLEISTOTHECIUM An ascocarp which is generally spherical in shape, and which lacks an ostiole. A cleistothecium lacks a hymenial layer.

COENOBIC A condition common in some algae of producing a thallus comprised of a constant characteristic number of cells.

COENOCYTIC A multinucleate thallus not composed of individual cells. Most coenocytic organisms are filamentous and lack cross walls.

COLUMELLA A central sterile tissue found within a sporangium. The columella is an extension of the sporangiophore among fungi but is the central part of the capsule of bryophytes.

COMMENSALISM A symbiotic relationship between two organisms in which both derive benefit from the relationship.

COMPANION CELL A phloem cell of an angiosperm which is closely associated with and is thought to control a sieve tube element.

COMPLETE FLOWER An angiosperm flower in which all floral parts are present.

CONCEPTACLE A small chamber within the receptacle of some Phaeophyta. Conceptacles contain the gametangia.

CONE SCALE The ovule-bearing scale of a conifer. One or more ovules are produced on the upper surface of the cone scale.

CONIDIOPHORE The conidium-bearing hypha of a fungus. A conidiophore may be characteristically shaped or it may resemble a vegetative hypha.

CONIDIUM An asexual spore of many fungi, particularly Ascomycetes and Deuteromycetes (Fungi Imperfecti).

CONJUGATION An isogamous type of sexual reproduction wherein entire protoplasts of separate cells act as gametes.

COROLLA The collective term for the entire complement of petals in an angiosperm flower.

CRUSTOSE LICHEN An encrusting lichen which is attached to the substrate along its entire undersurface.

CRYOPHILIC Cold-loving organisms. For instance, algae which inhabit snow banks and glaciers are cryophilic.

CUTICLE A waxy covering which is secreted by the cells of the epidermis. The cuticle protects the plant against water loss.

CYANOPHYCEAN STARCH The storage product of Cyanophyta. This is a carbohydrate similar to starch stored by Chlorophyta and the higher plants.

CYTOKINESIS The process of cytoplasmic division of a cell which follows nuclear division.

CYTOPLASM The living portion of a cell containing the semi-liquid groundplasm and the organelles. The cell membrane is also part of the cytoplasm.

DIATOMACEOUS EARTH A deposit of fossil diatom cells (frustules) which is often commercially valuable.

DICHOTOMOUS BRANCHING A type of equal branching which occurs when the apical meristem or apical cell of an organ divides equally to produce two branches each with a new meristem or apical cell.

DIFFUSION The process of movement of any substance from an area of greater concentration to an area of lesser concentration.

DIKARYOTIC The condition of a cell which contains two nuclei each of which has been derived from a separate sexual strain.

DIOECIOUS (1) The condition of a gametophyte plant which bears gametangia on separate thalli. (2) The condition in angiosperms of producing unisexual flowers on separate plants so that an individual plant produces only male or female flowers.

DIPLOID The condition of a nucleus which contains two sets of homologous chromosomes. For instance, a diploid human cell contains two sets of 23, or a total of 46 chromosomes. A diploid corn cell contains two sets of 10, or a total of 20 chromosomes.

DIVISION The highest taxonomic grouping of similar plants except kingdom. The plant kingdom is composed of several separate divisions. Each division contains one or more plant classes.

DOCTRINE OF SIGNATURES The practice of using plants medicinally for the treatment of bodily parts they resemble. For instance, liverworts were so named since the thallus of some species resembles a human liver.

DOLIPORE The pore in a basidiomycete septum which has a thickened margin.

DOUBLE FERTILIZATION The process characteristic of angiosperms which occurs when two sperm are released into the embryo sac and one fertilizes the egg and the second unites with the polar nuclei.

ECOLOGY The study of the relationship between an organism or population and its environment.

ELATER A structure produced in the capsule of many bryophytes and attached to the spores of *Equisetum*. Elaters aid in breaking up the spore mass within a sporangium and disseminating spores.

ELIOPLAST A colorless plastid which is responsible for storing oils.

EMBRYO A juvenile sporophyte plant produced by higher plants which results from the germination of a zygote.

EMBRYO SAC The female gametophyte of an angiosperm. This plant is composed from four to sixteen nuclei (often 8 in 7 cells) and produces the egg.

ENDODERMIS The innermost layer of cells of the cortex which surrounds the stele of a root or stem or the leaf bundles of many vascular plants.

ENDOPHYTE An organism which lives within a plant. Several plants and animals are endophytic.

ENDOPLASMIC RETICULUM A double membrane network in the cytoplasm of a cell which is responsible for the synthesis of certain materials and for intracellular communication.

ENDOSPERM The tissue which nourishes the developing embryo of a seed plant. The endosperm is synonymous with the megagametophyte in gymnosperms. However, it is a separate, often 3N tissue in angiosperms.

ENDOSPORE A spore produced within a blue-green algal cell which is non-motile and lacks a cell wall.

ENDOSPORIC GAMETOPHYTE A gametophyte plant which is produced entirely within the confines of the meiospore. For example, the megagametophyte of *Selaginella* is endosporic.

ENERGY CHAIN The chain of energy transfer from one organism to another as organisms are consumed by those higher on the food chain. Energy chain is often used interchangably with food chain.

EPIDERMIS The outermost layer of cells in the organs of higher plants. This tissue functions in protecting against water loss.

EPIPHYTE An organism which grows upon the surface of a plant.

EPITHECA The larger half of the diatom cell wall or frustule.

EQUATORIAL PLATE The imaginary plate within a cell upon which the chromosomes become aligned at nuclear division.

EVOLUTION The process of change through time caused

by natural selection acting upon variations in organisms.

EXOSPORIC GAMETOPHYTE A gametophyte plant which develops outside the confines of the meiospore. All lower plants are exosporic in development.

EYE SPOT A light sensitive spot in some motile algae or zoospores which functions to direct the organism to a region of optimum illumination for photosynthesis or germination.

FALSE BRANCH An apparent branch of some Cyanophyta which arises from stresses produced when a hormogonium begins growth within the same sheath as the parent trichome.

FAMILY A taxonomic grouping of related genera with similar characteristics. From one to several genera may comprise a family.

FERTILIZATION TUBE A tube which allows the entrance of a male gamete into a female gametangium for fertilization.

FIBER An elongate, thick-walled cell which functions in adding support and strength to some plants.

FILAMENT (1) A group of cells arranged end to end to form an elongate thallus. (2) The portion of a stamen which bears the anther at its tip.

FISSION Cell division without the formation and orderly separation of chromatids. Blue-green algal cells divide by fission.

FLAGELLUM An organelle functioning in locomotion of the vegetative cells of some bacteria, algae and fungi and the gametes of many plants.

FLORIDEAN STARCH The carbohydrate storage product of Rhodophyta. This substance stains red rather than blue-black when treated with iodine solution.

FOLIOSE LICHEN A rather leaf-like lichen which is attached to the substrate by rhizoids.

FOOD CHAIN The chain of utilization of organisms for food from primary producers (green plants) through various consumers. Energy chain is another term used to describe a food chain.

FOOT The basal region of the sporophyte plant of a bryophyte or the embryo of a vascular plant which absorbs water and nutrients from the gametophyte plant.

FRAGMENTATION A type of asexual reproduction wherein the thallus physically breaks and each fragment may become a new plant.

FROND The large, often compound leaf of a fern or cycad. The term is also occasionally applied to certain large angiosperm leaves.

FRUIT The matured ovary of an angiosperm. A fruit generally contains seeds.

FRUSTULE The complete cell of a diatom. The frustule is composed of two overlapping valves and is highly siliceous.

FRUTICOSE LICHEN A highly divided or dissected, often lace-like lichen. Fruticose lichens are often epiphytes on vascular plants.

FUCOXANTHIN A unique xanthophyll pigment found in Phaeophyta.

GAMETANGIUM A gamete bearing cell or structure. The gametangium is unicellular in lower plants and multicellular among higher plants. Antheridia, oogonia and archegonia are examples of gametangia.

GAMETE A haploid cell which is capable of fusing with another haploid cell to produce a zygote.

GAMETOPHORE The gametangium bearing stalk common in some Bryophyta.

GAMETOPHYTE GENERATION The haploid generation which produces gametes.

GEMMAE An asexual reproductive structure produced by some bryophytes.

GENERATIVE CELL The cell in an angiosperm pollen grain which divides to produce two sperm.

GENUS A taxonomic grouping of similar species. The name of a plant always includes the generic name first. For example, the common dandelion, *Taraxicum officinale,* is placed in the genus *Taraxicum.*

GOLGI BODY An organelle found in the cytoplasm of plant cells which is thought to function in cell wall formation.

GROUND PLASM The semi-liquid portion of the cytoplasm in which organelles are imbedded.

GULLET An invagination of the apex of some Euglenophyta into which particles of food may be ingested under certain circumstances.

GYMNOSPERM A seed plant which produces its seeds nakedly on megasporophylls or ovuliferous scales.

GYNOECIUM The collective term used for the entire complement of female reproductive structures in an angiosperm flower.

HETEROECIOUS The term applied to a rust (Uredinales) which completes its life cycle on two separate host plants.

HETEROMORPHIC ALTERNATION A type of alternation of generations wherein the sporophyte and the gametophyte plants are different in appearance.

HETEROSPOROUS The condition of producing two kinds of meiospores known as megaspores and microspores. All seed plants are heterosporous.

HETEROTHALLIC The condition when gametes from two separate thalli are required for syngamy. Thus, gametes from the same thallus will not unite with each other.

HETEROTROPHIC The term applied to an organism which does not manufacture its own food supply.

HOLDFAST A cell or multicellular structure which attaches the thallus to the substrate.

HOMOLOGOUS CHROMOSOME A chromosome which is a structural duplicate of another chromosome in a diploid cell. Each chromosome in a diploid nucleus has a homolog.

HOMOSPOROUS The condition of producing a single kind of meiospore.

HOMOTHALLIC The condition wherein gametes produced by the same thallus will unite with each other to produce a zygote.

HORMOGONIUM A fragment of a trichome produced by some Cyanophyta. Hormogonia are often formed by separation disks.

HYMENIAL LAYER (HYMENIUM) The layer of meiospore bearing cells and paraphyses produced by some fungi.

HYPHA An individual filament of a fungal thallus.

HYPODERMIS A layer of thick-walled cells just beneath the epidermis of some plants. The hypodermis functions in protection and support.

HYPOTHECA The inner half of the diatom cell wall or frustule.

IMPERFECT FLOWER An angiosperm flower which lacks either male or female reproductive structures.

INCOMPLETE FLOWER An angiosperm flower which lacks any floral part including sepals and petals as well as reproductive structures.

INDUSIUM A protective flap of tissue covering the sorus of some ferns.

INTEGUMENT The outer protective layer of an ovule. The integument matures to become the seed coat.

INTERNODE That portion of a stem occurring between two successive nodes.

INTERPHASE The stage of a nucleus when not actively engaged in division.

ISOGAMETE A gamete which is alike in size and shape with the gamete it fuses with at fertilization.

ISOGAMY The process of fusion of gametes which appear exactly alike. Isogamy is thought to be the primitive type of syngamy.

ISOMORPHIC The type of alternation of generations when the gametophyte and sporophyte plants appear alike.

ISTHMUS The central, often constricted portion of a desmid cell where the nucleus generally lies.

LAMELLAE The thin ribs of tissue (gills) lining the underside of the basidiocarp of some Basidiomycetes (mushrooms).

LAMINA The thin, flattened portion of a thallus or leaf which is highly efficient in photosynthesis.

LATERAL BRANCH A branch which develops some distance behind an apical meristem. Lateral branches often originate from buds in the axils of leaves.

LEAF GAP A region in the stele of a higher vascular plant filled with parenchyma tissue located at the point of departure of a leaf trace.

LEAF SHEATH A leaf or group of fused leaves which encircles or ensheathes a stem. A leaf sheath often protects buds.

LEAF TRACE A vascular connection between the stem and leaf of a vascular plant.

LEUCOPLAST A non-colored plastid which often stores food products.

LIFE CYCLE The pattern of growth and reproduction of a plant including especially sexual reproduction and meiosis.

LONG SHOOT A normal branch of *Ginkgo* and some other gymnosperms. Long shoots produce short or spur shoots which bear leaves and reproductive structures.

LYSOSOME An organelle which produces enzymes that are apparently responsible for the degradation of the protoplast upon death of the cell.

MACROPHYLL A leaf normally containing many veins which form leaf gaps in the stele of the stem.

MEGAGAMETOPHYTE The gametophyte produced by the germination of a megaspore. The megagametophyte is responsible for the production of female gametes.

MEGASPORANGIUM A sporangium responsible for the production of megaspores.

MEGASPORE A relatively large meiospore responsible for the production of the megagametophyte plant. Megaspores are produced by heterosporous plants.

MEIOSIS The process of nuclear division wherein a diploid nucleus divides to produce four haploid nuclei. Meiosis is often called reduction division.

MEIOSPORE A spore resulting from meiosis. The sporophyte generation is responsible for producing meiospores.

MERISTEM A region in a plant composed of cells which carry on continued cell division. Meristems terminate roots, stems and leaves and are responsible for the elongation of those organs.

MESOPHYLL The tissue comprising most of the interior of many leaf laminae. Mesophyll is responsible for most of the photosynthesis of many vascular plants.

METAPHASE The stage of nuclear division when chromosomes become aligned at the equatorial plate. Chromosomes align independently at mitosis but with their homolog in metaphase I of meiosis.

METAPHASE PLATE The term used by some botanists to describe the equatorial plate.

MICROFIBRIL A fiber-like bundle of cellulose molecules which is the fundamental building unit of the cell wall of many plants.

MICROGAMETOPHYTE The gametophyte plant which originates from the germination of a microspore and is responsible for the production of male gametes.

MICROPHYLL A leaf produced by lower vascular plants which is usually small and contains a single vascular bundle that does not form a leaf gap in the stem.

MICROPYLAR TUBE An elongate tube formed from the inner integument in some Gnetophyta.

MICROPYLE The opening at the apex of an ovule formed by the incomplete closure of the integument.

MICROSPORANGIUM The sporangium which produces microspores.

MICROSPORE A relatively small meiospore which germinates to produce the microgametophyte plant. Heterosporous plants produce microspores.

MICROSPOROCYTE A diploid cell which divides by meiosis to produce four microspores.

MITOCHONDRION An organelle in most plant cells responsible for respiration.

MITOSIS The process of nuclear division which proceeds by an orderly sequence of steps or phases. Two daughter cells each with the same chromosome number as the mother cell result from mitosis.

MONOECIOUS (1) The condition of a gametophyte plant

which produces both gametangia on the same thallus. (2) The condition of an angiosperm which produces unisexual flowers on the same plant.

MORPHOLOGY The study of form and reproduction of plants.

MULTINUCLEATE The condition of a cell which contains more than one nucleus.

MULTISERIATE A filament of cells which is more than one cell broad.

MYCELIAL CONNECTION A root-like projection of hyphae which attaches a basidiocarp to its subtending mycelium.

MYCELIUM The vegetative thallus of many fungi. A mycelium is composed of many hyphae growing closely together.

MYCORRHIZA The association between a fungus and a root, rhizome or gametophyte plant of a higher plant.

NATURAL SELECTION The evolutionary process by which those organisms most fit for their particular environment experience a greater rate of survival.

NECK The elongate apical portion of an archegonium. Sperm enter through the archegonial neck and migrate to the egg.

NECRIDIUM A cell in the trichome of some Cyanophyta which dies and causes the fragmentation of the trichome into hormogonia.

NITROGEN FIXATION The process occurring in some plants of utilizing atmospheric nitrogen in their metabolism. This nitrogen usually becomes available to other organisms upon the death of the nitrogen fixing plant.

NODE The point on a stem at which a leaf is produced. Buds often occur in the axils of these leaves and later develop into lateral branches.

NUCELLUS The tissue within the integument of an ovule. The nucellus produces the megasporocyte and is used to nourish the developing megagametophyte.

NUCLEOPLASM The semi-liquid portion of the nucleus in which the chromatin material is imbedded.

NUCLEUS A large, prominent organelle containing chromosomes found in most cells. The nucleus is responsible for most cellular processes by controlling protein synthesis.

OOGAMY A type of syngamy in which the female gamete is a non-motile egg and the male gamete is a smaller sperm.

OOGONIUM The unicellular, egg-bearing female gametangium common in many lower plants.

OPERCULUM The covering or lid at the apex of the capsule of the sporophyte plant of a moss.

ORDER A taxonomic grouping of similar plant families. The order is above the rank of family and below the rank of class.

ORGANELLE A membrane limited structure with a particular function found in the cytoplasm of a cell. Examples of organelles include the nucleus, mitochondrion, golgi body, and chloroplast.

OSMOSIS The process of diffusion through a differentially permeable membrane.

OSTIOLE A small opening in an otherwise enclosed structure.

OVARY The enlarged basal portion of the female reproductive structure (pistil) of an angiosperm which is composed of one or more carpels enclosing ovules. The ovary matures to become a fruit.

OVULE The specialized megasporangium and integument/integuments of a seed plant. An ovule matures to become a seed.

PALISADE MESOPHYLL A layer of cells normally found immediately beneath the upper epidermis of a leaf lamina. Palisade cells are oriented perpendicular to the epidermal layer and are responsible for much of the photosynthesis of the leaf.

PARAMYLUM A carbohydrate storage product produced by Euglenophyta.

PARAPHYSES Sterile, hair-like filaments which occur at the hymenium of many fungi and associated with the gametangia or sporangia of many other plants.

PARASITE An organism which derives its nourishment from another living organism.

PARENCHYMA The fundamental cell type of all plants. Parenchyma cells are generally rather large and isodiametric and are thought to be the primitive cell type.

PERENNIAL A plant which lives for more than two growing seasons. For example, all large conifers are perennial.

PERFECT FLOWER A flower which contains both male and female reproductive structures.

PERFORATION PLATE The plate at the end of a vessel element. This plate may contain a large, simple hole or it may contain a series of smaller openings.

PERICENTRAL CELL A cell of an external filament of a polysiphonous thallus.

PERICYCLE The tissue immediately internal to the endodermis of most roots and some stems. This tissue may become meristematic and is responsible for branch root formation.

PERIDERM The outermost, secondary, corky or woody tissue produced by most perennial gymnosperms and angiosperms. Periderm is produced by a cork cambium to replace the epidermis and cortex.

PERIPLAST The outer rather rigid membrane of the cell of many organisms which lack a cell wall. *Euglena* and some dinoflagellates commonly produce periplasts.

PERISTOME A structure in a moss capsule composed of a ring of tooth-like structures which project into the capsule. The peristome aids spore dispersal.

PERITHECIUM An ascocarp which is enclosed except for an ostiole at its apex. Perithecia produce a definite hymenial layer.

PETAL A sterile, often highly colored, leaf-like structure common in most angiosperm flowers. Petals often function in insect attraction. They occur internal to sepals and external to stamens.

PHLOEM The vascular tissue responsible for transporting

photosynthetic products to non-photosynthetic parts of the plant.

PHOTOSYNTHESIS The process of manufacturing simple sugar compounds from carbon dioxide and water in the presence of chlorophyll utilizing the radiant energy of the sun.

PHYCOCYANIN A blue pigment found in Cyanophyta and Rhodophyta which absorbs light and aids in photosynthesis.

PHYCOERYTHRIN A red pigment found in Cyanophyta and Rhodophyta which absorbs light and aids in photosynthesis.

PHYLLIDIUM A leaf-like structure common in mosses and many liverworts. A phyllidium lacks vascular tissue.

PHYTOPLANKTON Small, often microscopic plants which float passively in the water. Phytoplankton are especially important in food chains of most aquatic organisms.

PILEUS The basidiocarp cap of many Basidiomycetes (mushrooms).

PINNA A segment of a frond or leaf which is composed of a central rachis with attached leaflets.

PINNULE The smallest segment of a fern or cycad frond. Pinnules are attached along the rachis of a pinna.

PISTIL The term used to describe an angiosperm ovary with its attached style and stigma.

PISTILLATE FLOWER A unisexual angiosperm flower which produces only female reproductive structure.

PITH The centermost tissue of a siphonostele. The pith is usually composed of parenchyma cells which often function in storage.

PLANKTER An individual planktonic organism.

PLANKTON Plants or animals which float passively in the water.

PLANT COMMUNITY An association of plants inhabiting the same geographical region and determined by a distinct set of environmental conditions.

PLANT SUCCESSION The replacement of one plant community by another caused by a change in environmental conditions.

PLASMALEMMA The term used by many botanists to describe the cell membrane.

PLASMODIUM The thallus of a slime mold which is coenocytic and moves by amoeboid motion.

PLASMOLYSIS The condition of losing water from the vacuole or cytoplasm of a cell so that the protoplast shrinks away from the cell wall.

PLASTID An organelle common in plant cells which functions in photosynthesis or in storage of food products.

POLAR NUCLEI Two generally haploid nuclei of an angiosperm megagametophyte plant. The polar nuclei are fertilized by one sperm to become the endosperm.

POLLEN DROPLET A sticky droplet of liquid which is extruded from the micropyle of some gymnosperm ovules to trap pollen grains and thus aid in pollination.

POLLEN GRAIN A specialized matured microspore produced by a seed plant. The pollen grain is often ornate and easily carried on wind currents or by pollinating organisms.

POLLEN TUBE A cytoplasmic tube produced by a pollen grain upon germination. The pollen tube grows toward the female gametophyte and eventually deposits sperm near the egg.

POLLINATION TUBE The term used to describe the tube which aids in pollination formed by the inner integument of some Gnetophyta.

PRIMARY CONSUMER An organism which is the first to consume a primary producer. The second level of a food chain.

PRIMARY HYPHA A haploid basidiomycete hypha which results from the germination of a basidiospore.

PRIMARY PRODUCER An organism which manufactures its own food by photosynthesis. All green plants are primary producers.

PRIMARY TISSUE A plant tissue resulting from the germination of an embryo or from the activity of an apical cell or meristem.

PROPHASE A stage in nuclear division wherein the chromatin material becomes organized to form recognizable chromosomes. The nuclear membrane also disappears at prophase.

PROTONEMA An algal-like filament resulting from the germination of a meiospore produced by most bryophytes.

PROTOPLAST A term used for the entire living unit of a plant cell. Functionally, the term is often used for an entire plant cell excluding the cell wall.

PROTOSTELE A stele composed of a solid core of vascular tissue lacking a pith.

PYCNOSPORE A spore formed within the pycnium of some rusts (Uredinales). A pycnospore fuses with a receptive hyphal cell of opposite mating strain to produce a dikaryotic cell.

PYRENOID A structure found in the chloroplast of many algae and some bryophytes which usually functions in producing and storing starch.

RAPHE A longitudinal furrow or opening in the frustule of some diatoms which allows motility of the organism.

RAY (1) A projection which develops from the female receptacle of some Marchantiales. (2) A region of parenchyma cells in vascular tissue which function in lateral conduction.

RECEPTACLE (1) The swollen branch terminus in some Phaeophyta which contains conceptacles. (2) The flattened disk of tissue in an angiosperm flower which produces the floral parts.

RECEPTIVE HYPHA A haploid hypha produced within the pycnium of many rusts (Uredinales) which fuses with a pycnospore of opposite mating strain to produce a dikaryotic cell.

REPLICATION The process of exact duplication of a strand of chromatin material. Replication occurs during interphase, prior to nuclear division.

RESIN DUCT A canal-like structure common in many conifers which stores and translocates resins.

RESPIRATION The process of converting the energy stored in food products to a form usable to the cell for its various metabolic activities.

RHIZOID A root-like structure produced by many lower plants which attaches the organism to the substrate and absorbs water and nutrients. Rhizoids lack vascular tissue.

RHIZOME A stem which grows beneath the surface of or along the ground.

RHIZOMORPH A group of hyphae of a basidiomycete which resembles a root and attaches a basidiocarp to a subtending mycelium. This term is often used synonymously with *mycelial connection*.

RIBOSOME An organelle responsible for synthesizing proteins in a cell as controlled by the nucleus.

ROOT CAP A protective structure covering the apical meristem of a root. Cells of the root cap are often mucilagenous and are sloughed off as the root grows through the soil.

SAPROPHYTE An organism which obtains its nourishment from non-living organic matter.

SCLERENCHYMA Thick-walled cells which function in lending strength and support to some plants.

SCLEROTIUM A hardened mass of fungal hyphae produced especially by some Ascomycetes which functions in overwintering the fungus.

SECONDARY CONSUMER An organism which consumes other consumer organisms. The top consumer is the top organism on the food chain.

SECONDARY HYPHA The dikaryotic hypha of a Basidiomycete which results from plasmogamy of two primary hyphal cells.

SECONDARY TISSUE A tissue which results from the activity of a lateral meristem such as the vascular cambium.

SEED A matured ovule. A seed has an external seed coat and contains an embryo and occasionally other tissues.

SEED COAT The outermost layer of a seed which develops from the integument.

SEMICELL One-half of a desmid cell which terminates at the isthumus.

SEMIPERMEABLE MEMBRANE A membrane which is permeable to certain substances but restrictive to others. Such a membrane is often referred to as a differentially permeable membrane.

SEPAL The outermost floral part of an angiosperm flower. A sepal may be green and leaf-like or colored and petal-like.

SEPARATION DISK The term used by some botanists for the necridium produced by some Cyanophyta.

SEPTUM A cross wall of a trichome which divides the filament into individual cells.

SETA The stalk-like structure of the sporophyte plant of a bryophyte which elevates the capsule for efficient spore dispersal.

SEXUAL REPRODUCTION Reproduction of an organism involving syngamy.

SIEVE CELL A conductive cell of the phloem of lower vascular plants including conifers.

SIEVE TUBE ELEMENT A conductive cell of angiosperm phloem. A sieve tube element lacks a nucleus and is closely associated with a companion cell.

SIPHONOSTELE A stele with a central pith. Several kinds of siphonosteles are recognized by plant anatomists.

SOREDIUM An asexual reproductive body in some lichens. A soredium is composed of some algal cells surrounded by fungal cells.

SORUS A cluster of sporangia normally occurring on the undersurface of a fern leaf. The sorus may be naked or covered by an indusium.

SPONGY MESOPHYLL A region of mesophyll tissue which generally occurs beneath the palisades layer in a leaf lamina. This tissue is composed of large parenchyma cells with intercellular spaces.

SPORANGIOPHORE A stalk which bears a sporangium.

SPORANGIUM Any spore producing cell or structure.

SPORE A structure which is often resistant germinates directly to produce a new plant. Spores may be either haploid or diploid.

SPOROCARP A specialized, highly modified sporophyll which contains the sori of some ferns.

SPOROCYTE A cell which divides by meiosis to produce meiospores.

SPOROPHYTE The plant generation which is normally diploid and is responsible for producing meiospores.

SPUR SHOOT (SHORT SHOOT) A specialized, short side branch produced by some gymnosperms. Leaves and reproductive organs are produced on spur shoots.

STAMEN The male reproductive structure of an angiosperm flower. Stamens are produced inside of the petals and external to the female reproductive structures.

STAMINATE FLOWER A unisexual flower which bears only stamens.

STELE The term applied to the entire vascular tissue system of a stem or root.

STERIGMA A small protrusion on a basidium upon which the basidiospores are produced.

STIGMA (1) A term used by some biologists for the eye spot of an organism. (2) The apex of the pistil of an angiosperm flower which receives the pollen grains at pollination.

STIPE A stalk-like structure which bears the lamina of some Phaeophyta or the pileus of some Basidiomycetes.

STOMA An opening in the epidermis of a higher plant which allows the exchange of gases. The stoma is generally bordered by specialized guard cells.

STROBILUS The term applied to a group of sporangia and sporophylls which are tightly arranged at the apex of a stem or branch. The term cone is often used interchangeably with strobilus.

STROMA A tissue-like fungal mycelium which produces fruiting bodies. For instance, the stroma of *Claviceps* produces perithecia.

STYLE An elongate appendage produced by the ovary of most angiosperms which elevates the stigma for pollination.

SYNERGID A vegetative cell of the angiosperm embryo sac. A single synergid is usually produced on each side of the egg.

SYNGAMY The process of fusion of gametes. Syngamy is often known as fertilization.

TAXONOMY The science of classifying organisms.

TELIOSPORE A dikaryotic spore produced by some Basidiomycetes (rusts and smuts) which is resistant to adverse conditions.

TELIUM A pustule-like structure produced on a host plant by a rust (Uredinales) which produces teliospores.

TELOPHASE The stage of nuclear division following anaphase when homologous chromosomes or chromatids have completely separated. A new nuclear membrane develops and a new cell wall begins to form during telophase.

TERTIARY HYPHA A dikaryotic hypha produced by some Basidiomycetes which is organized to become a portion of a basidiocarp.

TETRAD (TETRAD OF SPORES) Four meiospores produced from the same sporocyte and remaining in contact.

TETRASPORE This term is often used synonymously with meiospore. Some botanists reserve it for use among some Rhodophyta for the spores produced by the tetrasporophyte plant.

TETRASPOROPHYTE The true sporophyte generation of some Rhodophyta. The tetrasporophyte plant originates from a carpospore and produces tetraspores (meiospores).

THERMOPHILIC An organism that is able to live under conditions of elevated temperature is said to be thermophilic.

TRACHEID An elongate, tapering xylem cell of many vascular plants. These cells are dead at maturity and conduct water through pits in their side walls.

TRICHOGYNE An elongate structure produced by the carpogonium of most Rhodophyta and the ascogonium of many Ascomycetes. This structure receives the male nuclei at sexual reproduction.

TRICHOME A filament of cells.

TRUMPET CELLS Cells of the central portion of the stipe and lamina of some Phaeophyta which function in translocating photosynthetic products.

TUBE NUCLEUS The cell or nucleus within an angiosperm pollen grain which may be responsible for the formation of the pollen tube.

TUBER A swollen underground stem which functions in storing food materials, particularly starch.

TURGID The condition of a cell which contains ample water to cause it to swell and press the cell membrane against the cell wall.

TURGOR PRESSURE The pressure exerted by a turgid cell against the cell wall.

UNIVERSAL VEIL The tissue-like layer of tertiary hyphae which surrounds the entire immature basidiocarp in certain Basidiomycetes.

UREDIUM A pustule-like structure produced by some rusts (Uredinales) which ruptures through the host epidermis to release uredospores.

UREDOSPORE A single celled, dikaryotic reinfection spore produced by some rusts (Uredinales).

VALLECULAR CANAL The large canal-like air spaces in the cortex of *Equisetum* which are thought to function in aeration of submerged portions of the plant.

VALVE One-half of the frustule of a diatom cell. Thus, the hypotheca and epitheca are referred to as valves.

VASCULAR BUNDLE A distinct, individual bundle of vascular tissue.

VEIN A vascular bundle in a leaf. Veins are often characteristically arranged.

VENTER The basal swollen portion of an archegonium which contains the egg.

VESSEL An end to end association of vessel elements to form long, hollow conductive tubes in the xylem of some plants.

VESSEL ELEMENT A cell of the xylem of some vascular plants. These cells are more or less barrel-shaped and have porose ends to facilitate conduction.

VULVA The cup-shaped structure at the base of the basidiocarp stipe of some Basidiomycetes.

XANTHOPHYLL A pigment common in the plastids of most green plants. Xanthophyll pigments are generally brownish or yellowish in color.

XYLEM The vascular tissue responsible for conducting water and mineral nutrients throughout the plant.

ZOOPLANKTON Microscopic animals which are free-floating or only slightly motile in the water.

ZOOSPORE A motile spore.

ZYGOSPORE An encysted zygote which is resistant to adverse conditions.

ZYGOTE A diploid cell which results from the fusion of two gametes.

Index